AF356600

Impact Welding of Materials

Editors
Ivan Galvão
ISEL, Department of Mechanical Engineering,
Polytechnic Institute of Lisbon
Portugal
University of Coimbra
Portugal

Altino Loureiro
University of Coimbra
Portugal

Ricardo Mendes
University of Coimbra
Portugal

Editorial Office
MDPI
St. Alban-Anlage 66
4052 Basel, Switzerland

This is a reprint of articles from the Special Issue published online in the open access journal *Metals* (ISSN 2075-4701) (available at: https://www.mdpi.com/journal/metals/special_issues/impact_welding).

For citation purposes, cite each article independently as indicated on the article page online and as indicated below:

LastName, A.A.; LastName, B.B.; LastName, C.C. Article Title. *Journal Name* **Year**, *Volume Number*, Page Range.

ISBN 978-3-0365-0696-8 (Hbk)
ISBN 978-3-0365-0697-5 (PDF)

Cover image courtesy of Gustavo H.S.F.L. Carvalho, Ivan Galvão, Rui M. Leal, Ricardo Mendes and Altino Loureiro.

Contents

About the Editors vii

Ivan Galvão, Altino Loureiro and Ricardo Mendes
Impact Welding of Materials
Reprinted from: *Metals* **2020**, *10*, 1668, doi:10.3390/met10121668 1

Huimin Wang and Yuliang Wang
High-Velocity Impact Welding Process: A Review
Reprinted from: *Metals* **2019**, *9*, 144, doi:10.3390/met9020144 5

Yulia Yu. Émurlaeva, Ivan A. Bataev, Qiang Zhou, Daria V. Lazurenko, Ivan V. Ivanov, Polina A. Riabinkina, Shigeru Tanaka and Pengwan Chen
Welding Window: Comparison of Deribas' and Wittman's Approaches and SPH
Simulation Results
Reprinted from: *Metals* **2019**, *9*, 1323, doi:10.3390/met9121323 23

Yasir Mahmood, Kaida Dai, Pengwan Chen, Qiang Zhou, Ashfaq Ahmad Bhatti and Ali Arab
Experimental and Numerical Study on Microstructure and Mechanical Properties of
Ti-6Al-4V/Al-1060 Explosive Welding
Reprinted from: *Metals* **2019**, *9*, 1189, doi:10.3390/met9111189 41

Henryk Paul, Robert Chulist and Izabela Mania
Structural Properties of Interfacial Layers in Tantalum to Stainless Steel Clad with Copper
Interlayer Produced by Explosive Welding
Reprinted from: *Metals* **2020**, *10*, 969, doi:10.3390/met10070969 59

Gustavo H. S. F. L. Carvalho, Ivan Galvão, Ricardo Mendes, Rui M. Leal and Altino Loureiro
Aluminum-to-Steel Cladding by Explosive Welding
Reprinted from: *Metals* **2020**, *10*, 1062, doi:10.3390/met10081062 77

Daisuke Inao, Akihisa Mori, Shigeru Tanaka and Kazuyuki Hokamoto
Explosive Welding of Thin Aluminum Plate onto Magnesium Alloy Plate Using a Gelatin Layer
as a Pressure-Transmitting Medium
Reprinted from: *Metals* **2020**, *10*, 106, doi:10.3390/met10010106 95

Masatoshi Nishi, Shigeru Tanaka, Matej Vesenjak, Zoran Ren and Kazuyuki Hokamoto
Fabrication of Composite Unidirectional Cellular Metals by Using Explosive Compaction
Reprinted from: *Metals* **2020**, *10*, 193, doi:10.3390/met10020193 109

Shan Su, Shujun Chen, Yu Mao, Jun Xiao, Anupam Vivek and Glenn Daehn
Joining Aluminium Alloy 5A06 to Stainless Steel 321 by Vaporizing Foil Actuators Welding with
an Interlayer
Reprinted from: *Metals* **2019**, *9*, 43, doi:10.3390/met9010043 119

Huimin Wang and Yuliang Wang
Characteristics of Flyer Velocity in Laser Impact Welding
Reprinted from: *Metals* **2019**, *9*, 281, doi:10.3390/met9030281 131

Omid Emadinia, Alexandra Martins Ramalho, Inês Vieira de Oliveira, Geoffrey A. Taber and Ana Reis
Influence of Surface Preparation on the Interface of Al-Cu Joints Produced by Magnetic Pulse Welding
Reprinted from: *Metals* **2020**, *10*, 997, doi:10.3390/met10080997 . . . . **143**

Jörg Bellmann, Jörn Lueg-Althoff, Benedikt Niessen, Marcus Böhme, Eugen Schumacher, Eckhard Beyer, Christoph Leyens, A. Erman Tekkaya, Peter Groche, Martin Franz-Xaver Wagner and Stefan Böhm
Particle Ejection by Jetting and Related Effects in Impact Welding Processes
Reprinted from: *Metals* **2020**, *10*, 1108, doi:10.3390/met10081108 . . . . **155**

Benedikt Niessen, Eugen Schumacher, Jörn Lueg-Althoff, Jörg Bellmann, Marcus Böhme, Stefan Böhm, A. Erman Tekkaya, Eckhard Beyer, Christoph Leyens, Martin Franz-Xaver Wagner and Peter Groche
Interface Formation during Collision Welding of Aluminum
Reprinted from: *Metals* **2020**, *10*, 1202, doi:10.3390/met10091202 . . . . **179**

About the Editors

Ivan Galvão is an assistant professor at the Department of Mechanical Engineering of ISEL—Polytechnic Institute of Lisbon (Portugal) and an integrated researcher in CEMMPRE—University of Coimbra (Portugal). He completed a Ph.D. in Mechanical Engineering (Production Technology field) at the University of Coimbra. His research field includes the study of materials joining and processing, specifically welding technology and metallurgy, the mechanical behavior of welded joints, the solid-state welding of similar and dissimilar materials and the solid-state processing of materials. He has published many scientific papers in high impact journals and conference proceedings, and he has participated in several international conferences. He has also been involved in research projects, collaborated with international journals and engineering associations, as well as supervised research works and theses.

Altino Loureiro is a full professor at the Faculty of Sciences and Technology, University of Coimbra, Portugal. He has a Ph.D. in Mechanical Engineering, and an M.Sc. in Materials and Manufacturing Processes, Mechanical Engineering and Welding Engineering. He is the co-author of an internationally published technical book and of more than one hundred and twenty (120) articles published in international scientific journals, with more than two thousand eight hundred (2800) citations and an h-index of 31, according to the SCOPUS citation database. He has supervised or co-supervised ten (10) doctoral theses and more than sixty (60) master's theses.

Ricardo Mendes was born in Tomar (Portugal) in 1967. He is an assistant professor at the Faculty of Sciences and Technology, University of Coimbra, Portugal. He graduated in Physics Engineering and obtained his Ph.D. in Mechanical Engineering. His main scientific activity is within the area of high-pressure, high-temperature ultra-fast phenomena associated with the detonation physics of ideal and non-ideal explosives. Recently, he has also worked on the shock wave processing of materials like the detonation synthesis of nanopowders and explosion welding of metals. He is also involved in research projects and has published many scientific papers in high impact journals. He has participated in several international conferences, and supervised or co-supervised four (4) Ph.D. theses and more than fifty (50+) master's theses.

Editorial

Impact Welding of Materials

Ivan Galvão [1,2,*], Altino Loureiro [1] and Ricardo Mendes [3]

[1]   University of Coimbra, CEMMPRE, Department of Mechanical Engineering, Rua Luís Reis Santos,
      3030-788 Coimbra, Portugal; altino.loureiro@dem.uc.pt
[2]   ISEL, Department of Mechanical Engineering, Polytechnic Institute of Lisbon,
      Rua Conselheiro Emídio Navarro 1, 1959-007 Lisboa, Portugal
[3]   University of Coimbra, ADAI, LEDAP, Department of Mechanical Engineering, Rua Luís Reis Santos,
      3030-788 Coimbra, Portugal; ricardo.mendes@dem.uc.pt
*   Correspondence: ivan.galvao@dem.uc.pt; Tel.: +351-239-790-700

Received: 10 December 2020; Accepted: 12 December 2020; Published: 14 December 2020

1. Introduction and Scope

Recent industrial criteria, focused on obtaining increasingly efficient structures, require the production of multimaterial components. However, the manufacturing requirements of these components are not met by conventional welding techniques. Alternative solid-state technologies, such as friction or impact-based processes, must be considered. Impact welding processes have the advantage of presenting a very short cycle time, which minimises the interaction of the materials under high temperature. This fact strongly contributes to reducing the formation of brittle intermetallic compounds (IMCs), i.e., one of the main concerns of welding dissimilar materials. Moreover, as the influence of the welding process is confined to a very narrow band around the materials interface, similar and dissimilar welds with high-strength bonding and a minimal heat-affected zone can be produced.

The impact welding family encompasses different welding processes, such as explosion welding, magnetic pulse welding, vaporising foil actuator welding, and laser impact welding. Although these processes share the main operating principle, consisting of a high-velocity collision between a flyer and a target, they differ in the way the flyer is accelerated. These processes also present very different length scales, providing the impact welding family with a broad applicability range. The technical interest of impact welding is driving the ongoing development of many scientific studies, which are essential to optimise the current manufacturing processes by developing new welding strategies and solutions. The present special issue presents a sample of the cutting-edge research that is being conducted on the multidisciplinary field of impact welding.

2. Contributions

Eleven research papers and one review paper have been published in the present special issue. Different research approaches, i.e., experimental, numerical, and coupled experimental/numerical, are presented in the published papers. These studies, which were developed by some of the most relevant research groups working in impact welding, encompass a large spectrum of welding processes, such as explosion welding, magnetic pulse welding, vaporising foil actuator welding, and laser impact welding. Detailed analyses on the impact welding macro and microscopic phenomena, metallurgy, mechanical behaviour, numerical modelling and simulation, process developments, and industrial applications are presented in this group of works.

Six research papers on explosion welding have been published. Émurlaeva et al. [1] conducted a Smooth Particle Hydrodynamics (SPH) simulation work to further analyse the suitability of the classical approaches defining the boundaries of the weldability window. The numerical simulation results, which were applied to 6061-T6 aluminium alloy, were found to reproduce the basic phenomena typical of high-velocity impact welding, such as the jet and wave formation and the material deformation near the interface. The left, right, and lower boundaries of the weldability window found by numerical simulation were in good agreement with those resulting from Wittman's and Deribas's approaches. However, significant differences were found for the upper limit. In turn, a coupled experimental/numerical work focused on studying the microstructure and the mechanical properties of titanium/aluminium explosion welds before and after being submitted to heat treatment was developed by Mahmood et al. [2]. According to the authors, aluminium–titanium IMCs were formed at the weld interface, but these phases did not affect the weld strength. The authors also reported that the weld strength decreases after heat treatment due to changes in the structure of the weld interface. The SPH-based numerical simulation analysis conducted in this research was found to meet the experimental results. For example, the predicted temperatures for the weld interface were higher than the melting temperature of the alloys, which agreed well with the formation of aluminium–titanium IMCs.

Paul et al. [3] studied the microstructural and mechanical properties of tantalum/copper/stainless steel explosion-welded composites. An exhaustive experimental work, in which several characterisation techniques were used, such as scanning electron microscopy, energy dispersive spectroscopy, transmission electron microscopy, electron backscatter diffraction, microhardness testing, and bending testing, was conducted to characterise the tantalum/copper and the copper/stainless steel interfaces. The conditions experienced at the weld interface, such as severe plastic deformation, localised melting, and materials interaction, were deeply discussed based on the microstructural features and the mechanical properties of this region. An experimental work was also conducted by Carvalho et al. [4], which was focused on two different material combinations. Specifically, these authors analysed the coupled effect of two strategies for optimising the production of aluminium/carbon steel and aluminium/stainless steel explosion-welded clads, i.e., the use of a low-density interlayer and the use of a low-density and low-detonation velocity explosive mixture. The low values of collision point and impact velocities achieved for the welding tests provided the production of joints with sound microstructure and good mechanical behaviour, making the differences in weldability of both material combinations less significant. The low-detonation velocity, low-density, and ability to detonate in small thicknesses of the explosive mixture make it very suitable to be used for welding very thin flyers and dissimilar material couples that easily form IMCs.

Inao et al. [5] joined a very thin aluminium plate to two different magnesium alloys (AZ31, $Mg_{96}Zn_2Y_2$) by explosion welding, using a gelatine layer as a pressure-transmitting medium. The lower energetic conditions associated with the use of a very thin flyer and the gelatine medium promoted better bonding quality than in conventional explosion welding. The welds achieved under these conditions presented a uniform, smooth surface and an interface without intermetallic-rich layers. In turn, Nishi et al. [6] conducted a work highly focused on the development of new heat-exchanger solutions. Specifically, the authors addressed the manufacturing procedure of two types of copper/stainless steel composite UniPore structures. Two specific arrangements of thin copper and stainless steel pipes were successfully compacted/welded by explosion. They were found to present excellent interfacial bonding between the component walls.

Using a different welding process from the previous works, i.e., vaporising foil actuator welding, Su et al. [7] produced aluminium/stainless steel joints with an aluminium interlayer. The authors tested different input energy values and reported that higher energy values promoted welds with higher tensile and shear strengths because of the larger weld areas obtained under these conditions. Regarding the interface morphology, it was found to be different for both interfaces. While the flyer/interlayer interface (aluminium/aluminium) was wavy, the interface between the interlayer and the baseplate (aluminium/stainless steel) was composed of continuously or intermittently distributed IMCs.

The work developed by Wang and Wang [8] was focused on characterising the flyer velocity in laser impact welding, which was measured with Photon Doppler Velocimetry (PDV) under different experimental conditions. The influence of the laser energy and the flyer thickness on laser energy efficiency was analysed. Other aspects, such as the standoff working window, the rebound behaviour of the flyer, and the effect of the flyer size and confinement layer on the flyer velocity, were also addressed in this research.

A study focused on magnetic pulse welding was conducted by Emadinia et al. [9]. These authors studied the influence of the surface preparation on the interface properties of aluminium/copper tubular joints. They reported that the weld quality was influenced by the surface preparation of the copper target. From the microstructural and mechanical study of welds produced with different surface preparations, it was observed that the highest weld strength values were registered for a threaded copper target, being achieved without the interfacial waviness or IMCs.

Bellmann et al. [10] studied the formation of a cloud of particles during impact welding, its characteristics, and its influence on weld formation. A detailed experimental plan was implemented by the authors, who carried out impact welding tests on different setups. These tests were monitored using a high-speed camera, accompanied by long-term exposures, recordings of the emission spectrum, and an evaluation of the interaction of the cloud of particles with witness pins made of different materials. The interfacial phenomena were discussed based on the pressure conditions, the welding parameters, the base material properties, and the surface state of the welded materials. The authors provided an estimation of the temperature reached in the joining gap. In the sequence of this work, the influence of the process parameters on the main phenomena and mechanisms of bond formation in impact welding was addressed by Niessen et al. [11]. An in-depth experimental research was presented, in which different welding setups were studied. The processes were monitored by several techniques, such as high-speed imaging, PDV, and light emission measurements. The welds were experimentally characterised by ultrasonic inspection, metallographic and microstructural analyses, and mechanical testing. This study made it possible to better understand the influence of different process parameters on the weldability window, and consequently, on the weld characteristics and properties, and to predict different bonding mechanisms.

As reported above, a review work on high-velocity impact welding was also published in the present special issue. This study, which was developed by Wang and Wang [12], presents a global overview of different impact welding processes, such as explosion welding/gas gun welding, magnetic pulse welding, laser impact welding, and vaporising foil actuator welding, and explains the formation of the jet phenomenon during these processes. The macro and micro characteristics of the bonding interface of the impact welds were also addressed at the second stage of this research. Finally, the authors studied the welding parameters, specifically, their selection and their effect on wave formation and weld mechanical properties.

3. Conclusions and Outlook

An encompassing view of the advances that are being made in impact welding was presented in this special issue. This was a consequence of the quality and diversity of the published papers, in which highly supported experimental, numerical simulation, and review researches on a broad spectrum of impact welding processes were presented. However, the research in impact welding is far from being saturated, and strong research is still necessary so that the industrial applicability of this family of processes can grow in the future.

As the Editors of the present special issue, we consider that it was a very successful project, which made us especially happy. However, it would be impossible for this special issue to succeed without the valuable contributions of all the authors. They were the main actors of this project and we are profoundly grateful to all of them. We also want to thank all the reviewers, whose work was crucial for ensuring the quality, scientific relevance, and actuality of the published works. Finally, we acknowledge all support provided by the Metals editorial team, especially by Kinsee Guo, during the development of this special issue.

Conflicts of Interest: The authors declare no conflict of interest.

References

1. Émurlaeva, Y.Y.; Bataev, I.A.; Zhou, Q.; Lazurenko, D.V.; Ivanov, I.V.; Riabinkina, P.A.; Tanaka, S.; Chen, P. Welding Window: Comparison of Deribas' and Wittman's Approaches and SPH Simulation Results. *Metals* **2019**, *9*, 1323. [CrossRef]
2. Mahmood, Y.; Dai, K.; Chen, P.; Zhou, Q.; Bhatti, A.A.; Arab, A. Experimental and Numerical Study on Microstructure and Mechanical Properties of Ti-6Al-4V/Al-1060 Explosive Welding. *Metals* **2019**, *9*, 1189. [CrossRef]
3. Paul, H.; Chulist, R.; Mania, I. Structural Properties of Interfacial Layers in Tantalum to Stainless Steel Clad with Copper Interlayer Produced by Explosive Welding. *Metals* **2020**, *10*, 969. [CrossRef]
4. Carvalho, G.H.S.F.L.; Galvão, I.; Mendes, R.; Leal, R.M.; Loureiro, A. Aluminum-to-Steel Cladding by Explosive Welding. *Metals* **2020**, *10*, 1062. [CrossRef]
5. Inao, D.; Mori, A.; Tanaka, S.; Hokamoto, K. Explosive Welding of Thin Aluminum Plate onto Magnesium Alloy Plate Using a Gelatin Layer as a Pressure-Transmitting Medium. *Metals* **2020**, *10*, 106. [CrossRef]
6. Nishi, M.; Tanaka, S.; Vesenjak, M.; Ren, Z.; Hokamoto, K. Fabrication of Composite Unidirectional Cellular Metals by Using Explosive Compaction. *Metals* **2020**, *10*, 193. [CrossRef]
7. Su, S.; Chen, S.; Mao, Y.; Xiao, J.; Vivek, A.; Daehn, G. Joining Aluminium Alloy 5A06 to Stainless Steel 321 by Vaporizing Foil Actuators Welding with an Interlayer. *Metals* **2019**, *9*, 43. [CrossRef]
8. Wang, H.; Wang, Y. Characteristics of Flyer Velocity in Laser Impact Welding. *Metals* **2019**, *9*, 281. [CrossRef]
9. Emadinia, O.; Ramalho, A.M.; de Oliveira, I.V.; Taber, G.A.; Reis, A. Influence of Surface Preparation on the Interface of Al-Cu Joints Produced by Magnetic Pulse Welding. *Metals* **2020**, *10*, 997. [CrossRef]
10. Bellmann, J.; Lueg-Althoff, J.; Niessen, B.; Böhme, M.; Schumacher, E.; Beyer, E.; Leyens, C.; Tekkaya, A.E.; Groche, P.; Wagner, M.-X.; et al. Particle Ejection by Jetting and Related Effects in Impact Welding Processes. *Metals* **2020**, *10*, 1108. [CrossRef]
11. Niessen, B.; Schumacher, E.; Lueg-Althoff, J.; Bellmann, J.; Böhme, M.; Böhm, S.; Tekkaya, A.E.; Beyer, E.; Leyens, C.; Wagner, M.-X.; et al. Interface Formation during Collision Welding of Aluminum. *Metals* **2020**, *10*, 1202. [CrossRef]
12. Wang, H.; Wang, Y. High-Velocity Impact Welding Process: A Review. *Metals* **2019**, *9*, 144. [CrossRef]

Publisher's Note: MDPI stays neutral with regard to jurisdictional claims in published maps and institutional affiliations.

Review

High-Velocity Impact Welding Process: A Review

Huimin Wang [1] and Yuliang Wang [2,3,*]

[1] National Center for Materials Service Safety, University of Science and Technology Beijing, Beijing 100083, China; wanghuimin@ustb.edu.cn
[2] School of Mechanical Engineering and Automation, Beihang University, Beijing 100191, China
[3] Beijing Advanced Innovation Center for Biomedical Engineering, Beihang University, Beijing 100191, China
* Correspondence: wangyuliang@buaa.edu.cn; Tel.: +86-1861-252-5756

Received: 26 December 2018; Accepted: 27 January 2019; Published: 28 January 2019

Abstract: High-velocity impact welding is a kind of solid-state welding process that is one of the solutions for the joining of dissimilar materials that avoids intermetallics. Five main methods have been developed to date. These are gas gun welding (GGW), explosive welding (EXW), magnetic pulse welding (MPW), vaporizing foil actuator welding (VFAW), and laser impact welding (LIW). They all share a similar welding mechanism, but they also have different energy sources and different applications. This review mainly focuses on research related to the experimental setups of various welding methods, jet phenomenon, welding interface characteristics, and welding parameters. The introduction states the importance of high-velocity impact welding in the joining of dissimilar materials. The review of experimental setups provides the current situation and limitations of various welding processes. Jet phenomenon, welding interface characteristics, and welding parameters are all related to the welding mechanism. The conclusion and future work are summarized.

Keywords: impact welding; impact velocity; impact angle; welding interface

1. Introduction

Welding technique has wide applications in the areas of aerospace, automobiles, shipbuilding, pressure vessels, and bridges. It is one of the important manufacturing methods in industries. Based on the state of materials during the welding process, welding techniques are categorized as solid-state welding and fusion welding [1]. In fusion welding, metallurgical bonding occurs during the solidification of materials. In solid-state welding, metallurgical bonding occurs below the melting point of materials. Therefore, defects such as solidification cracking, distortion, and porosity [2], which appear in fusion welding as a result of the liquid phase, can be avoided in solid-state welding. Solid-state welding has a long history that predates the invention of arc welding. Ancients used hammers to weld gold earlier than 1000 B.C., which today is called forge welding. Since the early 20th century, the development of solid-state welding has been limited due to arc welding being easier than forge welding and having a higher efficiency.

The advantages of joining dissimilar and other specific materials, such as Al 7075 alloy, titanium, and zinc-coated sheet steels [3], brought solid-state welding back onto the stage. For example, the joining of steel and copper provides good electric conductivity and mechanical properties; the joining of steel and aluminum reduces the weight of automobile [4]. Dissimilar materials usually have different thermal conductivity, thermal expansion coefficients, and melting points, which may result in defects in fusion welding [5–7]. Furthermore, some specific alloys are sensitive to heat. Both Al 7075 alloy and titanium are important materials in aerospace and aircraft. However, Al 7075 alloy is susceptible to hot cracking [8,9], and titanium is chemically active at high temperature [10–12]. Zinc-coated steel is an important structural material in automobiles. However, during the fusion-welding process, zinc vaporizes, and porosity forms when zinc vapor is trapped.

Various solid-state welding methods have been developed recently, for example, friction stir welding, explosive welding (EXW), friction welding, magnetic pulse welding (MPW), cold welding, ultrasonic welding, roll welding, pressure gas welding, resistance welding, vaporizing foil actuator welding (VFAW), gas gun welding (GGW), and laser impact welding (LIW). Among those solid-state welding methods—GGW, EXW, MPW, VFAW, and LIW—are five kinds of high-velocity impact-welding methods. They shared the same welding mechanism, but have different welding energy sources indicated by their names: high-speed gas [13], explosive [14,15], capacitor bank energy (MPW and VFAW) [16], and laser [17], respectively.

High-velocity impact welding is characterized by a low welding temperature and fast welding speed. The process is conducted at room temperature. Furthermore, there is no external heat input during the welding process. Figure 1 is a schematic transient state in the welding process. After the transient force is applied on the external surface of the flyer, the flyer moves toward the target at the velocity of several hundred meters per second [18]. When the flyer collides on the target, a jet is generated at the collision point, which contains contaminants, oxide layers, and a thin layer of metals. As a result, the "clean" metals (no contaminants, oxide layers) are exposed to each other. With the transient force, they are brought within atomic distance, where the atomic bond is formed. This process is usually takes several to dozens of microseconds based on different welding processes. For example, in LIW, it usually takes less than one microsecond. In other high-velocity impact-welding processes, it takes longer than that.

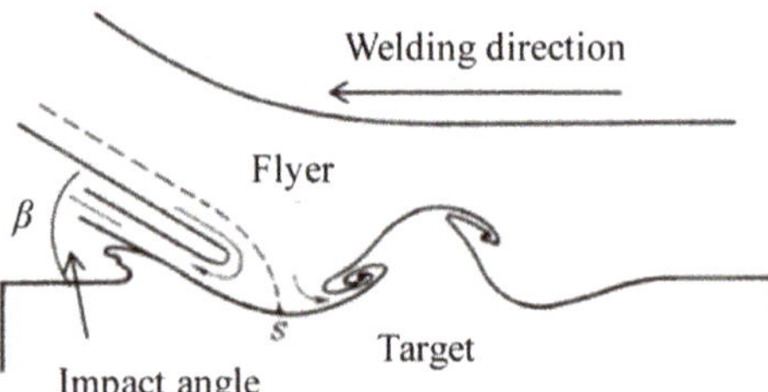

Figure 1. A transient state of bonding interface in high-velocity impact welding, reproduced from [19], with permission from Roral Society, 1934.

In this review, five high-velocity impact-welding methods and the corresponding welding mechanisms are reviewed in Section 2. In Section 3, the macro-characteristics and micro-characteristics of the bonding interface associated with weld quality are summarized. In Section 4, the welding parameters that may affect the welding quality are discussed, such as for example, the material properties, impact velocity, impact angle, and surface preparation. At the end, conclusions and future work are addressed based on the discussion of welding mechanisms, bonding interfaces, and welding parameters.

2. Welding Methods and Jet Phenomenon

2.1. Overview of High-Velocity Impact-Welding Methods

Nowadays, five high-velocity impact-welding methods have been developed, which are GGW, EXW, MPW, VFAW, and LIW. GGW was limited to the lab research on welding parameters and welding mechanisms. EXW experienced decent studies, and has been widely applied in manufacturing all over the world. MPW has limited application in automobiles, and is under development all over the world. VFAW studies have been mainly conducted by researchers from Ohio State University, and are in the transition state from lab research to industrial application. LIW was proposed recently, and is under lab research.

2.1.1. Explosive Welding and Gas Gun Welding

EXW is the first high-velocity impact-welding method. It was observed when a projectile was fired by explosives and collided on another metal surface [20]. However, there was no record of EXW until the First World War. Its rapid development and wide application in industries was in the middle of the 20th century. The first United States (U.S.) patent filed by Philipchuk et al. in 1962 [21]. As its name suggests, explosives detonation is implemented to provide the impact force for the flyer. It was reported that more than 200 material combinations have been successfully welded by EXW [22], for example, Fe/Ti [23–25], Zr/Fe [26], Al/Ti [27], Ti/Ni [28], and Al/Fe [29,30]. For plate welding, two generally used experimental setups are the inclined mode and parallel mode, as shown in Figure 2. Buffer was used to avoid severe damage to the flyer by explosives. The inclined setup came from the idea of hollow charge (which is introduced later); it was used earlier than parallel setup. α was the initial angle in the inclined mode. Since it was not capable of handling large sheet metal, the parallel setup was developed later, and was mainly used to weld large plates with a pre-determined standoff distance. In the parallel setup, the impact angle varies with different standoffs. EXW was also used for the cladding of tubes [15]. From a two-dimensional observation, the experimental setup for tube cladding or ring to tube welding involves the parallel mode for plate welding curved into a circle. In EXW, it is difficult to measure the impact velocity directly, if it is at all possible. Usually, the explosive detonation velocity is measured [31].

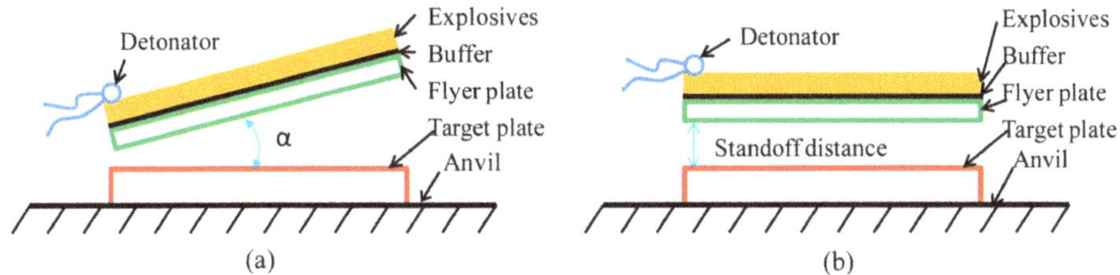

Figure 2. Schematic experimental setups of explosive welding (EXW): (**a**) inclined; (**b**) parallel.

Since EXW is not easy to conduct within the laboratory, GGW is usually used to study the welding parameters and welding mechanism for EXW. In GGW, the flyer is accelerated by high speed and high-pressure gas to collide on the target. Botros and Groves provided an impact force for the flyer with high-pressure gas from burning gunpowder in a 76-mm powder cannon gun [32]. Later on, compressed helium was used since it is clean and easy to control. In GGW, the initial angle is preset either on the flyer or target. The impact velocity can be measured with a high-speed camera [32] and electrical circuit [33]. The resolution of the high-speed camera was not enough to record the welding process accurately, because the welding process is usually done within several microseconds. In the electrical circuit method, there was a corresponding voltage change in this electrical circuit when wires were cut successively by the flyer. The time between the voltage change was then recorded to calculate the impact velocity [34]. The effect of the wires' transient block on impact velocity and impact angle has not yet been investigated.

EXW has been applied in industries all over the world. The Dynamic Materials Corporation is a world-leading explosive metalworking business. EXW has been successfully applied in the energy area. Its main application includes heat exchangers, as well as upstream and downstream products, in the oil and gas industries.

2.1.2. Magnetic Pulse Welding

MPW was used in Russia in the 1960s for the first time to weld an end closure for nuclear fuel rod holders [35,36]. The application of MPW is mainly tubular, for example, joining between tubes or tubes and cylinders. Currently, the most common application of MPW in industry is driveshaft production in the company of Pulsar and Dana [37]. MPW didn't experience a rapid development and wide application as EXW until recently due to the slow development of equipment [37]. Metal combinations of Cu/Cu [38], Cu/brass [39], Cu/Al [40], Cu/Fe [41], Al/Al [18], Al/brass [39], Al/Mg [42–44], Al/Fe [36,45–48], Al/ bulk metallic glass [49], and plastic/Al [50] etc. have been investigated. The general weld configurations of MPW are the lap joint and butt joint. MPW doesn't require a skilled operator. In addition, the welding parameters of MPW are controlled separately to permit fully exploring the effect of welding parameters over a wide range.

In MPW, the flyer is driven by electromagnetic force to collide on the target. The MPW system includes a capacitor bank, coil actuator, the flyer, and the target. The capacitor bank is the energy source of MPW. It is an open electrical circuit of inductance, resistance, and capacitance. The closed circuit was formed by the connection between the capacitor bank and coil actuator, which is made of electrical conductive materials. With the high-speed switch on, the capacitors begin to charge. The primary current passes through the actuator; thus, a changing magnetic field is around the actuator that interacts with any metals within it. Consequently, a secondary current with the opposite direction of the primary current is induced in the flyer. Simultaneously, the flyer is expelled by the electromagnetic force to collide on the target. A typical primary current and the flyer velocity, as measured by a Rogowski coil and photon Doppler velocimetry, are shown in Figure 3. The rise time is an important factor to relate impact velocity. The shorter the rise time, the higher the impact velocity will be.

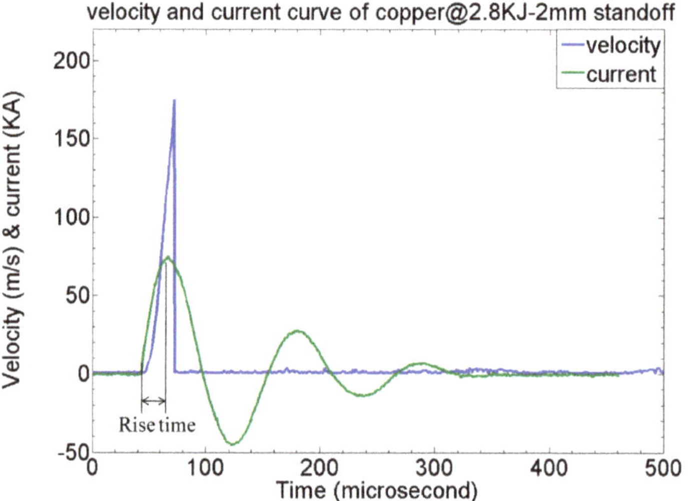

Figure 3. Typical primary current and flyer velocity in magnetic pulse welding (MPW).

Figure 4 shows schematic experimental setups for MPW with different coil actuators. A solenoid coil actuator is used for tube-to-tube welding (Figure 4a). Figure 4b shows the experimental setup with a uniform pressure actuator (UPA) for plate-to-plate welding. The UPA was initially designed for magnetic pulse forming at Ohio State University [51]. The wires between the flyer and the target form an impact angle during the welding process. The three types of bar actuators are I-shaped, U-shaped, and E-shaped (Figure 4c–e); they were developed by Kore et al. in India [52,53], Zhang et al. in the U.S. [54] and Aizawa et al. in Japan [45], separately. The thinner bar is used to push the flyer for U-shaped and E-shaped actuators. If an I-shaped bar and the thinner bars have the same dimension, an E-shaped actuator brings more inductance into the circuit. The rise time becomes longer. Currently, MPW is limited to lap joints. Experimental setups for other types of welding should be developed in the future.

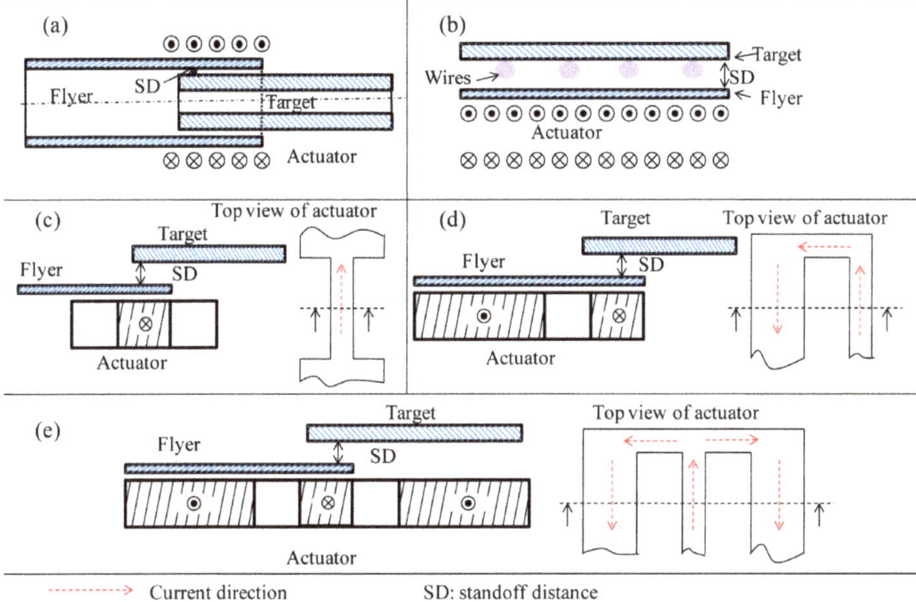

Figure 4. Schematic experimental setups for magnetic pulse welding (MPW) with various actuators: (**a**) solenoid coil actuator for tube-to-tube welding; (**b**) uniform pressure actuator for plate-to-plate welding; (**c–e**) I-shaped actuator, U-shaped actuator, and E-shaped actuator for plate-to-plate welding.

2.1.3. Laser Impact Welding

The first U.S. LIW patent was filed in 2009 by Daehn and Lippold from Ohio State University [55]. The welding system consists of a laser system, confinement layer, ablative layer, flyer, and target [17]. A laser beam ablates the ablative layer into plasma. With the confinement layer, the expansion of the plasma pushes the flyer to collide on the target. Materials such as copper, aluminum, steel, and titanium were welded with the LIW method [18,56,57]. The current experimental setup is similar to that for EXW, as shown in Figure 2. The possible industry application setup was proposed by Wang et al. [17]. In their study, the effect of the confinement layer, ablative layer, and the connection between the flyer and confinement layer were investigated for industrial application. The current weld configuration is limited to spot welding. In the study of Liu et al., they proposed the flyer movement with a Gaussian laser beam for laser spot welding. The central part of the spot was not welded, while welding occurred around the circumferential direction. The same phenomenon was also observed by Wang et al. [58], who also simulated the welding process with the smooth particle hydrodynamics (SPH) method [59] and reproduced the spalling and rebound phenomena by the simulation. For LIW, the spalling and rebound phenomena should be further studied in order to propose ways for the procedures to avoid those two behaviors. The welding area within the laser spot should be further increased. One of the possible future applications of LIW is the welding between Al and Ti in the heart pacemaker.

2.1.4. Vaporizing Foil Actuator Welding

VFAW utilizes the expanding plasma from the vaporization of thin foils to push the flyer to move toward the target. The development of VFAW before 2013 was summarized in [60]. In recent studies, the energy for the vaporization of thin foils is from the capacitor bank. The foil was designed in a dog-bone shape in order to concentrate the transient current from the capacitor bank to vaporize the foil. Vivek pointed out that the aluminum foil provided better mechanical impulse than copper foil through its rapid vaporization. Materials such as steel, aluminum, titanium, copper, magnesium, bulk metallic glass, etc., have been successfully welded with VFAW [61–65]. The original experimental setups in VFAW were similar to those used in EXW. The issue with those experimental setups was that the central part of the impacted region was not welded due to the rebound behavior of the flyer. The welds are very narrow. With this experimental setup, the impact angle was formed due to the standoff between the flyer and the target. Upon collision, the edge of the flyer was kept static, while the central part of the flyer moved toward the target. The impact angle was formed when the flyer collided on the target. However, unlike in EXW, in which the flyer gradually contacted with the target from one end to the other, in VFAW, the collision between the flyer and the target were at the same time. Thus, non-continuous metallurgical bonding resulted on the collision interface. By studying the effect of the impact angle on metallurgical bonding, the target was designed with grooves, which have different angles. The angle range for wave formation along the collision interface was proposed as 8–24° for 3003 Al and 4130 steel [64], and 16–24° for 3003 Al and pure Ti [61]. The recorded maximum impact velocity for VFAW was 900 m/s [66]. As for the numerical simulation, the finite element method with Eulerian formalism is a suitable way to predict the morphology of the collision interface in high-velocity impact welding [67]. The main possible application of the VFAW process is in the automobile industries. Researchers at Ohio State University are investigating its application in the car frame.

2.2. Jet Phenomenon

Jet is the essential reason for welding to occur in the high-velocity impact-welding process. In high-velocity impact welding, a jet is generated by the high-velocity oblique impact of the flyer and target, and then rapidly flies away. Without the impact angle, the jet would be trapped between the collision interfaces. Therefore, the impact angle is one of the significant parameters in order for metallurgical bonding to occur. It is commonly believed that the jet consists of thin metal layers, oxide layers, and other contaminants from the colliding surfaces of the flyer and the target [1]. After the jet flies away, clean virgin metal surfaces are generated. The surfaces are then brought together to a distance within the atomic scale by a transient high-impact pressure. As a result, the atomic bond is produced between the contact area of the flyer and the target. The jet phenomenon has been studied by several researchers [68,69]. In EXW, some welding parameters are applied to predict the jet generation. There is continuous jet formation in EXW due to the collision between the flyer and the target moving from one end to the other end continuously with the detonation of explosives. In other high-velocity impact-welding processes, the continuous generation of the jet along the collision interface is an issue to be resolved by the optimization of an experimental setup design.

The phenomenon in which explosives with a cavity in contact with a steel plate can make a deeper hole in the steel ahead of it than explosives without a cavity has been known for more than 200 years. This phenomenon is called hollow charge. Regarding the study of hollow charge, the phenomenon that explosives with a cavity lined with thin metal layers have enormous penetration ability was discovered around 1948. Birkhoff et al. studied the powerful penetrating and cutting phenomenon of hollow conical liners and hollow wedge-shaped liners (cross-section shown in Figure 5), respectively. As shown in Figure 5b, a high pressure from the detonation of explosives was applied on the outer wall of the wedge, causing it to collapse inward. Figure 5c shows the geometry of the collapse process of the metal liner: α is the initial angle that was set before the experiment; β ($\beta \neq 0$) is the dynamic angle (impact angle) that was formed during the experiment; V_0 is the velocity of the metal liner (impact velocity); V_1 (V_w) is the welding velocity in explosive welding; V_d is the detonation velocity of the explosives. The metal liner became two parts at the collision: slug and jet. The jet was from the inside of the metal liners, and the slug was from the outside of the metal liners. Birkhoff et al. [68] proved that the jet was the one with strong penetration ability. This is the first time that the jet phenomenon was studied in detail. It proved that the oblique collision between materials resulted in jet formation on the collision interfaces.

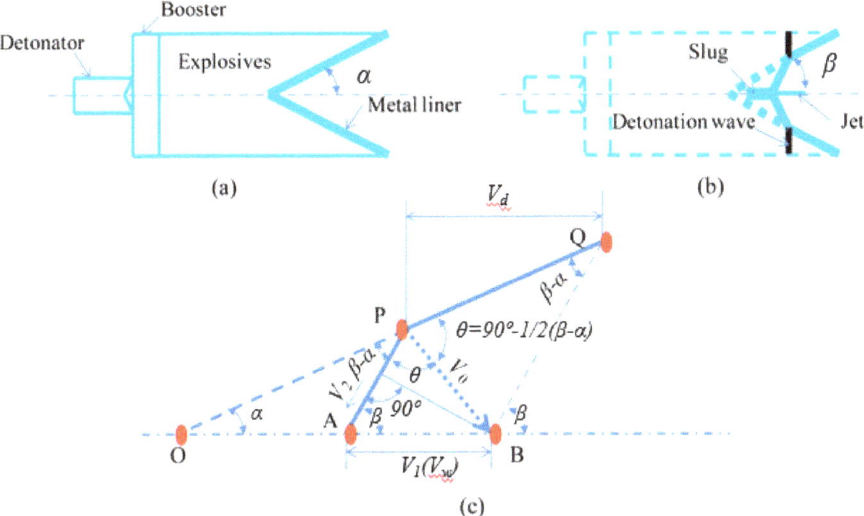

Figure 5. Hollow charge with wedge-shaped metal liner: (**a**) prior to the detonation, and (**b**) during the detonation; and (**c**) the geometry of the collapse process of the metal liner.

The geometry of the collapse process of the wedge-shaped metal line was adopted in EXW to build the math relationship between the welding parameters later on. The mathematic model to predict the velocity and the mass of the jet was built when V_2 is subsonic [19,70]. Cowan and Holtzman studied the limiting conditions for jet formation when V_2 is supersonic, and pointed out that a minimum impact angle should be satisfied for jet formation when V_2 is supersonic [69]. V_2 is the main entrance jet velocity under the assumption that the metal collision is the fluid-like flow.

However, the limitation of their model should be discussed. The pre-condition for their model is the fluid-like flow metal behavior. In their study, the metal was treated as fluid, but they didn't quantitatively clarify what the requirement was for that pre-condition. They assumed that the shear strength of the metals was negligible compared with the impact pressure from the explosives. Furthermore, in the study of Birkhoff et al. [68], a jet was always generated regardless of the other conditions. Actually, the pressure at the collision point should be high enough for the deformation of materials into a jet [22]. In other words, their model couldn't predict whether a jet will happen or not. Although the mechanism of jet generation has not been fully understood, the existence of a jet and the role of a jet in high-velocity impact welding have been generally accepted. As stated earlier, the penetration phenomenon of explosives with a cavity with a metal liner on the armor plate or concrete walls is direct evidence for the existence of a jet. The existence of a jet was also proved using radiographs according to Birkhoff et al. [68]. The existence of a jet was also verified by gas gun welding in which mass of the flyer and target can be measured accurately. Mass comparison before and after the welding has shown that the loss of mass should equal the mass of the jet. In the recent studies regarding the welding mechanism, the research has focused on the effect of the welding interface characteristics, weldability window, and welding parameters on the weld quality. However, jet formation should be studied further due to the continuous metallurgical bonding along the collision interface relying on the continuous generation of the jet, especially for MPW, VFAW, and LIW.

3. Bonding Interface Characteristics

In welding engineering, a welded zone with high strength, good toughness, and enough hardness is desirable. However, in reality, the welded zone is usually weakest welded part, since the microstructure at the welding zone keeps evolving during the welding process. Some defects may appear at the welded zone during the welding process. Therefore, it is necessary to study the bonding interface characteristics in high-velocity impact welding.

3.1. Macrocharacteristics

In high-velocity impact welding, regarding the geometric shape of the bonding interface, straight interface and wave interface (with or without vortices) have been observed [4,71,72], as shown in Figure 6. The wave interface was thought to be not only one of the typical characteristics, but also the sign of metallurgical bonding in EXW for a long time: since the invention of EXW. Recently, Behcet Gulenc [73] also believed that metallurgical bonding should be with a wave interface based on experimental results in which welded samples were characterized by a wave interface, while welded samples with a straight interface were not. However, the observation of a straight interface in some of the explosive-welded materials [71,72,74] demonstrated that a wave interface was not necessary in order for metallurgical bonding to occur. Furthermore, it has been established that the morphology of the bonding interface changed in the order of a straight interface, wave interface without vortices, and wave interface with vortices with the increase of impact pressure/impact velocity [4,14,74–77] in high-velocity impact welding. In addition, the wavelength and amplitude of the wave interface increased with the increase of the impact pressure/impact velocity once the wave interface formed [73,78,79]. The wavelength and amplitude of the wave interface were also related to the impact angle [62]. They also reported that the wavelength and amplitude of the wave interface was affected the flyer thickness. The bonding interface characteristics were also related to the base material properties [31]. In high-velocity impact welding, a collision interface characterized by a wave with vortices is usually accompanied by a melting phenomenon [80]. Previous studies have concluded that a wave is not essential in order for metallurgical bonding to occur in high-velocity impact welding.

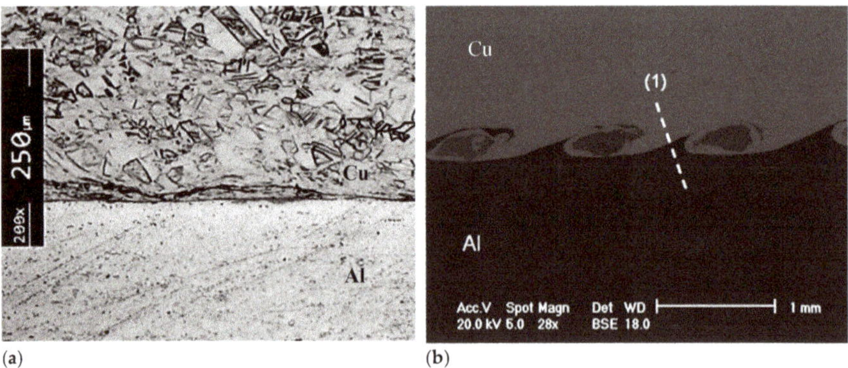

(a)   (b)

Figure 6. Bonding interface in high-velocity impact welding: (**a**) straight interface, reproduced from [81], with permission from Elsevier, 2016; (**b**) wave interface with vortices, reproduced from [18], with permission from Elsevier, 2011.

Wave interface and straight interface were also studied in relation to aspects of the mechanical properties, such as hardness and tensile strength. In one study that compared a straight interface and wave interface (Al/mild steel bonding interface), a higher hardness was detected near the bonding interface for both types of interfaces. By tensile test, it was demonstrated that a straight interface was stronger than a wave interface [71]. For welded samples with straight interfaces, fracture occurred at the Al side, and for welded samples with wave interfaces, fracture occurred at the welded zone. However, for welded samples with wave interfaces, the fracture mode was brittle. The brittle fracture of the welded sample with a wave interface indicated in a tensile test that there might be continuously distributed brittle phases along the bonding interface, which caused the fracture at the welded zone. Furthermore, some other researchers believed that a wave interface was better than a straight interface because of interlocking or mechanical locking [82,83]. In order to figure out which one is better, experiments should be done very carefully to find out the transition from a straight interface to a wave interface. The maximum strength should be within the transition range in which there is no continuous melting along the collision interface and interlocking played a role to increase the weld strength.

3.2. Microstructure at Bonding Interface

In high-velocity impact welding, two other types of interfaces were observed based on the composition variation across the bonding interface: a sharp interface [4,18,74,84] and a transition layer interface [34,36,40,43,49,75,85–87]. These both appeared periodically along the same wave interface with vortices. The transition layer interface was present at vortices where there was melting, and the sharp interface was present at other places along the bonding interface. It is generally accepted that a sharp interface is the result of a solid-state bond. However, it is not clear whether the transition layer interface is a solid-state bond or fusion bond. At the transition layer interface, defects (microvoids) and a crystal structure change (amorphous material) were also observed in some of the welded samples. Therefore, some researchers stated that a transition layer interface was the result of melting and solidification. However, others have argued that welding occurs by a solid-state bond at the transition layer interface where only compositional change happened, such as in cold-pressure welding. Previous studies have indicated that the transition layer interface could be the result of a solid-state bond and a fusion bond, or a mixture of both along the same collision interface, which relies on the welding parameters.

3.2.1. Grain Refinement

Grain refinement was reported along both types of metallurgical bonding interface [74,75,88,89]. In the study of the Al/Al metallurgical bonding interface, both fine and elongated grains were observed by Zhang et al. with the electron backscattered diffraction method (EBSD) [83]. They believed that grain refinement was the result of plastic deformation. They argued if it resulted from melting and solidification, grains with epitaxial orientation should be observed. However, Grignon et al. observed microvoids at the bonding interface of Al/Al beside grain refinement. They stated that grain refinement was the result of melting and solidification [90]. Therefore, both mechanisms for grain refinement are possible in high-velocity impact welding. In this specific experiment, the grain refinement mechanism should be based on the conversion of the kinetic energy of the flyer to heat. However, different mechanisms resulted in different mechanical properties. Severe plastic resulted in grain refinement that brought a high-stress concentration along the welding interface, whereas melting resulted in grain refinement that relieved the stress concentration.

3.2.2. Intermetallics, Microvoids, and Amorphous Materials

The characteristics of the Ti/steel metallurgical bonding interface were studied by Inal et al., Nishida et al., and Kahraman et al. [76,82,87]. Higher hardness was detected at the bonding interface by Inal et al. and Kahraman et al. [76,82]. Regarding the tensile test, welded samples failed out of the welded zone in the study of Inal et al. and Kahraman et al., while in the study of Nishida et al., failure occurred within the welded zone [87]. In the studies of Inal et al. and Kahramen et al., neither melting voids nor intermetallics were found along the bonding interface. However, the existence of an amorphous material was proved by a TEM diffraction pattern in the study of Nishida et al. The authors stated that the amorphous materials were the result of melting and rapid solidification. Therefore, those study indicated that melting in the bonding interface lowered the mechanical properties.

Ben-Artzy et al. and Kore et al. studied the bonding interface of Al/Mg [42,43]. Ben-Artzy et al. believed that bonding between aluminum and magnesium was the result of melting and solidification, while Kore et al. thought the bonding between aluminum and magnesium was pure solid state. Both observed a transition layer. In the study of Ben-Artzy et al., extensive microvoids were observed at the bonding interface within the transition layer. Microvoids were also observed by Marya M. and Marya S. in their study of the interfacial microstructures of magnetic pulse-welded Cu and Al [40]. Kore et al. did not observe defects such as intermetallics and microvoids by SEM and XRD. The above observations indicated that the metallurgical bonding as a result of melting and solidification can be avoided by a lower impact pressure input in high-velocity impact welding.

Liu et al. and Watanabe et al. investigated explosively welded Al/metallic glass and magnetic pulse welded Al/metallic glass. respectively [49,84]. In the study of Liu et al., hardness increased at both the Al and metallic glass sides close to the bonding interface, and a TEM bright field image showed a sharp transition at the bonding interface. The TEM diffraction pattern verified that the metallic glass kept its amorphous structure. However, the authors believed that melting happened and the cooling rate was high enough to allow the metallic glass to retain the amorphous structure based on the simulation result. In the study of Watanabe et al., hardness increased only at the Al side, and close to the bonding interface, the hardness of the metallic glass decreased, which was thought to be caused by the crystallization of metallic glass. An SEM backscattered electron image showed a transition layer at the bonding interface. However, TEM did not detect any crystal structure at the metallic glass side. From the above studies, it is hard to tell whether melting and solidification happened during the welding process. Two phenomena in their study could not be explained. Liu et al. thought that melting and solidification happened, but the hardness of the metallic glass increased at the bonding interface. In the study of Watanabe et al., a lower hardness of metallic glass was observed at the bonding interface, but no crystal structure was found at the metallic glass side.

In their investigation of magnetic pulse welded similar and dissimilar materials, Stern and Aizenshtein [86] believed that the flyer and target were bonded by melting and solidification. For combinations of dissimilar materials, compositions that were similar to some intermetallics were detected by EDS at the transition layer interface, such as in the study of Marya M. and Marya S. [40]. However, the determination of intermetallics needs further investigation, such as what Lee et al. did in their study of magnetic pulse welded low-carbon steel to aluminum [36]. In their study, varied composition was also detected by EDS at the bonding interface. However, the new composition, which is different from the flyer and the target, is not consistent with any composition of intermetallics in the Al–Fe phase diagram. So, they did further research using TEM. From the TEM image and diffraction patterns, fine aluminum grains, as well as fine Al–Fe intermetallics grains, were observed within the transition layer. The authors atttributed the higher hardness at the bonding interface to fine grains and possible intermetallics. Continuous intermetallics should be avoided by adjusting the welding parameters [91,92].

3.3. Summary

The straight interface and wave interface (with or without vortices) are two types of bonding interfaces in high-velocity impact welding based on the geometric shape of the bonding interface. Grain refinement and higher hardness were observed for both of them. There are two different explanations regarding the mechanism of grain refinement. One is melting and solidification, and the other is severe plastic deformation at the bonding interface. Both of them are possible mechanisms for grain refinement. The mechanism could depend on the welding parameters. However, melting along the bonding interface lowers the mechanical properties.

At the transition layer interface, compositional change, microvoids, and amorphous materials were observed, although they may not appear at the same time. The mechanism of compositional change with microvoids and amorphous materials at the bonding interface was the result of melting and solidification. The mechanism of compositional change at the bonding interface without defects may be due to solid-state diffusion. The existence of a sharp interface demonstrates that a bonding interface without defects could be produced in high-velocity impact welding. The formation of intermetallics caused compositional change, but this should be confirmed through detecting its crystal structure.

4. Welding Parameters

4.1. Welding Parameters Selection

A welded zone with acceptable weld quality should be as strong as the weaker part of the two welded parts, as determined by a mechanical test such as the tensile test, bending test, or hardness test. Additionally, damage to parent materials, such as spalling, should be avoided. To obtain acceptable weld quality, proper welding parameters should be selected.

It is generally accepted that a jet is essential in order for welding to occur in high-velocity impact welding. It has been shown from the "jet phenomenon" section that jet formation was related to the impact angle, impact velocity, and impact pressure. A large impact pressure, such as spalling, caused damage to the flyer and target [83]. The impact pressure at the collision point should have a maximum magnitude. The excessive kinetic energy of the flyer results in melting and continuous intermetallics at the bonding interface; thus, there is an upper limit for kinetic energy. The following parameters can be used to describe high-velocity impact welding: kinetic energy (E_k), impact pressure (P), impact velocity (V_p), and impact angle (β). Certainly, the properties of materials also affect the weld quality [93], density (ρ), and thickness of the flyer plate (t)). For convenience, the kinetic energy, impact pressure, impact velocity, impact angle, and materials properties are regarded as basic parameters, and others—such as for example the standoff distance (L), explosive properties [15,81] (explosive ratio, thickness, and detonation velocity (V_d)), and initial angle (α) in explosive welding, capacitor bank energy in magnetic pulse welding, and laser properties in laser impact welding—are process parameters. Basic parameters are determined by the process parameters. In this literature review, the discussion of welding parameters was confined to the basic parameters and standoff distance.

A map called the weldability window was proposed by Wittman et al. [1] in EXW, as shown in Figure 7. V_w is the welding velocity (see Figure 5c). The dynamic angle (impact angle) of obliquity is the angle between the flyer and the target at the collision point. On this map, the upper limit and lower limit of the welding velocity were included. On the right side of the map where the supersonic region is located, the critical angle for jet formation was defined. The transition velocity from a straight interface to a wave interface was also included. The experimental results demonstrated that there was a transition zone for the transition from a straight interface to a wave interface within the welding range [1]. Regarding this map, the following issues should be pointed out. Firstly, this map varies depending on materials' properties [94]. Secondly, welding velocity is not appropriate to be used in this map, except for EXW, as it is not one of the basic parameters that directly determines the weld quality. Thirdly, it is not appropriate for the transition velocity to be regarded as a constant. However, the weldability window guided people in the right direction in the research of welding parameters in high-velocity impact welding. Once it is built, it provides the manual for the application of high-velocity impact welding in industry.

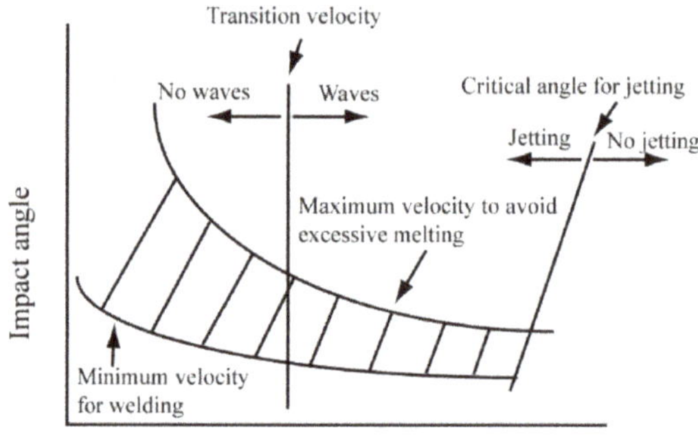

Figure 7. Weldability window. Velocity of welding (V_w) is represented by V_1 in Figure 5c. The dynamic angle of obliquity is the impact angle. When the combination of parameters is within the welding range, welding will take place.

4.2. Effect of Welding Parameters on Weld Quality

The welding parameters that have been investigated in different papers vary; for example, they have included the impact angle, detonation velocity, impact velocity, standoff distance, and discharge energy.

4.2.1. Effect of Welding Parameters on Wave Formation

In EXW, wave formation was believed to be a sign of strong metallurgical bonding between the flyer and the target. Therefore, early work on EXW focused on establishing critical parameters for wave formation. Deribas et al. studied the effect of detonation velocity, standoff distance, and initial angle on wave formation [95]. They pointed out that with a high detonation velocity, there was a critical angle below which there was no wave. With a low detonation velocity, there was no critical angle, but the wave dimension (amplitude and wavelength) would increase with the initial angle and standoff distance. Furthermore, for each fixed initial angle, there was a critical impact velocity for wave formation, and the impact velocity increased with the initial angle. However, they couldn't establish the direct relationship between the wave dimension and the initial angle, impact velocity, and standoff distance, since in each serial experiment, there were more than two parameters varying at the same time. Their conclusion was consistent with the jet formation regimes in which there was a critical initial angle that was required for supersonic flow, whereas there was no requirement for subsonic flow [19].

Acarer et al. [14] studied the effect of explosive loading and standoff distance on the wave dimension. The experimental results showed that the wave dimension increased with standoff distance and explosive loading. Durgutlu et al. [96] also found a similar effect of standoff on wave dimension; they also observed that the bonding interface was straight with a lower standoff distance, while with a higher standoff, the bonding interface had a wavy feature. However, they didn't build the quantitative relationship between the welding parameters and the wave dimension.

4.2.2. Effect of Welding Parameters on Mechanical Properties

The weld quality is usually evaluated with a tensile test and a peeling test; the weld quality is compared by measuring the fracture strength and observing the fracture location. When the fracture location is at the weld interface, the weld quality is ordered by the fracture strength. In some other cases, the fracture location is on the base metals, which indicated that the weld quality is better than the base metal strength. Several researchers investigated the effect of the parameters on the weld quality.

Kore et al. studied the welding parameters in MPW. Al with a thickness of one mm was welded with a discharge energy of 3.6 kJ. The bonding strength increased with increasing discharge energy.

They also found that there was an optimum standoff distance in magnetic pulse welding, which was also concluded by Hokari et al. [39]. This observation was not applicable to EXW. In MPW, the highest impact velocity occurs at the peak primary current. Therefore, before the impact velocity gets to its highest value, the flyer travels a specific distance, which is determined by the impact velocity and the rise time of the primary current. So, a specific standoff distance is needed for a specific MPW process.

In a study of the bonding interface of explosively welded steel to steel, microhardness increased at the bonding interface, but decreased far from the bonding interface [14]. However, Gulenc [73] observed that microhardness increased both at the bonding interface and far from the bonding interface. Microhardness should increase both at the bonding interface and the outer surface of the welded samples, which is caused by work hardening. The outer surface of the flyer is work-hardened by the explosives, and the outer surface of the target is work-hardened by the interaction with anvil. The decrease of hardness far from the bonding interface may be caused by softening.

Chizari and Barrett studied impact velocity with an aluminum flyer that was two-mm thick [33]. They found that there was no bonding when the impact velocity was lower than 250 m/s, and that a wave interface began to appear when the impact velocity was 340 m/s. Besides the impact velocity, they also studied the effect of two flyers on the weld quality. In their experiment, two pieces of parallel flyers with a separate distance were used. Their experimental results showed that the weld quality between the first flyer and the target was better than that between the two flyers. They attributed the poor bond between the two flyers to the increased roughness of the back surface of the first flyer during the colliding process. However, it should be noted that during the colliding process, the roughness of the target also increased. So, roughness was not causing the poor bond between the two flyers.

To summarize the research about welding parameters, welding parameters vary in different research papers, and some of them are dependent on specific techniques, and even specific experimental setups. For example, the flyer will be applied with a different impact force despite using the same capacitor bank energy and different actuators. Therefore, the basic parameters should be studied extensively. In former studies, the impact angle and impact velocity were chosen as the factors that can determine the weld quality. However, the jet formation is much more reliant on the impact pressure and impact angle. Kinetic energy is also very critical in determining the weld quality in order to avoid melting and solidification. Therefore, welding parameters need to be carefully investigated in order to figure out which ones could be selected to build the weldability window.

In explosive welding, the fatigue life of welded joints was also studied. Karolczuk et al. investigated the fatigue phenomena of explosively welded Fe/Ti [24]. They studied two types of bonding interface: flat interface and wavy interface. They concluded that a flat interface has a higher fatigue life than a wavy interface. Through the study of Szachogluchowicz et al. [27], it was found that the lower fatigue life of the wavy interface may be due to the stress concentration. In their study, heat treatment improved the fatigue life by stress relaxation and the elimination of microvoids. Prażmowski et al. [26] studied the fatigue life of explosively welded Zr/Fe, and concluded that the remelted layer in the bonding zone increased the fatigue life. This phenomena was also due to the stress relaxation. The corrosion behavior of explosively welded Al/Fe was studied by Kaya [29] with neutral salt spray (NSS) tests. Their results showed that the corrosion occurred on the steel side, while no corrosion behavior was observed on the Al side. Therefore, a cladded Al layer played a significant role in the protection of steel.

5. Conclusion and Future Work

High-velocity impact welding is suitable to join dissimilar materials and other specific materials, and has potential applications in industries. The conclusions are listed below.

1. In high-velocity impact welding, the weld configuration was limited by the experimental setup. The current general welding configurations are lap joint (cladding), spot weld, and tubular weld. More welding configurations should be developed for wider applications of high-velocity impact welding.

2. EXW and MPW have been applied in industries for lap weld/cladding. The current application is limited by weld configurations. The application of VFAW and LIW is under investigation.

3. In EXW, jet formation theory was built based on the assumption of a fluid-like flow when the materials are under high-impact pressure. However, the minimum impact pressure that is required for the fluid-like flow at the collision point was not determined. For the further optimization of this theory, this point should be stated.

4. Two types of bonding interface were observed in high-velocity impact welding based on the geometric shape of the bonding interface: a straight interface or a wave interface (with or without vortices). Generally, a straight interface has lower mechanical properties than a wave interface without vortices due to the interlocking or greater metallurgical bonding area between the flyer and the target. A wave interface with vortices is accompanied by a melting phenomenon, which resulted in the low mechanical properties. Therefore, for each material combination, there should be a transition range from a straight interface to a wave interface with vortices, which might be the strongest metallurgical bonding. The welding parameters should be optimized for each material combination for this transition range.

5. Regarding the compositional change across the collision interface, there are sharp interfaces and transition layer interfaces. The sharp interface is the result of solid-state welding, while the transition layer interface may be the result of solid-state bonding or melting. The melting also resulted in intermetallics, microvoids, and amorphous materials.

6. For the application of LIW in manufacturing, welding parameters need to be studied extensively in order to build the relationship between the welding parameters and weld quality systematically and avoid the spallation of the target and rebound behavior of the flyer.

7. Furthermore, the current weldability window should be re-evaluated for varied welding processes and material combinations. Whether bonding happens or not is not only related to impact angle and impact velocity, but also to the hardness, thickness, and density of the flyer.

Author Contributions: Conceptualization, H.W. and Y.W.; Literature search, H.W.; Literature summarization, H.W. and Y.W.; writing—original draft preparation, H.W.; writing—review and editing, Y.W.

Funding: This research was funded by the Fundamental Research Funds for the Central Universities, grant number 06500107 and the APC was funded by 06500107.

Acknowledgments: Thanks for lab members from The Ohio State University for their valuable discussions and instructions.

References

1. Crossland, B. *Explosive Welding of Metals and Its Application*; Oxford University Press: New York, NY, USA, 1982; pp. 10–38, 84–106.

2. Debroy, T.; David, S.A. Physical processes in fusion-welding. *Rev. Mod. Phys.* **1995**, *67*, 85–112. [CrossRef]

3. Li, X.; Lawson, S.; Zhou, Y.; Goodwin, F. Novel technique for laser lap welding of zinc coated sheet steels. *J. Laser Appl.* **2007**, *19*, 259–264. [CrossRef]

4. Durgutlu, A.; Gulenc, B.; Findik, F. Examination of copper/stainless steel joints formed by explosive welding. *Mater. Des.* **2005**, *26*, 497–507. [CrossRef]

5. Taban, E.; Gould, J.E.; Lippold, J.C. Characterization of 6061-T6 aluminum alloy to AISI 1018 steel interfaces during joining and thermo-mechanical conditioning. *Mater. Sci. Eng. A Struct. Mater. Prop. Microstruct. Process.* **2009**, *527*, 1704–1708. [CrossRef]

6. Sun, Z.; Ion, J.C. Laser-welding of dissimilar metal combinations. *J. Mater. Sci.* **1995**, *30*, 4205–4214. [CrossRef]

7. Torkamany, M.J.; Tahamtan, S.; Sabbaghzadeh, J. Dissimilar welding of carbon steel to 5754 aluminum alloy by Nd:YAG pulsed laser. *Mater. Des.* **2009**, *31*, 458–465. [CrossRef]

8. Yan, Y.B.; Zhang, Z.W.; Shen, W.; Wang, J.H.; Zhang, L.K.; Chin, B.A. Microstructure and properties of magnesium AZ31B-aluminum 7075 explosively welded composite plate. *Mater. Sci. Eng. A Struct. Mater. Prop. Microstruct. Process.* **2009**, *527*, 2241–2245. [CrossRef]

9. Fuller, C.B.; Mahoney, M.W.; Calabrese, M.; Micona, L. Evolution of microstructure and mechanical properties in naturally aged 7050 and 7075 Al friction stir welds. *Mater. Sci. Eng. A Struct. Mater. Prop. Microstruct. Process.* **2009**, *527*, 2233–2240. [CrossRef]

10. Atasoy, E.; Kahraman, N. Diffusion bonding of commercially pure titanium to low carbon steel using a silver interlayer. *Mater. Charact.* **2008**, *59*, 1481–1490. [CrossRef]

11. Liu, J.; Watanabe, I.; Yoshida, K.; Atsuta, M. Joint strength of laser-welded titanium. *Dent. Mater.* **2002**, *18*, 143–148. [CrossRef]

12. Qi, Y.L.; Deng, J.; Hong, Q.; Zeng, L.Y. Electron beam welding, laser beam welding and gas tungsten arc welding of titanium sheet. *Mater. Sci. Eng. A Struct. Mater. Prop. Microstruct. Process.* **2000**, *280*, 177–181.

13. Strand, O.T.; Goosman, D.R.; Martinez, C.; Whitworth, T.L.; Kuhlow, W.W. Compact system for high-speed velocimetry using heterodyne techniques. *Rev. Sci. Instrum.* **2006**, *77*, 083108. [CrossRef]

14. Acarer, M.; Gulenc, B.; Findik, F. Investigation of explosive welding parameters and their effects on microhardness and shear strength. *Mater. Des.* **2003**, *24*, 659–664. [CrossRef]

15. Mendes, R.; Ribeiro, J.B.; Loureiro, A. Effect of explosive characteristics on the explosive welding of stainless steel to carbon steel in cylindrical configuration. *Mater. Des.* **2013**, *51*, 182–192. [CrossRef]

16. Vivek, A.; Hansen, S.R.; Liu, B.C.; Daehn, G.S. Vaporizing foil actuator: A tool for collision welding. *J. Mater. Process. Technol.* **2013**, *213*, 2304–2311. [CrossRef]

17. Wang, H.M.; Taber, G.; Liu, D.J.; Hansen, S.; Chowdhury, E.; Terry, S.; Lippold, J.C.; Daehn, G.S. Laser impact welding: Design of apparatus and parametric optimization. *J. Manuf. Process.* **2015**, *19*, 118–124. [CrossRef]

18. Zhang, Y.; Babu, S.S.; Prothe, C.; Blakely, M.; Kwasegroch, J.; LaHa, M.; Daehn, G.S. Application of high velocity impact welding at varied different length scales. *J. Mater. Process. Technol.* **2011**, *211*, 944–952. [CrossRef]

19. Bahrani, A.S.; Black, T.J.; Crosslan, B. Mechanics of wave formation in explosive welding. *Proc. R. Soc. Lond. Ser. A Math. Phys. Sci.* **1967**, *296*, 123–136.

20. Young, G. Explosive welding, technical growth and commercial history. In Proceedings of the Stainless Steel World 2004, Houston, TX, USA, 20–22 October 2004; pp. 1–2.

21. Philipchuk, V.; Scituate, N.; Roy, L.F. Explosive Welding. U.S. patent 3,024,526, 13 March 1962.

22. Carpenter, S.H.; Wittman, R.H. Explosion welding. *Annu. Rev. Mater. Sci.* **1975**, *5*, 177–199. [CrossRef]

23. Rozumek, D.; Bański, R. Crack growth rate under cyclic bending in the explosively welded steel/titanium bimetals. *Mater. Des.* **2012**, *38*, 139–146. [CrossRef]

24. Karolczuk, A.; Kowalski, M.; Bański, R.; Żok, F. Fatigue phenomena in explosively welded steel–titanium clad components subjected to push–pull loading. *Int. J. Fatigue* **2013**, *48*, 101–108. [CrossRef]

25. Xie, M.-X.; Shang, X.-T.; Zhang, L.-J.; Bai, Q.-L.; Xu, T.-T. Interface Characteristic of Explosive-Welded and Hot-Rolled TA1/X65 Bimetallic Plate. *Metals* **2018**, *8*, 159. [CrossRef]

26. Prażmowski, M.; Rozumek, D.; Paul, H. Static and fatigue tests of bimetal Zr-steel made by explosive welding. *Eng. Fail. Anal.* **2017**, *75*, 71–81. [CrossRef]

27. Szachogluchowicz, I.; Sniezek, L.; Hutsaylyuk, V. Low cycle fatigue properties of AA2519–Ti6Al4V laminate bonded by explosion welding. *Eng. Fail. Anal.* **2016**, *69*, 77–87. [CrossRef]

28. Topolski, K.; Szulc, Z.; Garbacz, H. Microstructure and Properties of the Ti6Al4V/Inconel 625 Bimetal Obtained by Explosive Joining. *J. Mater. Eng. Perform.* **2016**, *25*, 3231–3237. [CrossRef]

29. Kaya, Y. Microstructural, Mechanical and Corrosion Investigations of Ship Steel-Aluminum Bimetal Composites Produced by Explosive Welding. *Metals* **2018**, *8*, 544. [CrossRef]

30. Findik, F. Recent developments in explosive welding. *Mater. Des.* **2011**, *32*, 1081–1093. [CrossRef]

31. Carvalho, G.H.S.F.L.; Galvão, I.; Mendes, R.; Leal, R.M.; Loureiro, A. Influence of base material properties on copper and aluminium–copper explosive welds. *Sci. Technol. Weld. Join.* **2018**, *23*, 501–507. [CrossRef]

32. Botros, K.K.; Groves, T.K. Fundamental impact-welding parameters—An experimental investigation using a 76-Mm powder cannon. *J. Appl. Phys.* **1980**, *51*, 3706–3714. [CrossRef]

33. Chizari, M.; Barrett, L.M. Single and double plate impact welding: Experimental and numerical simulation. *Comput. Mater. Sci.* **2009**, *46*, 828–833. [CrossRef]

34. Mousavi, A.A.A.; Al-Hassani, S.T.S. Numerical and experimental studies of the mechanism of the wavy interface formations in explosive/impact welding. *J. Mech. Phys. Solids* **2005**, *53*, 2501–2528. [CrossRef]
35. Katzenstein, J. System and Method for Impact Welding by Magnetic Propulsion. U.S. patent 4,504,714, 12 March 1985.
36. Lee, K.J.; Kumai, S.; Arai, T.; Aizawa, T. Interfacial microstructure and strength of steel/aluminum alloy lap joint fabricated by magnetic pressure seam welding. *Mater. Sci. Eng. A Struct. Mater. Prop. Microstruct. Process.* **2007**, *471*, 95–101. [CrossRef]
37. Kochan, A. Magnetic pulse welding shows potential for automotive applications. *Assem. Autom.* **2000**, *20*, 129–131. [CrossRef]
38. Kore, S.; Date, P.; Kulkarni, S.; Kumar, S.; Rani, D.; Kulkarni, M.; Desai, S.; Rajawat, R.; Nagesh, K.; Chakravarty, D. Electromagnetic impact welding of copper-to-copper sheets. *Int. J. Mater. Form.* **2009**, *3*, 117–121. [CrossRef]
39. Hokari, H.; Sato, T.; Kawauchi, K.; Muto, A. Magnetic impulse welding of aluminium tube and copper tube with various core materials. *Weld. Int.* **1998**, *12*, 619–626. [CrossRef]
40. Marya, M.; Marya, S. Interfacial microstructures and temperatures in aluminium-copper electromagnetic pulse welds. *Sci. Technol. Weld. Join.* **2004**, *9*, 541–547. [CrossRef]
41. Patra, S.; Arora, K.S.; Shome, M.; Bysakh, S. Interface characteristics and performance of magnetic pulse welded copper-Steel tubes. *J. Mater. Process. Technol.* **2017**, *245*, 278–286. [CrossRef]
42. Kore, S.D.; Imbert, J.; Worswick, M.J.; Zhou, Y. Electromagnetic impact welding of Mg to Al sheets. *Sci. Technol. Weld. Join.* **2009**, *14*, 549–553. [CrossRef]
43. Ben-Artzy, A.; Stern, A.; Frage, N.; Shribman, V. Interface phenomena in aluminium-magnesium magnetic pulse welding. *Sci. Technol. Weld. Join.* **2008**, *13*, 402–408. [CrossRef]
44. Jiang, X.; Chen, S. Texture evolution and plastic deformation mechanism in magnetic pulse welding of dissimilar Al and Mg alloys. *Weld. World* **2018**, *62*, 1159–1171. [CrossRef]
45. Aizawa, T.; Kashani, M.; Okagawa, K. Application of magnetic pulse welding for aluminum alloys and SPCC steel sheet joints. *Weld. J.* **2007**, *86*, 119S–124S.
46. Geng, H.; Mao, J.; Zhang, X.; Li, G.; Cui, J. Strain rate sensitivity of Al-Fe magnetic pulse welds. *J. Mater. Process. Technol.* **2018**, *262*, 1–10. [CrossRef]
47. Deng, F.; Cao, Q.; Han, X.; Li, L. Electromagnetic pulse spot welding of aluminum to stainless steel sheets with a field shaper. *Int. J. Adv. Manuf. Technol.* **2018**, *98*, 1903–1911. [CrossRef]
48. Cui, J.; Sun, T.; Geng, H.; Yuan, W.; Li, G.; Zhang, X. Effect of surface treatment on the mechanical properties and microstructures of Al-Fe single-lap joint by magnetic pulse welding. *Int. J. Adv. Manuf. Technol.* **2018**, *98*, 1081–1092. [CrossRef]
49. Watanabe, M.; Kumai, S.; Hagimoto, G.; Zhang, Q.; Nakayama, K. Interfacial microstructure of aluminum/metallic glass lap joints fabricated by magnetic pulse welding. *Mater. Trans.* **2009**, *50*, 1279–1285. [CrossRef]
50. Cui, J.; Li, Y.; Liu, Q.; Zhang, X.; Xu, Z.; Li, G. Joining of tubular carbon fiber-reinforced plastic/aluminum by magnetic pulse welding. *J. Mater. Process. Technol.* **2019**, *264*, 273–282. [CrossRef]
51. Kamal, M.; Daehn, G.S. A uniform pressure electromagnetic actuator for forming flat sheets. *J. Manuf. Sci. Eng. Trans. ASME* **2007**, *129*, 369–379. [CrossRef]
52. Kore, S.D.; Date, P.P.; Kulkarni, S.V. Electromagnetic impact welding of aluminum to stainless steel sheets. *J. Mater. Process. Technol.* **2008**, *208*, 486–493. [CrossRef]
53. Kore, S.D.; Date, P.P.; Kulkarni, S.V.; Kumar, S.; Kulkarni, M.R.; Desai, S.V.; Rajawat, R.K.; Nagesh, K.V.; Chakravarty, D.P. Electromagnetic impact welding of Al-to-Al-Li sheets. *J. Manuf. Sci. Eng. Trans. ASME* **2009**, *131*, 1–4. [CrossRef]
54. Zhang, Y.; Babu, S.; Daehn, G.S. Impact welding in a variety of geometric configurations. In Proceedings of the 4th International Conference on High Speed Forming, Columbus, OH, USA, 9–10 March 2010; pp. 97–107.
55. Daehn, G.S.; Lippold, J.C. Low Temperature Spot Impact Welding Driven without Contact. U.S. patent 8084710B2, 27 December 2011.
56. Wang, H.; Vivek, A.; Wang, Y.; Taber, G.; Daehn, G.S. Laser impact welding application in joining aluminum to titanium. *J. Laser Appl.* **2016**, *28*, 032002. [CrossRef]
57. Liu, H.; Gao, S.; Yan, Z.; Li, L.; Li, C.; Sun, X.; Sha, C.; Shen, Z.; Ma, Y.; Wang, X. Investigation on a Novel Laser Impact Spot Welding. *Metals* **2016**, *6*, 179. [CrossRef]

58. Wang, H.; Wang, Y. Laser-driven flyer application in thin film dissimilar materials welding and spalling. *Opt. Lasers Eng.* **2017**, *97*, 1–8. [CrossRef]
59. Wang, X.; Shao, M.; Gao, S.; Gau, J.-T.; Tang, H.; Jin, H.; Liu, H. Numerical simulation of laser impact spot welding. *J. Manuf. Process.* **2018**, *35*, 396–406. [CrossRef]
60. Vivek, A. Rapid Vaporization of Thin Conductors Used for Impulse Metalworking. Ph.D. Thesis, The Ohio State University, Columbus, OH, USA, 2012.
61. Chen, S.; Huo, X.; Guo, C.; Wei, X.; Huang, J.; Yang, J.; Lin, S. Interfacial characteristics of Ti/Al joint by vaporizing foil actuator welding. *J. Mater. Process. Technol.* **2019**, *263*, 73–81. [CrossRef]
62. Lee, T.; Zhang, S.; Vivek, A.; Kinsey, B.; Daehn, G. Flyer thickness effect in the impact welding of aluminum to steel. *J. Manuf. Sci. Eng.* **2018**, *140*, 121002. [CrossRef]
63. Liu, B.; Vivek, A.; Daehn, G.S. Joining sheet aluminum AA6061-T4 to cast magnesium AM60B by vaporizing foil actuator welding: Input energy, interface, and strength. *J. Manuf. Process.* **2017**, *30*, 75–82. [CrossRef]
64. Chen, S.; Daehn, G.S.; Vivek, A.; Liu, B.; Hansen, S.R.; Huang, J.; Lin, S. Interfacial microstructures and mechanical property of vaporizing foil actuator welding of aluminum alloy to steel. *Mater. Sci. Eng. A* **2016**, *659*, 12–21. [CrossRef]
65. Vivek, A.; Presley, M.; Flores, K.M.; Hutchinson, N.H.; Daehn, G.S. Solid state impact welding of BMG and copper by vaporizing foil actuator welding. *Mater. Sci. Eng. A* **2015**, *634*, 14–19. [CrossRef]
66. Hahn, M.; Weddeling, C.; Taber, G.; Vivek, A.; Daehn, G.S.; Tekkaya, A.E. Vaporizing foil actuator welding as a competing technology to magnetic pulse welding. *J. Mater. Process. Technol.* **2016**, *230*, 8–20. [CrossRef]
67. Gupta, V.; Lee, T.; Vivek, A.; Choi, K.S.; Mao, Y.; Sun, X.; Daehn, G. A robust process-structure model for predicting the joint interface structure in impact welding. *J. Mater. Process. Technol.* **2019**, *264*, 107–118. [CrossRef]
68. Birkhoff, G.; Macdougall, P.D.; Pugh, M.E.; Taylor, G. Explosives with lined cavities. *J. Appl. Phys.* **1948**, *19*, 563–582. [CrossRef]
69. Cowan, R.G.; Holtzman, H.A. Flow configurations in colliding plates-explosive bonding. *J. Appl. Phys.* **1963**, *34*, 928–939. [CrossRef]
70. Bahrani, A.S.; Crossland, B. Explosive welding and cladding: An introductory survey and preliminary results. *Proc. Inst. Mech. Eng.* **1964**, *179*, 264–304. [CrossRef]
71. Szecket, A.; Inal, O.T.; Vigueras, D.J.; Rocco, J. A wavy versus straight interface in the explosive welding of aluminum to steel. *J. Vac. Sci. Technol. A* **1985**, *3*, 2588–2594. [CrossRef]
72. Jaramillo, V.D.; Szecket, A.; Inal, O.T. On the transition from a waveless to a wave interface in explosive welding. *Mater. Sci. Eng.* **1987**, *91*, 217–222. [CrossRef]
73. Gulenc, B. Investigation of interface properties and weldability of aluminum and copper plates by explosive welding method. *Mater. Des.* **2008**, *29*, 275–278. [CrossRef]
74. Acarer, M.; Demir, B. An investigation of mechanical and metallurgical properties of explosive welded aluminum-dual phase steel. *Mater. Lett.* **2008**, *62*, 4158–4160. [CrossRef]
75. Kaçar, R.; Acarer, M. Microstructure-property relationship in explosively welded duplex stainless steel-steel. *Mater. Sci. Eng. A* **2003**, *363*, 290–296. [CrossRef]
76. Kahraman, N.; Gulenc, B.; Findik, F. Joining of titanium/stainless steel by explosive welding and effect on interface. *J. Mater. Process. Technol.* **2005**, *169*, 127–133. [CrossRef]
77. Mousavi, S.; Sartangi, P.F. Experimental investigation of explosive welding of cp-titanium/AISI 304 stainless steel. *Mater. Des.* **2009**, *30*, 459–468. [CrossRef]
78. Kahraman, N.; Gulenc, B. Microstructural and mechanical properties of Cu-Ti plates bonded through explosive welding process. *J. Mater. Process. Technol.* **2005**, *169*, 67–71. [CrossRef]
79. Mousavi, S.; Al-Hassani, S.T.S.; Atkins, A.G. Bond strength of explosively welded specimens. *Mater. Des.* **2008**, *29*, 1334–1352. [CrossRef]
80. Nassiri, A.; Chini, G.; Vivek, A.; Daehn, G.; Kinsey, B. Arbitrary Lagrangian–Eulerian finite element simulation and experimental investigation of wavy interfacial morphology during high velocity impact welding. *Mater. Des.* **2015**, *88*, 345–358. [CrossRef]
81. Loureiro, A.; Mendes, R.; Ribeiro, J.B.; Leal, R.M.; Galvão, I. Effect of explosive mixture on quality of explosive welds of copper to aluminium. *Mater. Des.* **2016**, *95*, 256–267. [CrossRef]
82. Inal, O.T.; Szecket, A.; Vigueras, D.J.; Pak, H.R. Explosive welding of Ti-6al-4v to mild-steel substrates. *J. Vac. Sci. Technol. A Vac. Surf. Films* **1985**, *3*, 2605–2609. [CrossRef]

83. Zhang, Y.; Babu, S.S.; Zhang, P.; Kenik, E.A.; Daehn, G.S. Microstructure characterisation of magnetic pulse welded AA6061-T6 by electron backscattered diffraction. *Sci. Technol. Weld. Join.* **2008**, *13*, 467–471. [CrossRef]
84. Liu, K.X.; Liu, W.D.; Wang, J.T.; Yan, H.H.; Li, X.J.; Huang, Y.J.; Wei, X.S.; Shen, J. Atomic-scale bonding of bulk metallic glass to crystalline aluminum. *Appl. Phys. Lett.* **2008**, *93*, 1–3. [CrossRef]
85. Göbel, G.; Kaspar, J.; Herrmannsdörfer, T.; Brenner, B.; Beyer, E. Insights into intermetallic phases on pulse welded dissimilar metal joints. In Proceedings of the 4th International Conference on High Speed Forming, Columbus, OH, USA, 9–10 March 2010; pp. 127–136.
86. Stern, A.; Aizenshtein, M. Bonding zone formation in magnetic pulse welds. *Sci. Technol. Weld. Join.* **2002**, *7*, 339–342. [CrossRef]
87. Nishida, M.; Chiba, A.; Honda, Y.; Hirazumi, J.; Horikiri, K. Electron-microscopy studies of bonding interface in explosively welded Ti/Steel clads. *Isij Int.* **1995**, *35*, 217–219. [CrossRef]
88. Zhang, Y.; Babu, S.; Daehn, G. Interfacial ultrafine-grained structures on aluminum alloy 6061 joint and copper alloy 110 joint fabricated by magnetic pulse welding. *J. Mater. Sci.* **2010**, *45*, 4645–4651. [CrossRef]
89. Wronka, B. Testing of explosive welding and welded joints: Joint mechanism and properties of explosive welded joints. *J. Mater. Sci.* **2010**, *45*, 4078–4083. [CrossRef]
90. Grignon, F.; Benson, D.; Vecchio, K.S.; Meyers, M.A. Explosive welding of aluminum to aluminum: Analysis, computations and experiments. *Int. J. Impact Eng.* **2004**, *30*, 1333–1351. [CrossRef]
91. Carvalho, G.H.S.F.L.; Galvão, I.; Mendes, R.; Leal, R.M.; Loureiro, A. Formation of intermetallic structures at the interface of steel-to-aluminium explosive welds. *Mater. Charact.* **2018**, *142*, 432–442. [CrossRef]
92. Carvalho, G.H.S.F.L.; Galvão, I.; Mendes, R.; Leal, R.M.; Loureiro, A. Explosive welding of aluminium to stainless steel. *J. Mater. Process. Technol.* **2018**, *262*, 340–349. [CrossRef]
93. Lueg-Althoff, J.; Bellmann, J.; Gies, S.; Schulze, S.; Tekkaya, A.E.; Beyer, E. Influence of the flyer kinetics on magnetic pulse welding of tubes. *J. Mater. Process. Technol.* **2018**, *262*, 189–203. [CrossRef]
94. Carvalho, G.H.S.F.L.; Mendes, R.; Leal, R.M.; Galvão, I.; Loureiro, A. Effect of the flyer material on the interface phenomena in aluminium and copper explosive welds. *Mater. Des.* **2017**, *122*, 172–183. [CrossRef]
95. Deribas, A.A.; Kudinov, V.M.; Matveenk, F.I. Effect of initial parameters on process of wave formation in explosive welding. *Combust. Expl. Shock Waves* **1967**, *3*, 344–348. [CrossRef]
96. Durgutlu, A.; Okuyucu, H.; Gulenc, B. Investigation of effect of the stand-off distance on interface characteristics of explosively welded copper and stainless steel. *Mater. Des.* **2008**, *29*, 1480–1484. [CrossRef]

Article

Welding Window: Comparison of Deribas' and Wittman's Approaches and SPH Simulation Results

Yulia Yu. Émurlaeva [1], Ivan A. Bataev [1,*], Qiang Zhou [2], Daria V. Lazurenko [1], Ivan V. Ivanov [1], Polina A. Riabinkina [1], Shigeru Tanaka [3] and Pengwan Chen [2]

[1]   Department of Mechanical Engineering and Technologies, Novosibirsk State Technical University, Novosibirsk 630073, Russia; emurlaeva@corp.nstu.ru (Y.Y.É.); pavlyukova_87@mail.ru (D.V.L.); i.ivanov@corp.nstu.ru (I.V.I.); ryabinkina.2013@stud.nstu.ru (P.A.R.)

[2]   State Key Laboratory of Explosion Science and Technology, Beijing Institute of Technology, Beijing 100081, China; zqpcgm@gmail.com (Q.Z.); pwchen@bit.edu.cn (P.C.)

[3]   Institutes of Pulsed Power Science, Kumamoto University, Kumamoto 860-8555, Japan; tanaka@mech.kumamoto-u.ac.jp

*   Correspondence: ivanbataev@ngs.ru

Received: 14 November 2019; Accepted: 5 December 2019; Published: 7 December 2019

Abstract: A welding window is one of the key concepts used to select optimal regimes for high-velocity impact welding. In a number of recent studies, the method of smoothed particle hydrodynamics (SPH) was used to find the welding window. In this paper, an attempt is made to compare the results of SPH simulation and classical approaches to find the boundaries of a welding window. The experimental data on the welding of 6061-T6 alloy obtained by Wittman were used to verify the simulation results. Numerical simulation of high-velocity impact accompanied by deformation and heating was carried out by the SPH method in Ansys Autodyn software. To analyze the cooling process, the heat equation was solved using the finite difference method. Numerical simulation reproduced most of the explosion welding phenomena, in particular, the formation of waves, vortices, and jets. The left, right, and lower boundaries found using numerical simulations were in good agreement with those found using Wittman's and Deribas's approaches. At the same time, significant differences were found in the position of the upper limit. The results of this study improve understanding of the mechanism of joint formation during high-velocity impact welding.

Keywords: high-velocity impact welding; smoothed particle hydrodynamics simulation; welding window

1. Introduction

High-velocity impact welding is one of the best methods for joining dissimilar materials. This group of welding processes include explosive welding, magnetic pulse welding, laser impact welding, etc. Among multiple parameters of these processes, the collision point velocity (V_c) and the collision angle (α) are of the highest importance to select the regime of welding. Thus, the choice of the welding regime, as a rule, is associated with establishing the optimal combination of these two values. For most materials, the range of V_c and α is quite wide, so in the diagrams plotted in the coordinates (V_c–α), a large range of regimes providing reasonable welding quality can be distinguished. This area is called the welding window (or weldability window). An example of weldability window is shown in Figure 1.

Among the pioneering studies in which successful attempts were made to determine the boundaries of the weldability window, it is worth highlighting the studies of Wittman [1] and Deribas [2]. Their approaches are still widely used in practice to select welding regimes. The approaches to determining the lower and right boundaries used by both authors almost coincide. However, the position of the upper limit calculated according to the formulas stated in their studies differs (as a rule, the Deribas calculation predicts a wider range of acceptable welding regimes). Despite the fact that in many cases the classical approaches to constructing a weldability window work quite well, their application is difficult when working with dissimilar materials. In addition, adequately describing the boundaries of the weldability window, the expressions used do not give an idea of the mechanics of the deformation process near the interlayer boundary and practically do not discuss the formation of melt zones in explosively welded joints.

To address these issues, in recent studies [3–6] the weldability window was found using smoothed particle hydrodynamics (SPH) simulation. SPH technique was used extensively in recent years to analyze various aspects of high-velocity impact welding. Unlike other methods, for example, Euler or Lagrangian, SPH reproduces well the basic phenomena associated with high-velocity impact welding-the formation of jets, waves and vortex zones, and it is well suited for analyzing the pressures in the impact zone [4,7–14].

This study compares approaches to constructing a weldability window proposed by Deribas and Wittman with an approach based on SPH simulation. Aluminum alloy 6061-T6 was used as a material for analysis. A large amount of experimental data on explosive welding of this alloy presented by Wittman [1] was used to verify the simulation results. In this study, an attempt was made to determine the regimes of formation and disappearance of a jet, the area of wave formation and vortex zones using numerical simulation and "classical" approaches proposed by Deribas and Wittman. To assess the upper limit of the weldability window, the influence of melt zones and their cooling rates on the formation of joints was analyzed. In addition, the paper discusses the features of the plastic flow of material near the interface, which are compared with experimental data obtained in the works of Chugunov et al. [15].

The majority of recent studies devoted to numerical simulation of high-velocity impact welding consider only few collision regimes providing only a limited understanding of the bonding phenomena [16–18]. Besides, most of the currently existing SPH studies on high-velocity impact welding simply state the fact of good agreement between the simulation results and experimentally observed interface. The relationship between the flow of individual particles and the jetting and wave formation is usually not discussed. Thus, the understanding of some important phenomena is still limited. In the current study, we simulated a large number of welding regimes, varying the collision angle and collision velocity. Thus, a more complete understanding of the bonding phenomena was achieved.

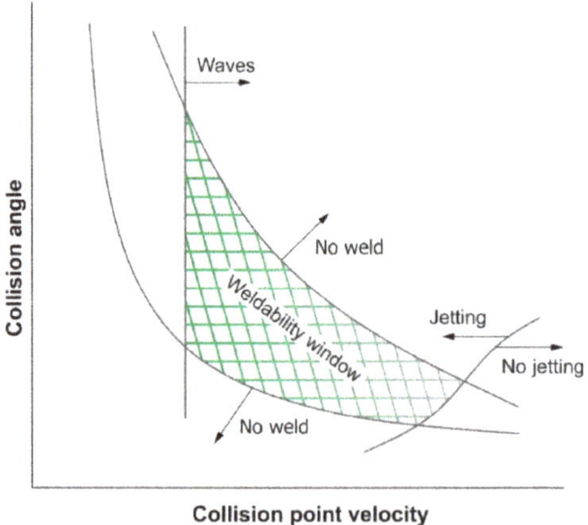

Figure 1. A scheme of the weldability window.

From the practical point of view, this study is interesting for engineers, who are responsible for selection of regimes of high-velocity impact welding as it provides better understanding of the weldability window concept. The application of SPH simulation will allow to predict the shape of the interface and minimize the amount of liquid phase which may appear when the collision velocity is too high. Understanding of the thermal situation near the interface allows predicting the phases which may appear due to the high cooling rates.

2. Description of Numerical Simulation

2.1. Simulation of Impact

Plates of 6061-T6 aluminum alloy with dimensions of 63.5 × 6 mm^2 were used for numerical simulation. This material was chosen due to the previous study of Wittman [1], who published a significant amount of experimental data on explosive welding of this alloy and described the morphology of the interlayer boundaries and the properties of the joints. Thus, the data obtained by Wittman were used for comparison with data obtained by simulation.

The computational domain is shown in Figure 2. At the initial moment of time, the flyer plate had a velocity V_p and it was positioned at an angle α with respect to the base plate. The collision point velocity (V_c) was determined in accordance with the Equation [1]:

$$V_p = 2V_c \sin \frac{\alpha}{2} \tag{1}$$

In this study, a series of simulations were carried out in which the angle varied from 5° to 40° with a step of 5°, and the collision point velocity varied from 500 m/s to 9300 m/s with a step of 800 m/s. After the approximate position of the lower limit was found, a series of additional simulations were carried out in order to specify the minimum welding parameters with greater accuracy.

The collision process was simulated in the Ansys Autodyn 19 software using the SPH method. This method is well suited for simulating phenomena associated with large strains and rapidly moving borders. In a number of previous studies, this method was successfully used to simulate high-velocity impact welding of similar [4,19–21] as well as dissimilar materials [17,22–24] showing a very good agreement between the simulation results and experiments.

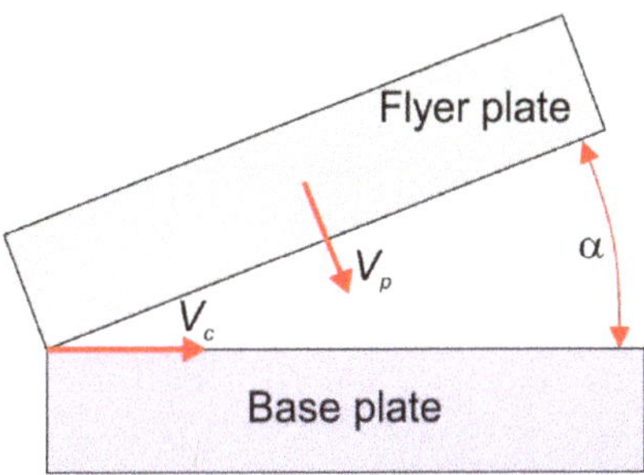

Figure 2. Scheme of the computational domain at the initial time.

High-velocity impact can lead to the formation of shock waves in the material. At a shock wave-front, discontinuity in materials properties is observed. For numerical analysis of discontinuity, it is required to set the properties before and after the shock front. For this reason, the modified Mie-Grüneisen equation of state based on the Hugoniot shock adiabat was used [25]:

$$p = p_H + \Gamma\rho(e - e_H), \tag{2}$$

where p is the pressure, p_H is the Hugoniot pressure, Γ is the Grüneisen gamma, e is the internal energy, e_H is the Hugoniot internal energy. For solid materials, the equation of the Hugoniot shock adiabat can be written in the following form:

$$U_s = C_1 + S_1 U_P, \tag{3}$$

where U_s is the shock velocity, U_P is the particle velocity, C_1, S_1 are empirically determined coefficients depending on the material.

The coefficients used in this study are shown in Table 1.

Table 1. Parameters of the modified Mie-Grüneisen equation of state for aluminum alloy 6061-T6.

Parameter	Value	Units
Reference density	2.7	g/cm^3
Grüneisen parameter	1.97	-
Parameter C_1	5.35	m/ms
Parameter S_1	1.34	-
Reference Temperature	293	K
Specific Heat	$8.85 \cdot 10^{-4}$	kJ/(g·K)

The dependence of the material's strength on the deformation conditions was taken into account by using the Johnson-Cook empirical model, which is widely used in the simulation of various phenomena associated with a high-velocity deformation [26,27]. This model allows considering the influence of strain hardening, thermal softening, and a strain rate on the yield stress of the material and it is described by the following Equation:

$$\sigma = \left(A + B\varepsilon^n\right)\left(1 + C\ln\dot{\varepsilon}^*\right)\left(1 - T^{*m}\right), \tag{4}$$

where σ is the current yield strain, ε is the effective plastic strain, $\dot{\varepsilon}^*$ is the dimensionless plastic strain rate, $T^* = \frac{T - T_r}{T_m - T_r}$, T is the current temperature, T_m is the melting temperature, T_r is the reference temperature. A, B, C, n and m are the material constants determined from an empirical fit of flow stress data, A, B and n are the yield stress, the hardening constant and the hardening exponent associated with quasi-static test respectively, C is the strain rate constant, m is the thermal softening exponent. The constants used in the Johnson-Cook model for Al 6061-T6 are shown in Table 2.

Table 2. Johnson-Cook model parameters [28,29].

Parameter	Value	Units
Shear Modulus	26	GPa
Yield Stress	0.324	GPa
Hardening Constant	0.114	GPa
Hardening Exponent	0.42	-
Strain Rate Constant	0.002	-
Thermal Softening Exponent	1.34	-
Melting Temperature	925	K
Ref. Strain Rate (/s)	1	-

2.2. Description of Cooling Model

The estimation of time which material spent in the molten state is of fundamental importance to assess the upper limit of welding. It is believed that, if this time is less than the duration of compressive stresses, a joint will form. Otherwise, the incoming tensile wave breaks the joint that did not have enough time to solidify. It should be noted that the system of equations used by Autodyn software does not consider the heat transfer. Therefore, it is believed that the deformation process is completely adiabatic. For this reason, the calculation of the cooling process was carried out separately using self-developed code in Python programming language. The temperatures obtained by simulation in

Autodyn were exported as a text file and were used as initial conditions for solving the heat equation, which in the 2D case can be written as follows:

$$\frac{\partial U}{\partial t} = a\left(\frac{\partial^2 U}{\partial x^2} + \frac{\partial^2 U}{\partial y^2}\right), \tag{5}$$

where $U = (x, y, t)$ is the function that describes the temperature at a point with x and y coordinates at the moment of time t, a is the thermal diffusivity.

The equation was solved by the finite difference method in a "Forward-Time Central-Space" explicit scheme. In this case, it was believed that the outer boundaries of the plates were isolated from the environment, and cooling occurred due to heat transfer from the interface to the inner volumes of the plates, which practically did not heat up during a collision.

3. Results and Discussion

Figure 3a shows the results of an experimental study of 6061-T6 alloy explosive welding performed by Wittman. Based on these studies, he developed approaches to calculate the limiting conditions that provided a satisfactory quality of the joint and built a weldability window also plotted in Figure 3a. For the upper limit of the weldability window shown in Figure 3a, two lines are plotted: one of them is based on the predictions of Wittman's model, and the other is based on the estimations of Deribas [30]. Figure 3b represents the simulation results showing the regions of modes characterized by the absence of a jet, the presence of a jet, and the simultaneous presence of a jet and waves at the interface. In addition, the limits of welding calculated based on approaches of Deribas and Wittman are also plotted in this Figure for comparison purpose. In the following sections of the paper, each of these limits is discussed in detail.

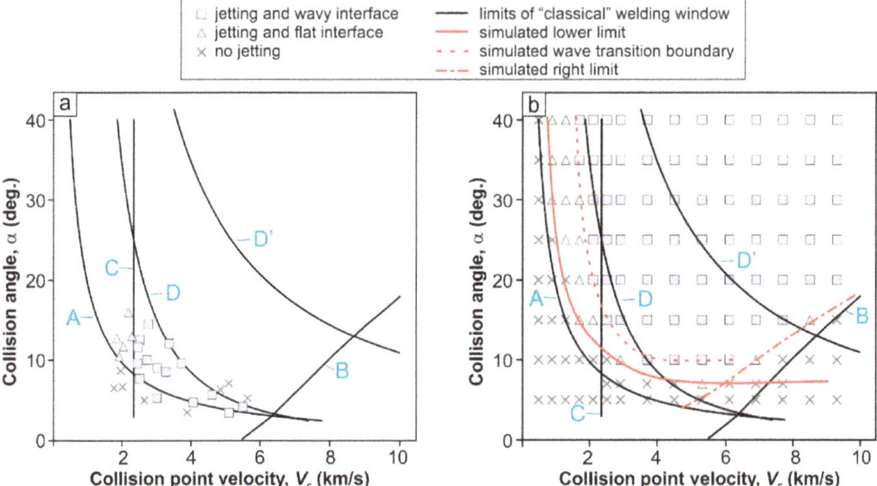

Figure 3. Experimental studies of welding regimes of 6061-T6 alloy carried out by Wittman [1] (**a**) and the results of SPH numerical simulations performed in this study (**b**). The limits of the weldability window calculated using the approaches developed by Deribas and Whitman are plotted on both diagrams for comparison purpose: A is the lower limit, B is the right limit, C is the limit of transition to the wave formation regimes, D is the upper limit according to Wittman, D' is the upper limit according to Efremov and Zakharenko [31].

3.1. The Lower Limit of the Weldability Window

The lower limit of the weldability window represents the locus of points that define the "softest" welding regimes, below which the collision does not lead to the formation of a strong bond. Following the reasoning typical for various solid-state welding methods, the impact should ensure the achievement of a certain critical degree of plastic strain and cooperative flow of the impacted plates, providing activation of the welded surfaces.

One of the conditions necessary for the joint formation in the process of high-velocity collision is considered to be the jet formation in front of the contact point [1,2,30]. This jet consists of the material of the surface layers of the impacted plates, which is ejected from the welding zone in the form of a peculiar cloud [4] providing contact to surfaces free from contamination. Deribas called this phenomenon self-cleaning [2]. In their studies, Deribas and Wittman believed that the critical regimes of jet formation determine the lower limit of the weldability window in the subsonic range. At the same time, both researchers developed an entirely empirical criterion which suggests that the pressure in the collision zone should be several times higher than the Hugoniot elastic limit (5 times according to Wittman [1]). In the same time, the study [1] does not give clear explanations for the choice of the empirical coefficient. It can be assumed that this coefficient was chosen to fit the calculated position of the lower boundary as closely as possible to the experimental data. In the absence of data on the Hugoniot elastic limit, the critical impact velocity that ensures jet formation according to Wittman can also be found using the following empirical expression:

$$V_p = \sqrt{\frac{\sigma_{ts}}{\rho_f}},\tag{6}$$

where σ_{ts} is the tensile strength and ρ_f is the material density.

This expression was used to plot the line A in Figure 3a.

It can be noted that jet formation was observed for most collision regimes analyzed by numerical simulation (Figure 3b). The lower limit predicted by the calculation has a similar trend to the line A, however, it is located more to the upper right part of the diagram, which may be due to the insufficient spatial resolution to simulate weak jets at relatively soft collision regimes.

To understand the reasons leading to the formation of a jet, it is interesting to analyze the differences in conditions near the contact point for some regimes between $V_c = 1300$ m/s and $V_c = 9300$ m/s at a constant collision angle $\alpha = 15°$ (Figure 4a–e). The collision point velocity equal to $V_c = 1300$ m/s corresponds to the regime at which the jet is not formed yet (Figure 4a), while collision point velocity of $V_c = 1700$ m/s provides the first signs of the weak jet formation (Figure 4b). At higher collision point velocities (Figure 4c–e), the jet becomes more noticeable until it finally disappears when crossing the so-called supersonic limit of the weldability window (Figure 4f), which occurs for the reasons described in the next section. One can observe the specific nature of the pressure distribution near the collision point in the modes corresponding to the jet formation. The pressure distribution lines are convex in the direction of collision point movement and at the same time, high pressures outpace it to some extent. This shape of the pressure distribution zone causes the surface layer material to rise above the surface forming an elevation when approaching the collision point. Thus, the collision angle of the plates in the immediate vicinity of the collision point increases compared to the initial collision angle α. Further, the material of the surface layers "flies" onto the area of extremely high pressures and is thrown forward in the form of a peculiar jet. It should be noted that the term "jet" as applied to explosive welding should be used cautiously. Experiments carried out in a number of studies indicate that due to the heating, the material moving in front of the collision point represents a cloud (probably a plasma cloud) rather than a liquid metal [2,4,32].

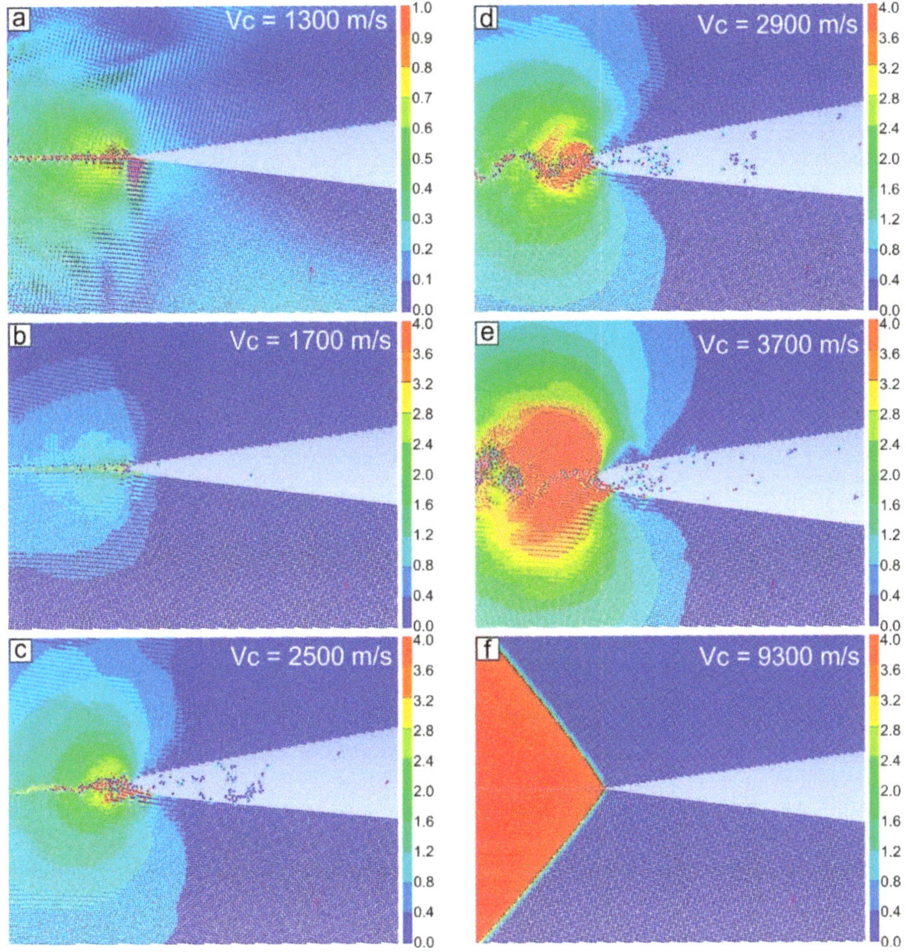

Figure 4. Pressure distribution near the collision point at different collision point velocities and constant collision angle ($\alpha = 15°$). The color bar represents the pressure in GPa. Notice, that the upper limit for colorbar in (**a**) is different from that for (**b**–**f**).

From the current simulation, it follows that Wittman's assumption that the pressure near the collision point should substantially exceed the strength of the material for the jet to form seems plausible. Taking into account the deformation conditions typical for the high-velocity impact welding (strain, strain rate and temperature), and inserting them in Equation (4) it can be found that the yield strength of 6061-T6 alloy can reach 0.45 GPa. In the current simulation, a weak jet was formed when the pressure near the collision point was approximately 2 GPa or more (that was about 5 times higher than the yield strength estimated by Equation (4)) (Figure 4b), and a stable jet could be formed when the pressure at the collision point exceeded 4 GPa (Figure 4c). Weaker pressures did not lead to the formation of a significant elevation and the material of the surface layer "entered" the collision point at an acuter angle. Moreover, according to the simulation, the strain near the interface exceeded $\varepsilon = 1.5$ for the regimes corresponding to the jet formation (for the regimes near the lower limit) and could be significantly larger for the regimes located near the right limit of the weldability window.

3.2. Supersonic Limit

At extremely high collision velocities, a jet does not form, and the surface self-cleaning phenomenon is not observed. In addition, tensile waves quickly come into the deformed area, which does not give sufficient time for solidification and leads to the destruction of the joint. In many studies (for instance, [33–35]), little attention is paid to the right limit of welding when constructing a weldability window. It is common to simplify its position by taking the maximum allowed collision velocity equal to the bulk sound velocity of welded materials. Although from a practical point of view this simplification is probably acceptable, since welding at such high velocities is quite rare, nevertheless, these simplifications do not correspond to real experiments. The concept of the critical angle for jet formation at supersonic velocities was considered in detail in the studies of Walsh et al. [11] and Cowan et al. [19], and an example of its practical application was considered in detail in [36]. According to this concept, at high collision point velocity, a jet is formed only if the collision angle exceeds a certain threshold value. In pursuance of the instructions given in [36–39], the following set of equations was obtained:

$$\tan \alpha = \frac{\left((\sin \beta - 5.35/V_c)/1.34\right)\sqrt{1 - \sin^2 \beta}}{(1 - \sin \beta(\sin \beta - 5.35/V_c)/1.34)}, \tag{7}$$

$$p = \rho(V_c \sin \beta)(V_c \sin \beta - 5.35)/1.34, \tag{8}$$

$$U_S = 5.35 + 1.34U_P, \tag{9}$$

$$U_S = V_c \sin \beta, \tag{10}$$

where β is the angle between the shock front and the vector of material flow into the collision point.

By varying β in the range from 0 to 90°, it is possible to develop a set of plots of β versus p for each V_c value. The obtained curves with a specific maximum provide the value of the critical angle of jet formation α for a given collision point velocity V_c (line B in Figure 3a).

One can note that there is a very good agreement between the position of the right limit, which was determined in accordance with [37,38] and in accordance with the results of SPH simulation (Figure 3b). With an increase of collision point velocity, the pressure front was getting more and more straightened (Figure 4f). Thus, the specific elevation of the surface layers in front of the contact point described in the previous section was barely observed and the plates collided at an acute angle without jet formation.

3.3. Wave Formation Limit

The wave formation at the interface is one of the features typical for high-velocity impact welding. However, it should be noted that a number of studies [40–43] showed that the mechanical properties of joints with a flat interface were as high as those with wavy ones. For this reason, the process of wave formation is primarily of fundamental interest. However, it is likely that a complete understanding of the explosive welding process is impossible without understanding the wave formation process.

One of the first sufficiently reliable criteria for the wave formation was proposed by Cowan et al. [8], who came to the conclusion that the transition from waveless to wavy mode of welding depends only on the collision point velocity. They found that for each combination of materials there is such a critical collision point velocity (V_T), above which the wave formation occurs, and they proposed the following Equation:

$$V_T = \sqrt{2R_T(H_f + H_b)/(\rho_f + \rho_b)}, \tag{11}$$

where ρ_f and ρ_b are the densities of flyer and base plates, respectively, H_f and H_b are the diamond pyramid hardness of flyer and base plates, respectively. R_T is an empirically determined parameter, introduced by analogy with the Reynolds number. The critical value of R_T leading to the wave formation varies in the range from 8.1 to 13.1 for different materials, which was established experimentally in [44].

This expression was used to construct line C in Figure 3a. In this case, following the data presented in [44], the value $R_T = 8$ was used to plot the line C.

It can be seen in Figure 3 that the critical value of the collision point velocity calculated using Equation (11), which describes the transition to wave formation, is in a good agreement with the results of Wittman's experimental studies. However, it should be noted that several studies showed that the transition to wave formation depends both on the collision point velocity and on the collision angle (for example, Szecket [45], or Lysak and Kuzmin [46]). This assumption is also supported by the results of the current numerical simulation (Figure 3b). One can note that at large collision angles, wave formation began at lower velocities than at small angles. Thus, the shape of the wave transition limit resembles line A more than line C, i.e., the transition to wave formation occurred at a certain critical flyer plate velocity, and not at some constant collision point velocity.

It is interesting to note that for regimes with very high V_c located slightly to the left of the jet formation limit, the wave formation terminated again. Thus, it can be assumed that there are two boundaries of wave formation-left and right. In principle, the termination of the wave formation process at high collision point velocities is described in a number of experimental works (e.g., [47,48]), however, the reasons of this phenomenon require additional study.

One can note a good agreement between the nature of material flow predicted by simulation and observed in experimental studies. For example, Figure 5a shows the deformation of the experimental grid after explosive welding of aluminum alloy plates. This image was obtained in the experiments of Chugunov et al. [15] using laminated inserts made of the material similar to the material of welded plates. The laminated nature of the inserts was used to track the peculiarities of plastic deformation near the interface. In Figure 5b an attempt to reproduce this experiment using SPH simulation is shown. To do this, the particles forming a square mesh were marked before the simulation was started. During the simulation, the positions of these particles were tracked, and the deformation of the mesh was observed. As it is seen from Figure 5, the flow pattern of the material observed in the simulation was in good agreement with that observed in the experiments. As an example, the flow patterns for different collision regimes are shown in Figure 6. It can be noted that the plastic strain substantially increased closer to the interface. Moreover, with an increase in the collision point velocity, the plastic strain also increased significantly. The proposed approach allows one to visualize the features of material deformation during explosion welding. In particular, it can be noted that the deformation of mesh cells in the flyer and base plates is different, which is probably due to the asymmetry of the collision scheme.

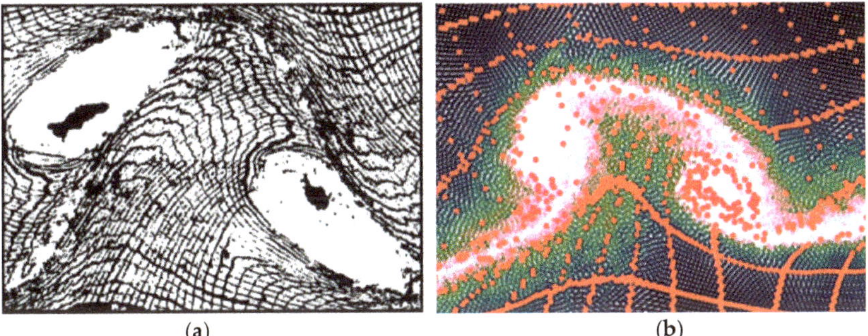

(a)   (b)

Figure 5. The grid deformation near interface observed in the experiments of Chugunov et al. [15], (reproduced from [15], with permission from V.E. Riecansky Technical Translations) (**a**) and in the simulation (**b**).

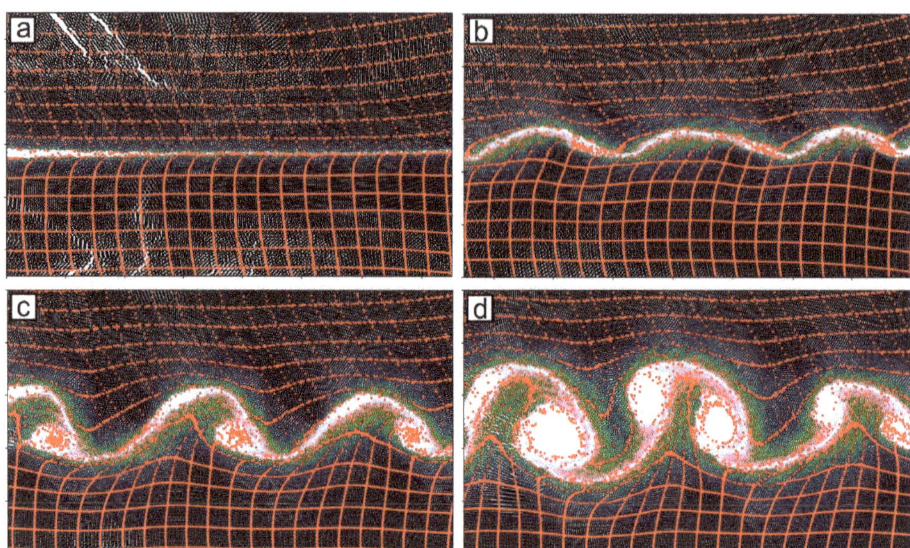

Figure 6. The calculated grid deformation at various collision regimes: (**a**) V_c = 1300 m/s, (**b**) V_c = 2100 m/s, (**c**) V_c = 2900 m/s, (**d**) V_c = 3700 m/s. The collision angle is 25° in all cases.

3.4. Upper Limit

The upper limit of the weldability window attracted much less attention of researchers than the lower one and the issues of transition from waveless to wavy modes of welding. This is probably due to the fact that the welding regimes near the lower limit are more preferable for industrial applications. However, the position of the upper limit is important from both industrial and fundamental points of view. It should be noted that welding using the regimes near the upper limit of the weldability window may lead to the formation of an excessive amount of molten zones leading to embrittlement of the welded joint and decrease in its strength.

So far, there are two main studies devoted to the analysis of the upper limit of the weldability window known from the literature—the study of Wittman [1] and the study of Efremov and Zakharenko [31]. The later became known due to the monographs of Deribas [2] and Zakharenko [49]. A common feature of both approaches is the comparison of the compressive stresses' duration in the deformed area with the time required to cool the material.

In his study, Wittman came to the following expression, limiting the highest allowable flyer plate velocity V_{max}:

$$V_{max} = (T_m C_B)^{1/2} (\alpha c C_B)^{1/4} / N V_c \left(\rho h_f\right)^{1/4}, \tag{12}$$

where T_m is the material melting temperature, C_B is the bulk sound velocity, c is the specific heat, h_f is the flyer plate thickness, N is empirically determined material-dependent coefficient. In the case of Al6061-T6 aluminum, alloy N is equal to 0.11. It should be also noted that all Wittman's calculations were performed in the CGS units, so when calculating in SI units the value of the coefficient N will be different.

In the work of Efremov and Zakharenko, the following equation was proposed, limiting the highest collision angle as a function of the collision point velocity and the thermophysical material properties:

$$\sin\frac{\alpha}{2} = 11.8 * V_c^{-5/4} \sqrt{\frac{T_m \frac{\chi}{a}}{\rho \frac{h_b}{h_f + h_b} \sqrt{h_f}}} \sqrt[4]{\xi}, \tag{13}$$

where χ is the thermal conductivity, h_b is the thickness of the base plate, $\xi = x^*/h_f$, x^* is the distance from the collision point to the point behind it where the tensile stresses are achieved. To simplify the calculations, it is common to consider that $\sqrt[4]{\xi} = 1$. The empirical coefficient equal to 11.8, which characterizes the amount of heat released and the irregularity of the melt location along the interface, was selected in [50] so that the calculated upper limit provides the best fit to the data obtained in practical studies. In the paper [51], it is proposed to use the coefficient 14.7 instead.

From Figure 3 it can be noted that Equation (12) gives a lower critical impact velocity, and, accordingly, a narrower weldability window, as compared to Equation (13).

To calculate the position of the upper limit using numerical simulation, it is reasonable to follow the same sequence of reasoning—firstly, it is necessary to calculate the duration of compressive stresses, and then compare the calculated time with that required to cool the weld below the melting temperature.

Typical pressure history in a point located near the interface is shown in Figure 7. It is possible to determine the time available for the material to cool, by measuring the time from the moment when the peak pressure is reached until the moment when the tensile wave arrived.

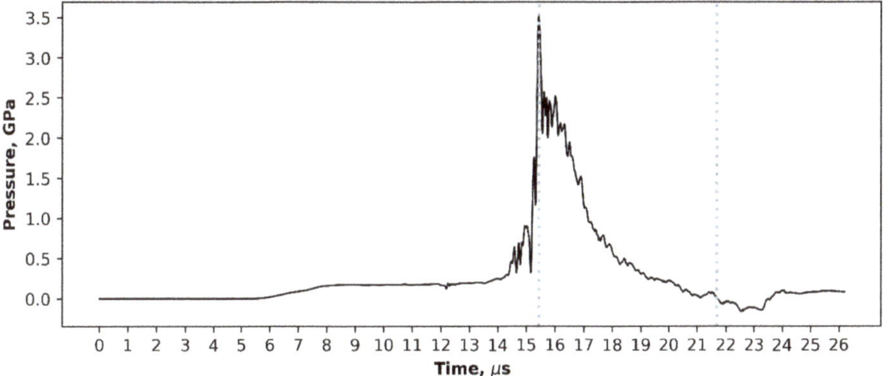

Figure 7. Change in pressure over time at a point near the interface (V_c = 2100 m/s, α = 20°).

The duration of compressive stresses on the collision regimes is shown in Figure 8. To determine the average value and the confidence interval, five measurements were carried out at various points along the interface. It can be noted that in the presented range of collision angles, the collision point velocity had the greatest impact on this parameter.

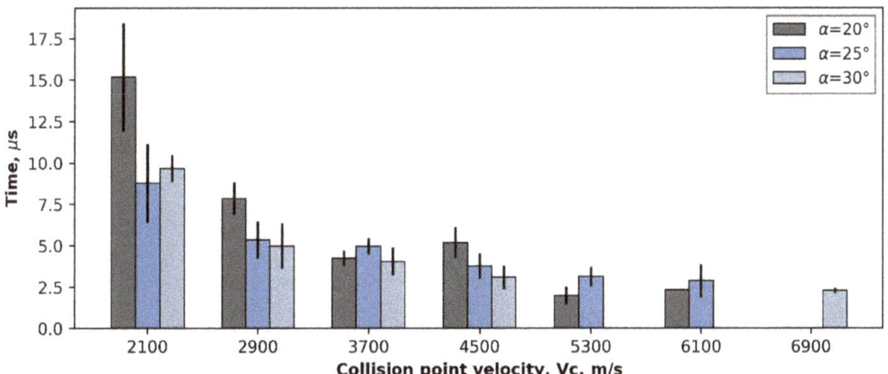

Figure 8. Relation between duration of compressive stresses in the deformed zone and the collision regimes.

Figure 9 shows a typical calculation of the cooling process in the vicinity of the contact point for V_c = 2100 m/s and α = 20°. The information about the temperature distribution obtained using SPH simulation in Ansys Autodyn was used as initial conditions for the heat equation. It should be noted that the time required for complete solidification of the material in the region of the interface exceeded the lifetime of compressive stresses in the weld. As already noted, the micro-volumes of the liquid phase were distributed unevenly along the weld, forming vortex zones on the sides of the waves. This process is explained by redistribution of the liquid phase behind the collision point in the process of wave growth, as it was previously shown in [4]. The current simulation shows, that these micro-volumes can remain in a partially molten state at the time of the arrival of the tensile wave. Due to this reason, they can't provide strong bonding and withstand tensile stresses. Thus, the bonding of the material near the upper limit entirely depends on the areas of interface free of liquid phase, where the direct contact of solid materials is possible, or on the areas where the liquid phase volume is extremely small. These areas allow the joint to withstand the passage of the tensile wave, while the final solidification of the melt clumps (i.e., vortex zones) probably occurs later.

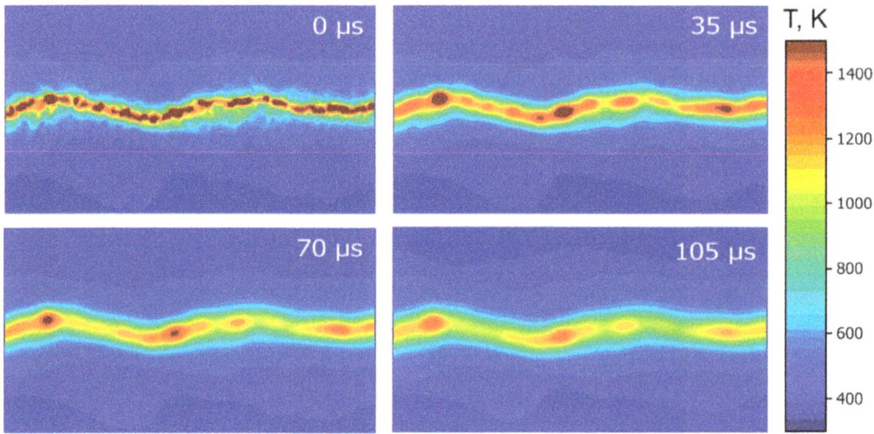

Figure 9. Temperature distribution near the interface after completion of plastic deformation.

According to Figure 10, the volume of molten metal at the interface raises almost linearly with the collision point velocity increase, and also, as a rule, slightly increases with an increase in the collision angle. Considering that the position of the upper welding limit described by Wittman is the most accurate (since it was received from the experiment), it is interesting to estimate the amount of the liquid phase near the interface defined by line D in Figure 3a. It can be noted that for all collision regimes close to line D, the liquid phase volume fraction was at the level of about 1% of the total plate volume. According to Efremov and Zakharenko (line D 'in Figure 3a), for the regimes located near the upper limit the estimated volume fraction of the liquid phase was at the level of 4–6%. It should be noted that determining the upper limit of the welding window, Wittman believed that the joint above this limit would have insufficient mechanical properties, while in Deribas's work it was supposed that the welding of the plates outside the upper limit would not occur in principle, since a tensile wave will lead to the destruction of the non-solidified weld joint. Thus, for the formation of a high-quality joint, the volume fraction of the liquid phase probably should not significantly exceed 1%. Nevertheless, this criterion requires more thorough experimental verification for materials of other compositions and other thicknesses.

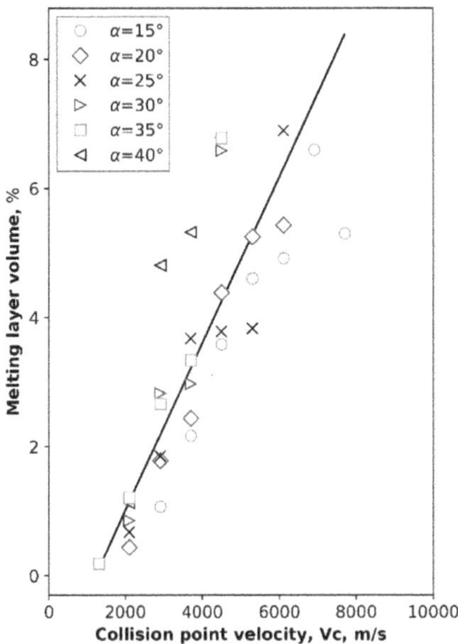

Figure 10. Relation between the volume of the melted material and the collision point velocity for different collision angles.

From the simulation, it follows that the volume fraction of molten metal is always slightly higher for a base plate, and the difference becomes more noticeable for high welding velocities (Figure 11).

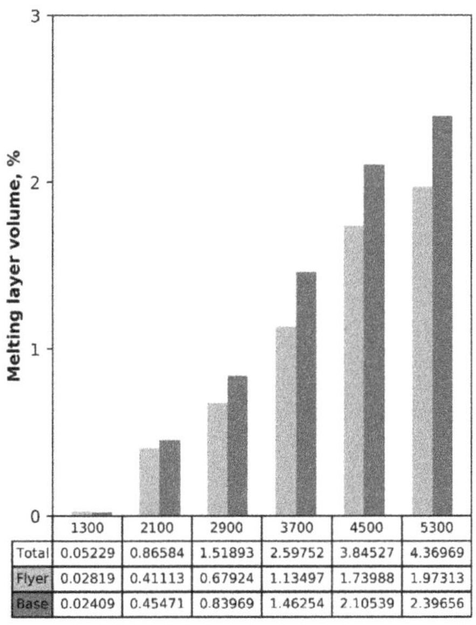

	1300	2100	2900	3700	4500	5300
Total	0.05229	0.86584	1.51893	2.59752	3.84527	4.36969
Flyer	0.02819	0.41113	0.67924	1.13497	1.73988	1.97313
Base	0.02409	0.45471	0.83969	1.46254	2.10539	2.39656

Figure 11. The volume fraction of molten metal in flyer and base plates as a function of welding conditions (the collision angle for all regimes was 25°).

4. Conclusions

Based on the results of numerical simulation and their comparison with experimental data of Wittman, we can draw the following conclusions:

1. SPH simulation reproduces all the basic phenomena typical for the high-velocity impact welding process: the jet (or cloud) formation in front of the collision point, the wave formation, as well as the material deformation mechanism near the interface. Flow patterns near the interface are in good agreement with the results of experimental studies obtained by other authors. The proposed approach allows one to build the weldability windows.

2. The simulated lower welding limit slightly differs from the results of Wittman's theoretical and experimental studies. While Wittman observed bonding (which implies the existence of a jet) even at very low collision angles (around 5 degrees), the simulation predicts jetting when the collision angle exceeds 7.5 or even 10 degrees. This may be due to inaccuracies in the material models used in the current simulation, as well as due to the insufficient resolution for observing weak jets.

3. The wave formation starts when the collision point velocity exceeds 1700 m/s. However, in comparison with most of the previous studies, the wave transition velocity turned out to be dependent on the collision angle. At low collision angles (e.g., 10 degrees), the transition to the wavy interface occurred around 4000 m/s.

4. The numerical simulation predicts the existence of the right limit of wave formation, which is consistent with several experimental studies. The transition to straight interface coincides with the supersonic limit of welding.

5. The position of the upper limit of the weldability window is difficult to determine using the model considered in the current study. The lifetime of compressive stresses is shorter than the time required for complete solidification of the molten areas. Thus, the joint formation near the upper limit is most likely occurs due to the presence of areas where direct contact in the solid phase is formed. Further refinement of the position of the upper limit requires the simultaneous solution of heat and deformation problems and is of interest for further studies. However, from a practical point of view, the welding regimes used in the industry should be closer to the lower limit of the weldability window. Thus, this approach can be used in practice.

Author Contributions: Methodology, S.T.; writing—original draft preparation, Y.Y.É.; writing—review and editing, I.A.B., D.V.L., Q.Z., P.C., I.V.I., P.A.R.; supervision, D.V.L.; project administration, I.A.B.; funding acquisition, Q.Z., P.C.

Funding: This paper is supported by the opening project of State Key Laboratory of Explosion Science and Technology (Beijing Institute of Technology). The opening project number is KFJJ18-05M.

Conflicts of Interest: The authors declare no conflicts of interest.

Nomenclature

V_c	collision point velocity (m/s)
α	collision angle (°)
V_p	flyer plate velocity (m/s)
p	pressure (GPa)
p_H	Hugoniot pressure
Γ	Grüneisen gamma
e	internal energy
e_H	Hugoniot internal energy
U_S	shock velocity
U_P	particle velocity
σ	current yield strain
ε	effective plastic strain
C_1, S_1	empirically determined coefficients depending on the material

$\dot{\varepsilon}^*$	dimensionless plastic strain rate
T	current temperature (K)
T_m	melting temperature (K)
T_r	reference temperature (K)
A	yield stress (GPa)
B	hardening constant (GPa)
C	strain rate constant
n	hardening exponent associated with quasi-static test
m	thermal softening exponent
U	function describing the temperature at a point
x, y	coordinates (mm)
t	moment of time (μs)
a	thermal diffusivity (mm^2/s)
σ_{ts}	tensile strength (Pa)
ρ	material density (kg/m^3)
β	angle between the shock front and the vector of material flow into the collision point (°)
C_B	bulk sound velocity (m/s)
c	specific heat (erg/(g×°C))
h_f	flyer plate thickness (m)
h_b	base plate thickness (m)
N	empirically determined material-dependent coefficient
χ	thermal conductivity (W/(m×K))
x^*	distance from the collision point to the point behind it where the tensile stresses appear (m)

References

1. Wittman, R.H. The influence of collision parameters of the strength and microstructure of an explosion welded aluminium alloy. In Proceedings of the Proc. 2nd Int. Sym. on Use of an Explosive Energy in Manufacturing Metallic Materials, Marianske Lazne, Czech Republic, 9–12 October 1973; pp. 153–168.
2. Deribas, A.A. физика упрочнения и сварки взрывом; Nauka: Novosibirsk, Russia, 1980.
3. Zhang, Z.L.; Ma, T.; Liu, M.B.; Feng, D. Numerical Study on High Velocity Impact Welding Using a Modified SPH Method. *Int. J. Comput. Methods* **2019**, *16*, 1–24. [CrossRef]
4. Bataev, I.A.; Tanaka, S.; Zhou, Q.; Lazurenko, D.V.; Junior, A.M.J.; Bataev, A.A.; Hokamoto, K.; Mori, A.; Chen, P. Towards better understanding of explosive welding by combination of numerical simulation and experimental study. *Mater. Des.* **2019**, *169*, 107649. [CrossRef]
5. Zhang, Z.L.; Liu, M.B. Numerical studies on explosive welding with ANFO by using a density adaptive SPH method. *J. Manuf. Process.* **2019**, *41*, 208–220. [CrossRef]
6. Zhang, Z.L.; Feng, D.L.; Liu, M.B. Investigation of explosive welding through whole process modeling using a density adaptive SPH method. *J. Manuf. Process.* **2018**, *35*, 169–189. [CrossRef]
7. Bataev, I. Structure of Explosively Welded Materials: Experimental Study and Numerical Simulation. *Met. Work. Mater. Sci.* **2017**, *4*, 55–67. [CrossRef]
8. Feng, J.; Chen, P.; Zhou, Q.; Dai, K.; An, E.; Yuan, Y. Numerical simulation of explosive welding using Smoothed Particle Hydrodynamics method. *Int. J. Multiphys.* **2017**, *11*, 315–325.
9. Nassiri, A.; Chini, G.; Vivek, A.; Daehn, G.; Kinsey, B. Arbitrary Lagrangian-Eulerian finite element simulation and experimental investigation of wavy interfacial morphology during high velocity impact welding. *Mater. Des.* **2015**, *88*, 345–358. [CrossRef]
10. Vivek, A.; Liu, B.C.; Hansen, S.R.; Daehn, G.S. Accessing collision welding process window for titanium/copper welds with vaporizing foil actuators and grooved targets. *J. Mater. Process. Technol.* **2014**, *214*, 1583–1589. [CrossRef]
11. Liu, C.B.; Palazzotto, A.N.; Nassiri, A.; Vivek, A.; Daehn, G.S. Experimental and numerical investigation of interfacial microstructure in fully age-hardened 15-5 PH stainless steel during impact welding. *J. Mater. Sci.* **2019**, *54*, 9824–9842. [CrossRef]
12. Lee, T.; Zhang, S.; Vivek, A.; Daehn, G.; Kinsey, B. Wave formation in impact welding: Study of the Cu–Ti system. *CIRP Ann.* **2019**, *68*, 261–264. [CrossRef]

13. Nassiri, A.; Vivek, A.; Abke, T.; Liu, B.; Lee, T.; Daehn, G. Depiction of interfacial morphology in impact welded Ti/Cu bimetallic systems using smoothed particle hydrodynamics. *Appl. Phys. Lett.* **2017**, *110*, 231601. [CrossRef]

14. Nassiri, A.; Zhang, S.; Lee, T.; Abke, T.; Vivek, A.; Kinsey, B.; Daehn, G. Numerical investigation of CP-Ti and Cu110 impact welding using smoothed particle hydrodynamics and arbitrary Lagrangian-Eulerian methods. *J. Manuf. Process.* **2017**, *28*, 558–564. [CrossRef]

15. Chugunov, E.A.; Kuzmin, S.V.; Lysak, V.I.; Peev, A.P. Основные закономерности деформирования металла околошовной зоны при сварке взрывомалюминия. *Phys. Chem. Mater. Treat.* **2001**, *3*, 39–44.

16. Mahmood, Y.; Dai, K.; Chen, P.; Zhou, Q.; Bhatti, A.A.; Arab, A. Experimental and Numerical Study on Microstructure and Mechanical Properties of Ti-6Al-4V/Al-1060 Explosive Welding. *Metals* **2019**, *9*, 1189. [CrossRef]

17. Li, Y.; Liu, C.; Yu, H.; Zhao, F.; Wu, Z. Numerical simulation of Ti/Al bimetal composite fabricated by explosive welding. *Metals* **2017**, *7*, 407. [CrossRef]

18. Wang, X.; Zheng, Y.; Liu, H.; Shen, Z.; Hu, Y.; Li, W.; Gao, Y.; Guo, C. Numerical study of the mechanism of explosive/impact welding using Smoothed Particle Hydrodynamics method. *Mater. Des.* **2012**, *35*, 210–219. [CrossRef]

19. Nassiri, A.; Kinsey, B. Numerical studies on high-velocity impact welding: Smoothed particle hydrodynamics (SPH) and arbitrary Lagrangian–Eulerian (ALE). *J. Manuf. Process.* **2016**, *24*, 376–381. [CrossRef]

20. Liu, M.B.; Zhang, Z.L.; Feng, D.L. A density-adaptive SPH method with kernel gradient correction for modeling explosive welding. *Comput. Mech.* **2017**, *60*, 513–529. [CrossRef]

21. Tanaka, K. Numerical studies on the explosive welding by smoothed particle hydrodynamics (SPH). *AIP Conf. Proc.* **2007**, *955*, 1301–1304.

22. Zhou, Q.; Feng, J.; Chen, P. Numerical and experimental studies on the explosive welding of tungsten foil to copper. *Materials* **2017**, *10*, 984. [CrossRef]

23. Nishiwaki, J.; Sawa, Y.; Harada, Y.; Kumai, S. SPH analysis on formation manner of wavy joint interface in impact welded Al/Cu dissimilar metal plates. *Mater. Sci. Forum* **2014**, *794–796*, 383–388. [CrossRef]

24. Chu, Q.; Zhang, M.; Li, J.; Yan, C. Experimental and numerical investigation of microstructure and mechanical behavior of titanium/steel interfaces prepared by explosive welding. *Mater. Sci. Eng. A* **2017**, *689*, 323–331. [CrossRef]

25. Meyers, M.A. *Dynamic Behavior of Materials*; John Wiley & Sons, Inc.: Hoboken, NJ, USA, 1994; ISBN 9780470172278.

26. Johnson, G.R.; Cook, W.H. A constitutive model and data for metals subjected to large strains, high strain rates and high temperatures. In Proceedings of the 7th International Symposium on Ballistics, the Hague, The Netherlands, 19–21 April 1983; pp. 541–547.

27. Johnson, G.R.; Cook, W.H. Fracture characteristics of three metals subjected to various strains, strain rates, temperatures and pressures. *Eng. Fract. Mech.* **1985**, *21*, 31–48. [CrossRef]

28. Corbett, B.M. Numerical simulations of target hole diameters for hypervelocity impacts into elevated and room temperature bumpers. *Int. J. Impact Eng.* **2006**, *33*, 431–440. [CrossRef]

29. Lesuer, D.R.; Kay, G.J.; LeBlanc, M.M. Modeling Large-Strain, High-Rate Deformation in Metals. In Proceedings of the Third Biennial Tri-Laboratory Engineering Conference Modeling and Simulation, Pleasanton, CA, USA, 3–5 November 1999.

30. Deribas, A.A. Classification of flows appearing on oblique collisions on metallic plates. In Proceedings of the Proc. 2nd Int. Sym. on Use of an Explosive Energy in Manufacturing Metallic Materials, Marianske Lazne, Czech Republic, 9–12 October 1973; pp. 31–44.

31. Efremov, V.V.; Zakharenko, I.D. Determination of the upper limit to explosive welding. *Combust. Explos. Shock Waves* **1977**, *12*, 226–230. [CrossRef]

32. Mali, V.I.; Simonov, V.A. Some effects appearing on interactions of shock waves with cavities in metals. In Proceedings of the Proc. 2nd Int. Sym. on Use of an Explosive Energy in Manufacturing Metallic Materials, Marianske Lazne, Czech Republic, 1974; pp. 83–96.

33. Carvalho, G.H.S.F.L.; Galvão, I.; Mendes, R.; Leal, R.M.; Loureiro, A. Explosive welding of aluminium to stainless steel. *J. Mater. Process. Technol.* **2018**, *262*, 340–349. [CrossRef]

34. Wu, Y.; Lu, J.; Tan, S.; Jiang, F.; Sun, J. Modified implementation strategy in explosive welding for joining between precipitate-hardened alloys. *J. Manuf. Process.* **2018**, *36*, 417–425. [CrossRef]

35. Saravanan, S.; Raghukandan, K. Influence of Interlayer in Explosive Cladding of Dissimilar Metals. *Mater. Manuf. Process.* **2013**, *28*, 589–594. [CrossRef]

36. de Rosset, W.S. Analysis of Explosive Bonding Parameters. *Mater. Manuf. Process.* **2006**, *21*, 634–638. [CrossRef]

37. Walsh, J.M.; Shreffler, R.G.; Willig, F.J. Limiting Conditions for Jet Formation in High Velocity Collisions. *J. Appl. Phys.* **1953**, *24*, 349–359. [CrossRef]

38. Cowan, G.R.; Holtzman, A.H. Flow Configurations in Colliding Plates: Explosive Bonding. *J. Appl. Phys.* **1963**, *34*, 928–939. [CrossRef]

39. Narsh, S.P. *LASL Shock Hugoniot Data*; University of California Press: Berkeley, CA, USA, 1980.

40. Lysak, V.I.; Kuzmin, S.V. Lower boundary in metal explosive welding. Evolution of ideas. *J. Mater. Process. Technol.* **2012**, *212*, 150–156. [CrossRef]

41. Chuvichilov, V.A.; Kuz'min, S.V.; Lysak, V.I.; Dolgiy, U.G.; Kokorin, A.V. Research of structure and properties of the composite materials received on battery scheme of explosion welding. *News Volgogr. State Tech. Univ.* **2010**, *5*, 34–43.

42. Lysak, V.I.; Kuzmin, S.V.; Dolgiy, U.G. Formation a welded joint by explosive spot welding. *News Volgogr. State Tech. Univ.* **2013**, *18*, 4–13.

43. Zlobin, B.S. *Development of the Scientific Basis for the Manufacturing Process of Bimetallic Bearing Blanks Using Explosion Welding*; Institute of Computational Technologies SB RAS: Novosibirsk Oblast, Russia, 2000.

44. Cowan, G.R.; Bergmann, O.R.; Holtzman, A.H. Mechanism of bond zone wave formation in explosion-clad metals. *Metall. Mater. Trans. B* **1971**, *2*, 3145–3155. [CrossRef]

45. Szecket, A. *An Experimental Study of the Explosive Welding Window*; Queen's University of Belfast: Belfast, Northern Ireland, 1979.

46. Lysak, V.I.; Kuzmin, S.V. *Сварка взрывом*; Mashinostroyeniye: Volgograd, Russia, 2005.

47. Deribas, A.A.; Kudinov, V.M. Влияние начальных параметров на процесс волнообразования на при сварке металлов взрывом. физика горения и взрыва. *Combust. Explos. Shock Waves* **1967**, *3*, 561–568.

48. Kuzmin, G.E.; Yakovlev, I.V. Исследование соударения металлических пластин со сверхзвуковой скоростью точки контакта. *Combust. Explos. Shock Waves* **1973**, *9*, 746–753.

49. Zakharenko, I.D. *Сварка металлов взрывом*; Навука i тэхнiка: Minsk, Russia, 1990; ISBN 5-343-00551-9.

50. Zakharenko, I.D. Критические режимы при сварке взрывом. *Combust. Explos. Shock Waves* **1972**, 422–427.

51. Efremov, V.V.; Zakharenko, I.D. К определению верхней границы области сварки взрывом. *Combust. Explos. Shock Waves* **1976**, 255–260.

Article

Experimental and Numerical Study on Microstructure and Mechanical Properties of Ti-6Al-4V/Al-1060 Explosive Welding

Yasir Mahmood, Kaida Dai *, Pengwan Chen *, Qiang Zhou, Ashfaq Ahmad Bhatti and Ali Arab

State Key Laboratory of Explosion Science and Technology, Beijing Institute of Technology, Beijing 100081, China; 3820160003@bit.edu.cn (Y.M.); zqpcgm@gmail.com (Q.Z.); ashfaqb@live.com (A.A.B.); arabali83@yahoo.com (A.A.)
* Correspondence: daikaida@bit.edu.cn (K.D.); pwchen@bit.edu.cn (P.C.); Tel.: +86-010-6891-8740 (K.D. & P.C.)

Received: 15 October 2019; Accepted: 2 November 2019; Published: 5 November 2019

Abstract: The aim of this paper is to study the microstructure and mechanical properties of the Ti6Al4V/Al-1060 plate by explosive welding before and after heat treatment. The welded interface is smooth and straight without any jet trapping. The disturbances near the interface, circular and random pores of Al-1060, and beta phase grains of Ti6Al4V have been observed by Scanning electron microscopy (SEM). Heat treatment reduces pores significantly and generates a titanium-island-like morphology. Energy dispersive spectroscopy (EDS) analysis results show that the maximum portion of the interfacial zone existed in the aluminium side, which is composed of three intermetallic phases: TiAl, $TiAl_2$ and $TiAl_3$. Heat treatment resulted in the enlargement of the interfacial zone and conversion of intermentallic phases. Tensile test, shear test, bending test and hardness test were performed to examine the mechanical properties including welding joint qualities. The results of mechanical tests show that the tensile strength and welding joint strength of the interfacial region are larger than one of its constituent material (Al-1060), the microhardness near the interface is maximum. Besides, tensile strength, shear strength and microhardness of heat treated samples are smaller than unheat treated. Smooth particle hydrodynamic (SPH) method is used to simulate the transient behaviour of both materials at the interface. Transient pressure, plastic deformation and temperature on the flyer and base side during the welding process were obtained and analyzed. Furthermore, the numerical simulation identified that almost straight bonding structure is formed on the interface, which is in agreement with experimental observation.

Keywords: explosive welding; Ti6Al4V/Al-1060; microstructure; mechanical properties; smooth particle hydrodynamic (SPH)

---

1. Introduction

Welding is a useful technique for joining two similar or different materials. Nowadays, various welding techniques i.e., gas welding, arc welding, laser welding, friction welding, explosive welding etc. have been widely used in many industries. The explosive welding is a solid state welding process in which a flyer plate is accelerated by an explosive and welded with other material in a short interval of time. During the welding process, a high velocity jet removes the impurities on the material surfaces [1]. Explosive welding was used in many mechanical related industries, especially in power plants and aerospace industry. Moreover, it was successfully applied to produce biocompatible materials in the medical related fields [2,3]. This method is advantageous to weld different kinds of metals and alloys that cannot easily be welded by some other means of welding. Materials having excellent mechanical properties (i.e., corrosive resistance, high strength to density ratio, good conductance and heat resistance, etc.) can be welded with materials having low mechanical properties. This type of welding makes the materials more reliable and cost effective.

Both aluminium and titanium are widely used in the transportation and aerospace industries. Al-1060 belongs to commercially pure aluminium. It is highly malleable and corrosion resistant. However, it is very reactive with air at high temperature and has low mechanical strength. Titanium and its alloys have the highest strength to mass ratio in all naturally existing materials. So their combination can be used as an excellent industrial application. Explosive welding is one of the best choice for bonding these two metal alloys because the two materials have a big difference in their melting points, and mechanical strengths [4]. Furthermore, this combination has been received special attention due to the formation of titanium aluminides at the interface. Titanium aluminides have low density, good oxidation ignition resistance and excellent mechanical properties at high temperatures. Due to these features titanium aluminides get special attention and researchers practised different techniques to develop titanium aluminides. Previously, Perusko et al. [5] investigated the intermetallic formation during Al/Ti welding by using Ar^+ ion implementation. Adeli et al. [6] used induction preheating process to synthesis the Ti/Al powder and prepared TiAl phase. Arakawa et al. [7] formed titanium aluminide foam with the help of self-propagating high-temperature synthesis (SHS). They created a foam which contains 60% to 70% porosity due to $TiAl_3$. Titanium aluminides percentage is increased by using diffusion and combustion [8]. DC magnetron sputtering method was opted by Ramos et al. [9] to prepare γ-TiAl. Furthermore, they controlled activation energy by silver foil. Romankov et al. [10] prepared $TiAl_3$ intermetallic by using thermal deposition technique.

Kahraman et al. [11] studied the complex microstructure of explosively welded Ti6Al4V and aluminium alloy using different charge to mass ratios. They found that hardness and corrosion were increased at the interface with the increase of the charge to mass ratios. E et al. [12] evaluated the tensile properties of the explosively welded Ti/Al sample with a load applied along parallel and perpendicular directions to the interface. Xia et al. [13] observed the micro grains and recrystallisation of titanium alloy in the interface. Fronczek et al. [14] and Bazarnik et al. [15] studied mechanical properties and microstructure of the explosively welded titanium alloy with aluminium alloy.

Additionally, the Ti/Al interface study is essential to check the quality of welded joints. According to Ege et al. [16], at the Ti/Al interface, the aluminium percentage content reinforces the Ti/Al joints, providing stability and increasing the strength up to 825 MPa. Inal et al. [17] found that the straight interface is more suitable than the wavy interface because it accompanied by excessive heat generation and could produce weak and brittle intermetallics. Ege et al. [18] presented that the heat treatment of the Ti/Al welded sample did not affect the stability, therefore could be used for high temperature environment. Multilayered combination for explosive welding of titanium and aluminium is one of the best options to fabricate titanium aluminides. Lazurenko et al. [19] welded forty layers and Mali et al. [20] joined 23 of Ti-Al by using explosive welding technique. They produced $TiAl_3$ stable intermetallics by using sintering at 640 °C temperature and 3MPa pressure. Furthermore, Bataev et al. [21] studied the thickness of $TiAl_3$ from top to bottom plates and compared the results with post heat treated samples. Likewise, Foaden et al. [22] analysed the $TiAl_3$ formation process after annealing of the explosively welded composite plate. Fan at al. [23] prepared multilayered TiAl foils by using vacuum hot pressing and studied the foil thickness effect. Similarly, Fan et al. [24] welded 5 Ti-Al plates by using an explosive welding technique to analyse the microstructure and mechanical properties of composite plates.

Simulation is an excellent tool to predict the explosive welding conditions, i.e., impact velocity, pressure, plastic strain, etc. Many authors applied various approaches to simulate explosive welding phenomenon. Recently, Mousavi et al. [25] used ANSYS AUTODYN to simulate the straight, wavy, jetting morphology and humps formation during the welding process. Nassiri et al. [26] replicated the jetting phenomena with the help of SPH method, and exercised Arbitrary Lagrangian-Eulerian (ALE) method to obtain other welding parameters, i.e., interface shape, temperature, the velocity of material, shear stress. Wang et al. [27] imitated the shear stress and effective plastic strain of materials by using the SPH approach. Wang et al. [28] reproduced the whole explosive welding phenomena with the help of material point method (MPM) in C++ program.

In this paper, microstructure and mechanical properties of the welded composite plate (Ti6Al4V and Al-1060) were investigated. The SPH approach with the help of LS-DYNA was used to understand the welding conditions i.e., interface morphology, transient pressure, temperature and plastic deformation.

2. Experimental Procedure

2.1. Materials and Explosive Welding Setup

Ti6Al4V was used as a flyer plate and Al-1060 as a base plate because titanium had lower thermal diffusivity than aluminium [29]. Both plates with the size of 200 mm × 150 mm × 3 mm were welded in air. The schematic experimental setup is shown in Figure 1. Both plates were arranged in parallel configuration with a standoff distance of 5 mm. Ammonium nitrate and fuel oil (ANFO) with 30 mm thickness (packing density is 670 Kg/m^3 and detonation velocity is 2600 m/s) was used to accelerate the flyer plate. The sand was employed as an anvil. The welded samples were heat treated at 525 °C for 4 h and then cooled in open air. Inal et al. [17] reported that the bond strength remains constant at this heating temperature and time duration.

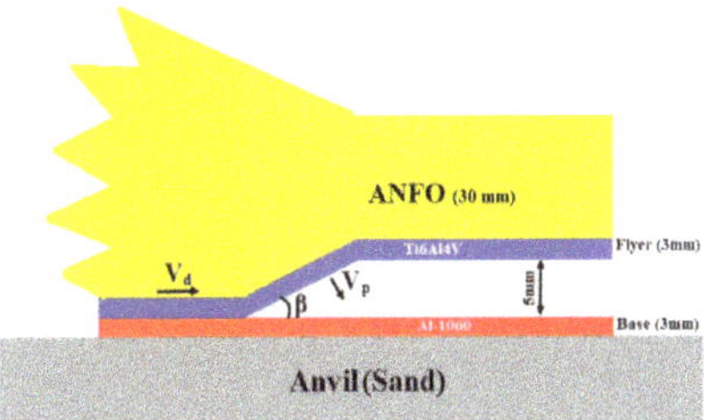

Figure 1. Schematic diagram of explosive welding.

2.2. Microstructure Test

For microstructure analysis, samples were cut from the middle of the welded plate approximately 100 mm away from the detonating point. The samples were mounted in an epoxy such that the welding interface was parallel to the detonation direction. The samples were cleaned with sandpapers (Grit 320, 1000, 1500, 2000), and then polished with ceramic powder. Finally, the samples were soaked in the Kroll reagent (distilled water 92 mL, nitric acid 6 mL, hydrofluoric acid 2 mL) for 20 s to examine the apparent microstructure.

The microstructure and elements distribution at the interface was observed by using scanning electron microscope (SEM, Hitachi S-4800, Tokyo, Japan) at an acceleration voltage of 15 kV. SEM elemental scan images were further edited to combine both elements (Ti-Al) using CoralDraw (Coral Corporation, Ottawa, ON, Canada).

2.3. Mechanical Test

The Vickers hardness tests were conducted along the perpendicular direction of the interface. The hardness tests were carried out on a microhardness machine using a 0.98 N load for 15 s.

For mechanical tests, stratified samples were obtained from the welded plate before and after heat treatment. Tensile test and shear test samples (as seen in Figure 2a,b) were performed on material testing machine (MTS) with a loading rate of 5 mm/min and 1.8 mm/min respectively to measure the tensile properties of the welded portion. For three point bending test, the sample with a thickness of 5 mm and a length of 100 mm was conducted along the perpendicular direction of the flyer-base interface.

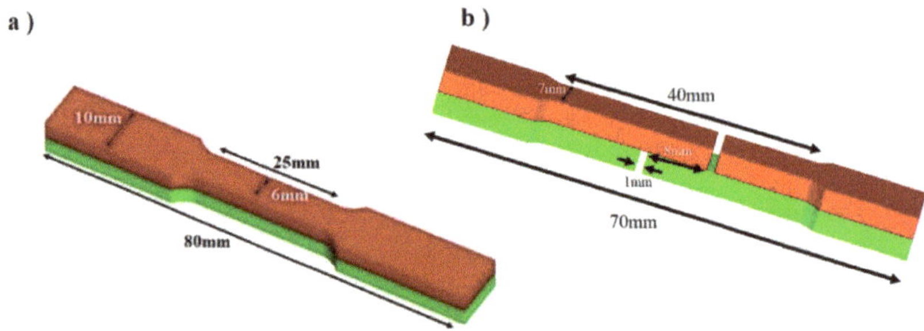

Figure 2. Schematic diagram for (**a**) tensile test (**b**) shear tensile test.

3. Simulation of Explosive Welding

ANFO has non-ideal detonation behavior [30]. It is difficult to simulate it with Jones-Wilkins-Lee (JWL) equation because it has a long reaction zone, so energy and momentum at Von-Neuman Spike cannot be neglected [31]. Therefore empirical formula was used to calculate the plate velocity and impact angle.

3.1. Calculation of Plate Velocity

Deribas et al. [32] and Manikandan et al. [33] developed a distance dependent empirical relation to calculate the impact angle between flyer and base plate. The impact angle is defined as,

$$\beta(rad) = \left(\sqrt{\frac{k+1}{k-1}} - 1 \right) \frac{\pi}{2} \frac{R}{\left(R + A + \frac{Bt_e}{S}\right)} \tag{1}$$

where $A = 2.71$, $B = 0.184$, t_e is the thickness of explosive, S is the standoff distance between the flyer and base plate. Polytropic exponent k can be measured with the by Gurney velocity [32].

$$k = \sqrt{\frac{D^2}{V_g^2} + 1} \tag{2}$$

where V_g represents the Gurney velocity, which can be calculated by using the following formula [34–36].

$$\sqrt{2E}(m/s) = 600 + 0.52 \frac{D}{\sqrt{\gamma + 1}} \tag{3}$$

where γ is a ratio of specific heat constant. For ANFO, the value of γ is 2.881 [37]. Plate velocity can be calculated by Crossland relation [38].

$$V_p = 2D sin\left(\frac{\beta}{2}\right) \tag{4}$$

Using the above equations for the current experimental conditions, plate velocity V_p with 5 mm standoff distance is 707 m/s and the impact angle β is 15.04°.

3.2. Equation of State and Constitutive Model

Mei Gruneisen equation of state is mostly used for shock wave propagation [27]. This equation gives us a relation between pressure and volume under shock conditions at a given temperature. Johnson-Cook material model was used for the simulation of explosive welding, because it can successfully predict the high deformation and Von-Misses yield stress of the material. Johnson-Cook equation can be written as

$$\sigma = (A + B\varepsilon^n)\left(1 + C\,ln\dot{\varepsilon}_p\right)(1 - T^{*m}) \tag{5}$$

where ε is plastic strain, C is strain rate constant, $\dot{\varepsilon}_p$ is plastic strain rate, n is hardening exponent, A is yield strength of the material, m is softening exponent, B is strain hardening coefficient and T^* is homologous temperature and equal to $T^* = \frac{(T - T_{room})}{(T_{melt} - T_{room})}$. The parameters of material models for Ti6Al4V and Al-1060 are listed in Table 1.

Table 1. Parameters of material model and equation of state.

Material	A (MPa)	B (MPa)	N	C	M	Density (Kg/m^3)	C_o (m/s)	G	S
Ti-6Al-4V [39]	1098	1092	0.93	0.014	1.1	4430	5130	1.23	1.028
Al-1060 [40]	66.56	108.8	0.23	0.029	1	2707	5386	1.97	1.339

3.3. Smooth Particle Hydrodynamics (SPH)

Numerical gridless lagrangian hydrodynamics simulation was carried out with the SPH method in ANSYS/LS-DYNA (Ansys 14.5, LSTC, Livermore, CA, USA). In SPH method, Kernel approximation of field variable at given points is applied to simplify the conservation equations. Discrete particle helps to find out field information while neighboring particles are used to solve the integrals. If neighboring particles are expressed by subscript j, then field variable for non zero Kernel approximation can be described as follow:

$$f(r) \cong \sum \frac{m_p}{\rho_p} f_j W(|r - r_j|,\ h|X|, h) \tag{6}$$

where m_p, ρ_p are particle mass and density respectively, $f(r)$ is a field variable, r represents the location of particle, h shows the length at which particle is effected by the neighboring particles. For the current simulation of the explosive welding, it is simplified with the collision of the flyer and base plate at a certain velocity and angle. These two parameters are calculated by Equation (1) and Equation (4), respectively. The schematic model for SPH simulation is shown in Figure 3, with a particle size of 0.5 μm.

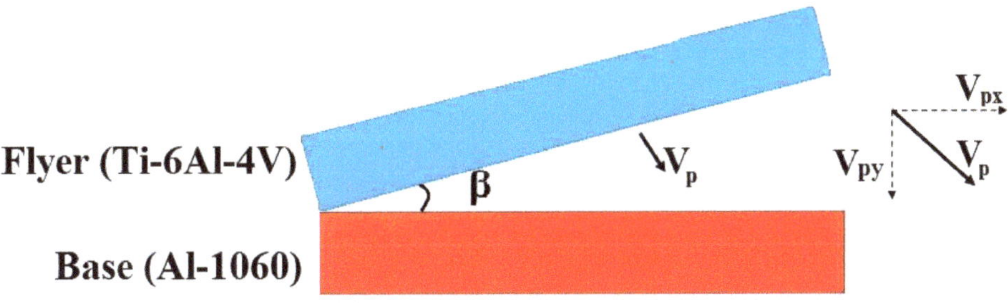

Figure 3. Model for SPH simulation.

4. Results and Discussion

4.1. Microstructure of Welding Interface before Heat Treatment

The SEM results of the welded sample before heat treatment are displayed in Figure 4. It shows that welding interface is smooth and flat without delamination. This type of smooth Ti/Al interface joint pattern was previously observed by Bazarnik et al. [15] for Ti6Al4V/Al2519 welding and Ege et al. [16] for Ti6Al4V/Al6061 multilayer welding.

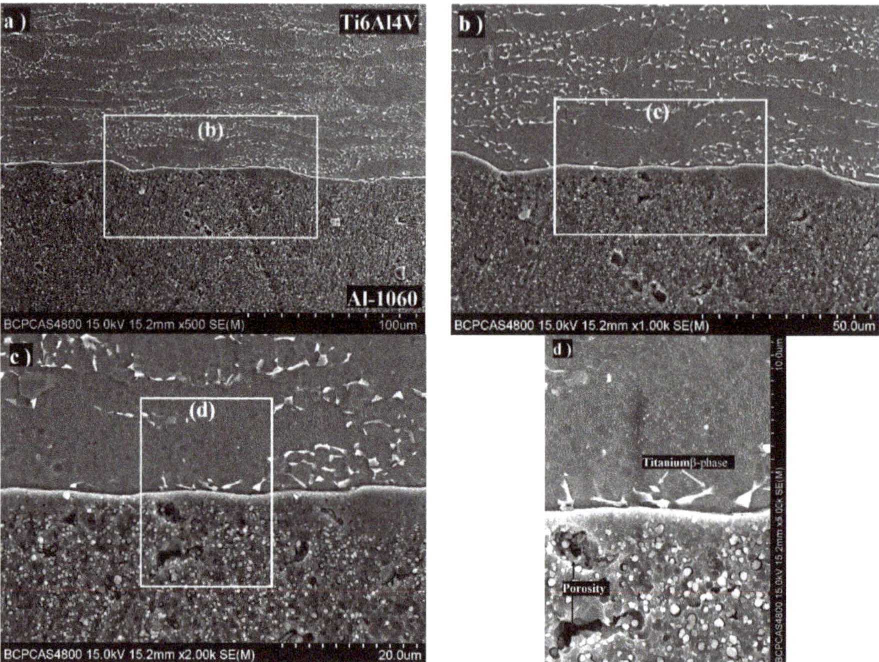

Figure 4. SEM image of explosively weldedTi6Al4V/Al1060 interface before heat treatment at magnifications of (**a**) 500×, (**b**) 1000×, (**c**) 2000×, (**d**) 5000×.

Seeing from Figure 4a–d, both metals have some disturbances in microstructures near the interface, especially aluminium side shows the maximum disturbances. High pressure and deformation during explosive welding process may generate recrystallization of microstructure. These phenomena near the interface formed under high pressure were previously observed by Zhang et al. [41] and Gloc et al. [42]. Furthermore, Figure 4d shows that Al-1060 contains some circular and random pores. These pores are formed during severe deformation and high temperature gradient, causing rupture of local surfaces. Raoelison et al. [43] analyzed the formation of pores during the welding process and stated that these pores might be the origin of the crack propagation. The same pores were observed by Su et al. [44] during the welding process of Fe-Al. As shown in Figure 4d, Ti6Al4V microstructure near the interface is also altered, especially beta phase grains that are elongated toward the detonation direction. During the explosive welding process, due to high strain and pressure, the temperature of the interfacial zone is abruptly raised. This temperature increment is high enough to change the phase of the plate [25]. Therefore, Ti6Al4V is most likely to be converted to the beta phase during the collision. However, the cooling rate is very high in this process, the titanium phase change process only takes a short interval and quickly turns back to the alpha phase [38]. According to Tomashchuk et al. [45] the titanium beta phase is more likely to react with aluminium than alpha phase to form titanium aluminide. The intense plastic deformation and elongation were also observed by Murr et al. [46] and Kacar et al. [47].

The interface between titanium and aluminiumhas a very complicated structure, so EDS technique is used to understand the microstructure. Figure 5a shows an elemental scan near the interface region, indicating that the Ti6Al4V/Al-1060 interface is smooth and flat. Figure 5b reveals that aluminium element counts in the base plate start decreasing about 0.86 µm away from the interface. This decrease in aluminium counts continues in flyer plate and gets normal counts after 0.24 µm thickness from the interface. So the total width of the interfacial zone is 1.1 µm with maximum portion existing in the base side (aluminium). Slope difference shows that aluminium counts near interface decline faster in the base plate as compared to the flyer plate side.

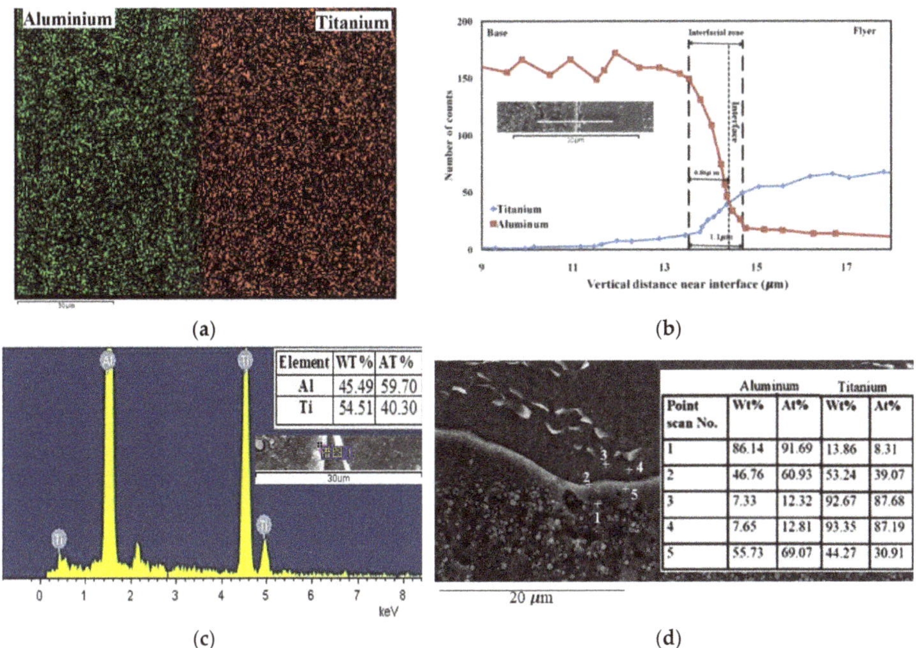

Element	WT%	AT%
Al	45.49	59.70
Ti	54.51	40.30

Point scan No.	Aluminum		Titanium	
	Wt%	At%	Wt%	At%
1	86.14	91.69	13.86	8.31
2	46.76	60.93	53.24	39.07
3	7.33	12.32	92.67	87.68
4	7.65	12.81	93.35	87.19
5	55.73	69.07	44.27	30.91

Figure 5. (**a**) Elemental scan of aluminium and titanium elements along with welding interface, (**b**) graphical representation of linear element analysis at the interface, (**c**) EDS point spectrum at the interface, (**d**) Ti-Al atomic distribution near interface at selected points.

Parasithi et al. [48] reported that materials with higher thermal conductivity keep a maximum percentage in interfacial area. Since aluminium has almost more than 25 times higher thermal conductivity as compared to titanium, that is why maximum microstructural changes are observed in the base plate. Manikandan et al. [33] gave another demonstration about the interfacial zone. According to their study, at the interface, the atomic concentration of base material is relatively higher than the flyer material.

Due to the extreme conditions in explosive welding process, chemical equilibrium was not achieved. Consequently, this may exhibit various kinds of intermetallics at different metastable equilibrium states [49]. Ti-Al has three main titanium aluminides TiAl, TiAl$_3$, and Ti$_3$Al. TiAl$_2$ is another equilibrium phase of titanium aluminide. However, this phase is formed under the overlap condition with the TiAl phase [14]. Figure 5c shows that the atomic percentage of aluminium in the interface is 60% and titanium is about 40%. It indicates that this region is formed by overlapping of two different phases of Ti-Al, i.e., TiAl + TiAl$_2$. Fronczek et al. [14] observed this phase with the help of X-ray synchrotron reflection method. Similarly, in Figure 5d, EDS point scan exhibits that point 2 has a chance of overlapped phase TiAl + TiAl$_2$ and point 5 may contains TiAl$_3$ phase. While the remaining points show no external interference to other elements.

4.2. Microstructure of Welding Interface after Heat Treatment

Figure 6a–d with different selected area magnifications show that the interfacial zone becomes wider after 4 h heat treatment. Pores are significantly decreased and grains are rearranged. High deformation can produce pores and cracks during explosive welding. Consequently, when the stress of the composite plate is relieved by heat treatment process, then these cracks can generate the island/peninsula-like morphology. Furthermore, the formation of this island/peninsula pattern is influenced by the detonation force and metal vortex flow. Previously, many researchers observed this type of morphology [16,48,49]. Figure 6d shows this titanium-island-like shape (marked by red line) in the aluminium rich area. These types of islands may affect the mechanical performance of the welded plate [50,51].

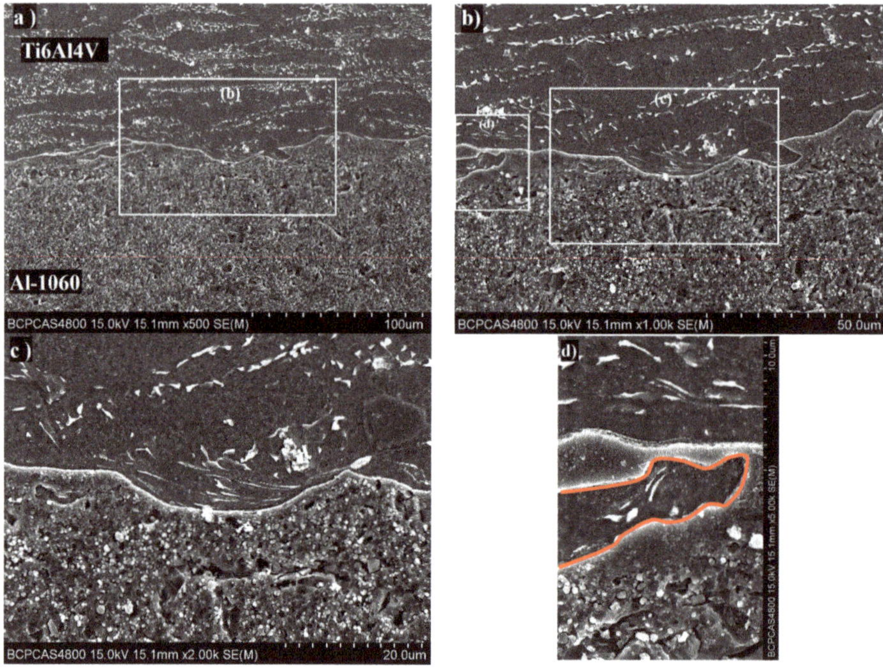

Figure 6. SEM images of explosively welded Ti6Al4V/Al1060 interfaces after heat treatment at different magnifications of selected regions (**a**) 500×, (**b**) 1000×, (**c**) 2000×, (**d**) 5000×.

The elemental scan of Ti-Al interface (Figure 7a) shows that at the interface, an island (as indicated by arrow) is formed with titanium as the dominant element while the surrounding area consists of titanium aluminides. Their detailed atomic percentage can be determined by using the EDS line and point scans (Figure 7b–d). As shown in Figure 7b, two positions are selected for line scan EDS. One line passes through islands and the neighboring area of interface (Figure 7c), while the second line passes through interface and surroundings (Figure 7d).

Figure 7b describes the point elemental scan near interface and island morphology. The point scans 3 and 5 show that the area near the interface contains titanium aluminides. Based on EDS atomic percentage analysis, there is a possibility of the existence of most stable titanium aluminide (TiAl$_3$). It can further be verified from line scan in Figure 7c, which shows that region 2 has higher titanium counts as compared to the base plate side. It indicates that the post heat treatment process influences the aluminides equilibrium state. Points 1, 2 and 4 scans show that there is no reasonable external involvement of any element in each side. Particularly point 4 is situated in the mid of island morphology. This point shows that the island is formed from the flyer plate.

EDS line scan (Figure 7c) shows that the interfacial zone is expended up to approximately 11.55 µm, which is almost 10 times larger than unheated sample (Figure 5b). In this interfacial area, the titanium island surrounded by Ti-Al intermetallics is observed, which has a width of 5 µm. Titanium aluminides of different equilibrium phases have been observed in the area between island and interface, which is expanded up to 4.6 µm. Heat treatment reduces the number of element counts in the unit area, increases the titanium penetration in aluminium and stretched the interfacial zone. Furthermore, the equilibrium phases of titanium aluminide are formed. Especially, TiAl$_3$ is dominated as compared to all other phases. This kind of variations were also observed by Gloc et al. [42] and Fronczek et al. [14].

Figure 7d shows that interfacial zone becomes more extensive than the sample before heat treatment. Although titanium is penetrated into aluminium up to 3 µm, the high level of Ti-Al mixing is in the range of 1.7 µm with a maximum area existed in the aluminium side. Comparison of the Figure 7c–d indicates that the interfacial area has a difference of about 9.8 µm. It shows that the post heat treatment process makes the interface irregular shape. Additionally, it is noticed that titanium aluminides have existed in the area between the titanium island and the flyer plate. This area did not directly expose during the collision of the plates. It implies that titanium aluminides was created during the heat treatment process.

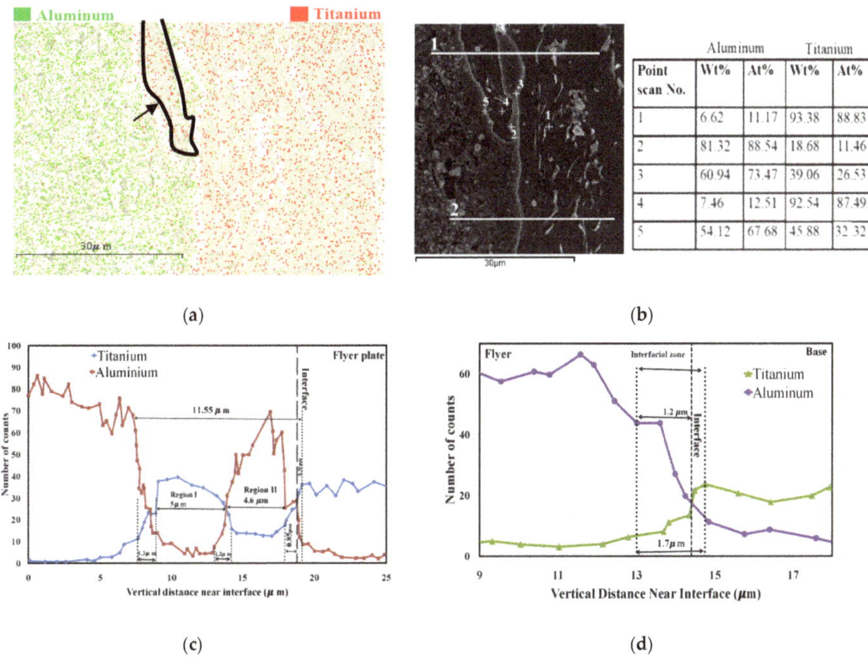

Point scan No.	Aluminum		Titanium	
	Wt%	At%	Wt%	At%
1	6.62	11.17	93.38	88.83
2	81.32	88.54	18.68	11.46
3	60.94	73.47	39.06	26.53
4	7.46	12.51	92.54	87.49
5	54.12	67.68	45.88	32.32

Figure 7. Pictures and EDS after heat treatment (**a**) Ti-Al mapping near interface, (**b**) Ti-Al atomic distribution near interface at selected points, (**c**) linear distribution of Ti-Al near interface (at position 1 in Figure 7b), (**d**) Linear distribution of Ti-Al near interface (at position 2 in Figure 7b).

4.3. Mechanical Tests

Tensile tests were conducted to study the mechanical response of the welded material. Two samples were prepared from normal and heat treated welded plates, as shown in Figure 8.

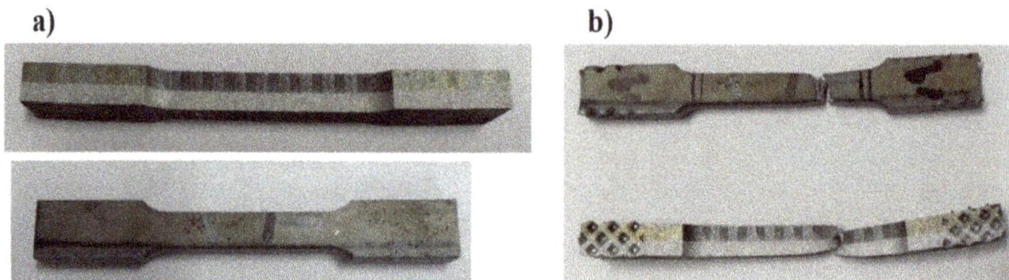

Figure 8. Tensile test samples (**a**) before the test; (**b**) after the test.

For successful bonding, the tensile strength of the welded sample should be higher than the weaker material used for welding [50]. Ege et al. [16] proposed that the yield strength of welded material depends on aluminium volumetric percentage and interfacial density. Figure 9 shows the stress-strain curves of the tested samples before and after heat treatment. After heat treatment, the tensile strength of the welded sample was decreased, while the elongation and plasticity were increased. Furthermore, Figure 9 shows that a small jerk occurred at the end of elastic limit of the unheated sample. This jerk state indicates that failure begins from the weak material (Al-1060) before the tensile strength is reached. The experimental results of the tensile tests were compared with the mechanical properties of the Al 1060 [51] and Ti6Al4V [52] based on their percentage thickness in the welded plate (Table 2). This approximation helps us to estimate the tensile properties of the welded sample [16,53]. Detailed results are shown in Table 2.

It is indicated that both samples have better results than the calculated values. Furthermore, experimental results exhibit that the tensile strength and yield stress of heat treated samples are reduced as compared to the unheated sample, but elongation is improved from 20% to 23%.

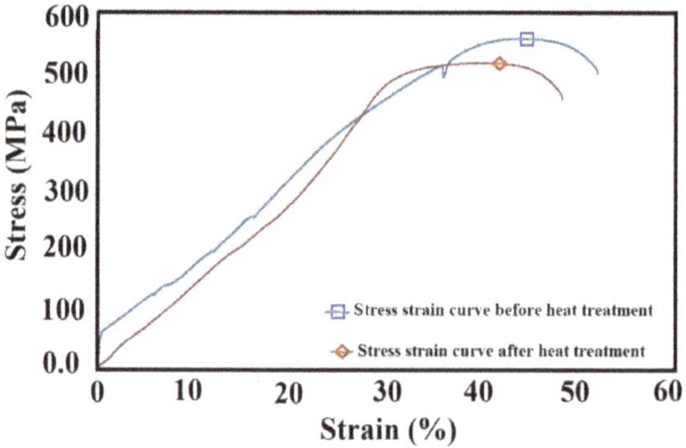

Figure 9. Engineering stress-strain profile before and after heat treatment.

Table 2. Tensile test results of the welded samples together with specific materials.

Samples	UTS (MPa)	Yield Stress (MPa)	Elongation (%)
Ti-6Al-4V [52]	947	872	13
Al-1060 [51]	97	110	28
Calculated value	509	446	20.7
Before heat treatment	560 ± 4	486 ± 8	20 ± 1
After heat treatment	525 ± 1	462.5 ± 8	23 ± 2

Three points bending test results show that no fractures or cracks were observed in any samples for both before and after heat treatment, as seen in Figure 10a,b.

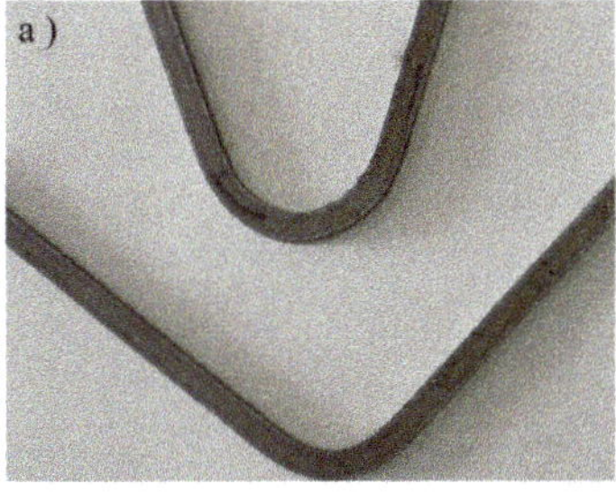

Figure 10. Three points bending test (**a**) side view of 90°and 180° bent sample; (**b**) bending interface.

In addition to the bending test and tensile test, the flat shear test was conducted to verify the welding joint strength for both samples before and after heat treatment. Figure 11a indicates that the tensile strength of aluminium decreases after the heat treatment. Figure 11b shows that the deformation starts in the aluminium and the joint displays no disturbance. This result reveals that the joint is stronger than the Al-1060 tensile strength ~40MPa.

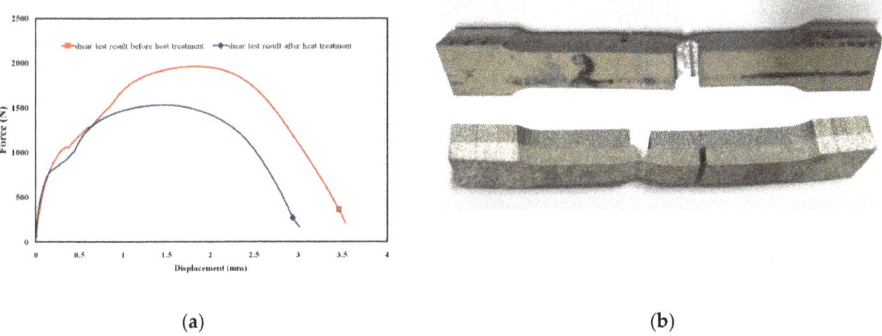

(**a**)   (**b**)

Figure 11. (**a**) Shear force versus displacement (**b**) fracture morphology after shear test.

Figure 12 shows that near the interface, the microhardness is maximum. Its value is about ~412, higher than the flyer and base standard values (flyer plate ~350, base plate ~30), which is due to high value of heat produced during explosive welding caused annealing at the contact point [47]. Heat treatment relaxed the interface and widened the interfacial zone, which affected the hardness value and caused it to decrease to ~390 at the interface. As moving away from the interface, hardness becomes approaching to the typical values of flyer and base plate

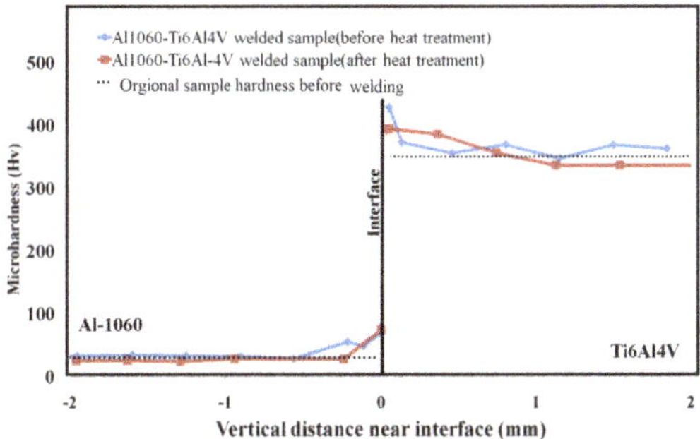

Figure 12. Microhardness (Hv) profiles before and after heat treatment.

4.4. Simulation Results

Figure 13a shows that smooth interface obtained by numerical simulation, which is in good agreement with the experimental results (see in Figure 4). Since Ti6Al4V is a very hard material and Al-1060 is soft, the flyer can easily penetrate into the base material. Furthermore, Figure 13a explains that some partial wavy shapes patterns appear, but due to high impact velocity and density difference, these waves are suppressed and became almost flat. Figure 13b exhibits that jet is formed during this process and the maximum portion of the jet consisted of base plate (Al-1060).

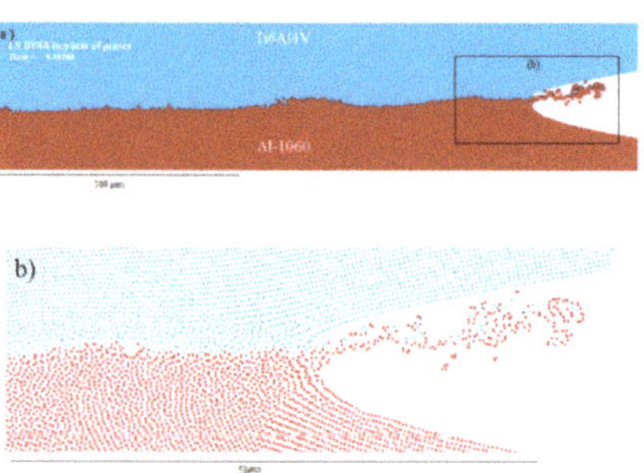

Figure 13. Simulation results for (**a**) Ti6Al4V/Al-1060 interface and (**b**) jet formation.

Figure 14a,b show that near the contact point, the pressure distribution is behind the jet. It means that the collision velocity is subsonic, which is an essential requirement for the bonding of materials. According to Blazynski et al. [4], the bonding contact pressure should be higher than material yield strength. Pressure contour plots (Figure 14a,b) illustrate that the transient pressure profile at all contact points during the collision is more than 10 GPa. Figure 14c supports that, at time 0.256 μs, the peak flyer transient pressure rises to 13.1 GPa, which is considerably higher than the yield strength of the flyer and base plate.

Additionally, it was observed (Figure 14c) that in the base plate, the pressure and impulse were slightly lower than that of the flyer. According to Holtzman et al. [54] investigation, if the base plate had a greater contribution to the jetting, there would be more transient pressure in the flyer side. Similarly, Mousavi et al. [37] simulated the same pressure difference between the flyer and base plates.

Furthermore, Figure 14c shows that there is a slight hump in the pressure-time curves before reaching the peak value. This hump indicates that the deformity had begun before the arrival of the shock wave. Cowen et al. [55] reported that deformation and jet formation were necessary for material bonding.

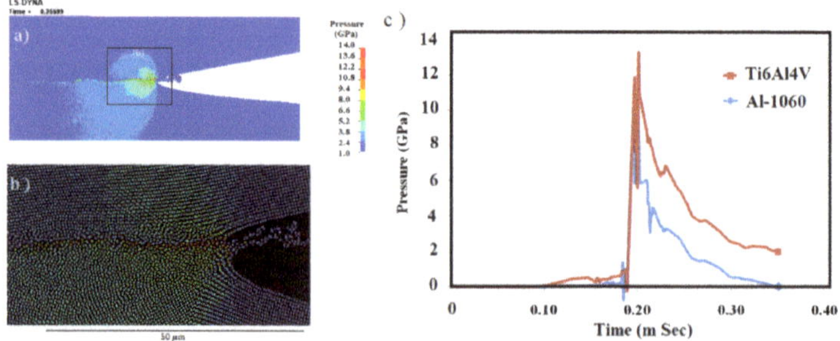

Figure 14. Numerica Simulation results for (**a**,**b**) pressure contour and (**c**) pressure-time graph at the interface.

The high impact creates shearing at the interface that causes to rise in heating and produce a bond between different metals. Kinetic energy during impact is converted into plastic work that causes the rise of temperature. Plastic deformation is not possible to witness experimentally, so simulation is an excellent way to explain the process.

Simulation results show that along with interface (Figure 15a–d), the maximum plastic strain is increased up to 7, while averagely its value is more than 5 throughout the interface. This strain value is enough to verify that at this impact velocity, the welding should be possible (the minimum plastic deformation required for welding should be more than 0.25 (Ti-MS) [37]). Figure 15a,b show that high plastic deformation expands up to 2 μm near the interface, which is similar to the experimentally measured interfacial area. It indicates that maximum deformation of the region generates the interfacial zone.

Figure 15c explains that this deformation is purely localized. As we move away from the interface, this deformation abruptly decreases to the minimum level. Furthermore, after 10 μm the flyer shows no plastic deformation, while the base plate has accommodated this deformation up to 30 μm.

Figure 15d illustrates that at the same impact point, the flyer and base plates have different values of plastic strain. The base plate has a plastic strain value almost double than that of the flyer plate. It is just because of the differences in their density and mechanical properties.

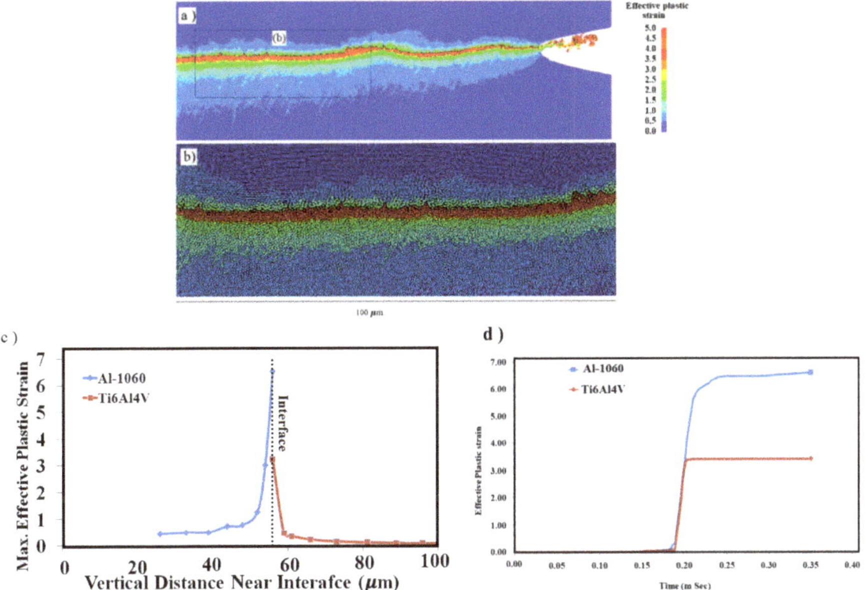

Figure 15. (**a**,**b**) Distribution of plastic deformation contour along with the interface, (**c**) graphical representation of plastic strain along the vertical distance of interface, and (**d**) plastic strain plot with respect to time at the interface.

During explosive welding, high pressure and plastic strain raise the temperature locally and abruptly because this phenomenon occurs in a very short interval of time and the cooling rate during this process is very high about 10^5–10^7 K/s [38].

In contour plots of temperature (Figure 16a–c), it is indicated that the temperature increment is enough to melt for both metals. However, both the temperature increments of flyer and base plates are not the same. At the flyer plate side, the average maximum temperature is 3000 K, while the base has an average of 1200 K. It is also observed that along with the interface, the peak temperatures have no fixed value. Figure 16d shows that the flyer has attained its melting point temperature in a very short time during the impact and then cools down immediately. While at the base plate side, the melting temperature sustains longer as compared to flyer. Obtained the melting point implies that the interfaces of both materials are more likely to form intermetallics. Furthermore, the interfacial zone behaves like the fluid flow and gets elongated toward detonation direction. Additionally, this process causes the refinement of grains near the interface.

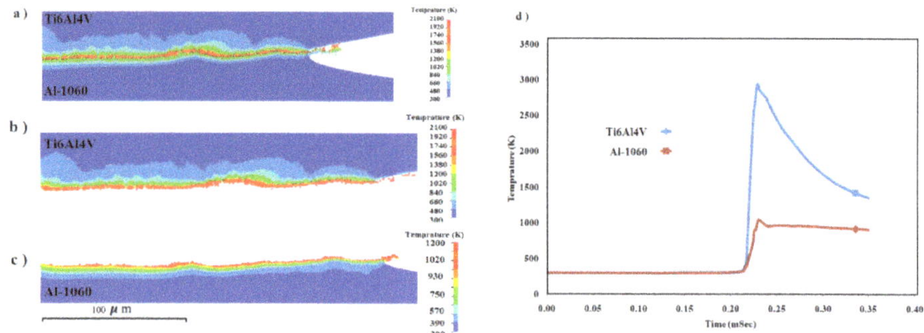

Figure 16. Plot of the temperature distribution at (**a**) the interface, (**b**) flyer side, (**c**) base side and (**d**) temperature profile of flyer and base plates at the interface.

5. Conclusions

Ti6Al4V was successfully welded with Al-1060 by explosive welding. The welded interface between Ti and Al was smooth and straight without any jet trapping. The maximum portion of the interfacial zone existed in the base side (Al-1060) where different phases of titanium aluminide were observed. Mechanical results, i.e., tensile test, bending test, shear test and Vickers hardness test, showed that welding quality was not highly affected by these titanium aluminides.

Heat treatment process stretched the interfacial zone with some titanium island/peninsula like shape. Due to this, strength of welded material was decreased as compared to the normal welded sample, but ductility was improved.

Numerical simulation depicted that impact pressure at all contact points had larger values than the yield strength of both welded materials, which is one of the basic requirements to meet the welding conditions. Furthermore, simulation results showed that in the interfacial zone, plastic deformation had values more than 5 and both materials obtained their melting points during impact. Melting of both materials provide a reason to form titanium aluminides. Since pressure, plastic deformation and temperature distribution for both materials (flyer and base) had different values, therefore, both materials had different interfacial thickness.

Author Contributions: Conceptualization, Y.M. and P.C.; explosive welding, Y.M. and Q.Z.; microstructure examination and mechanical tests, Y.M., A.A.B. and A.A.; numerical simulation, Y.M., K.D. and Q.Z.; writing–original draft manuscript, Y.M.; writing–review and editing, K.D. and P.C.

Funding: This research was funded by the National Natural Science Foundation of China (Grant No. 11521062) and State Key Laboratory of Explosion Science and Technology, Beijing Institute of Technology (Grant No. ZDKT18-01).

Conflicts of Interest: The authors declare no conflict of interest.

References

1. Findik, F. Recent developments in explosive welding. *Mater. Des.* **2011**, *32*, 1081–1093. [CrossRef]
2. Wang, B.F.; Luo, X.Z.; Wang, B.; Zhao, S.T.; Xie, F.Y. Microstructure and its formation mechanism in the interface of Ti/NiCr explosive cladding bar. *J. Mater. Eng. Perform.* **2014**, *24*, 1050–1058. [CrossRef]
3. Wang, B.F.; Xie, F.Y.; Wang, B.; Luo, X.Z. Microstructure and properties of the Ti/Al₂O₃/NiCr composites fabricated by explosive compaction/cladding. *Mater. Sci. Eng. C* **2015**, *50*, 324–331. [CrossRef] [PubMed]
4. Blazynski, T.Z. *Explosive Welding, Forming and Compact*; Springer: Dordrecht, The Netherlands, 1983.

5. Perusko, D.; Petrovic, S.; Stojianovic, M.; Mitric, M.; Cizmovic, M.; Panjan, M.; Milosavljevic, M. Formation of intermetallics by ion implantation of multilatered Al/Ti nano-structures. *Nucl. Instrum. Methods Phys. Res.* **2012**, *282*, 4–7. [CrossRef]

6. Adeli, M.; Seyedein, S.H.; Aboutalebi, M.R.; Kobashi, M.; Kanetake, N. A study on the combustion synthesis of titanium aluminide in the self-propagating mode. *J. Alloys Compd.* **2010**, *497*, 100–104. [CrossRef]

7. Arakawa, Y.Y.; Kobashi, M.; Kanetake, N. Foaming behavior of long-scale Al-Ti intermetallic foam by SHS mode combustion reaction. *Intermetallics* **2013**, *41*, 22–27. [CrossRef]

8. Goda, D.J.; Richards, N.L.; Caley, W.F.; Chaturvedi, M.C. The effect of processing variables on the structure and chemistry of Ti-aluminide based LMCS. *Mater. Sci. Eng. A* **2002**, *334*, 280–290. [CrossRef]

9. Ramos, A.S.; Calinas, R.; Vieira, M.T. The formation of γ-TiAl from Ti/Al multilayers with different periods. *Surf. Coatings Technol.* **2006**, *200*, 6196–6200. [CrossRef]

10. Romankov, S.E.; Mukashev, B.N.; Ermakov, E.L.; Muhamedshina, D.N. Structural formation of aluminide phases on titanium substrate. *Surf. Coat. Technol.* **2004**, *180*, 280–285. [CrossRef]

11. Kahraman, N.; Gulenc, B.; Findik, F. Corrosion and mechanical-microstructural aspects of dissimilar joints of Ti-6Al-4V and Al plates. *Int. J. Impact Eng.* **2007**, *34*, 1423–1432. [CrossRef]

12. E, J.C.; Huang, J.Y.; Bie, B.X.; Sun, T.; Fezzaa, K.; Xiao, X.H.; Sun, W.; Lou, S.N. Deformation and fracture of explosion-welded Ti/Al plates: A synchrotron-based study. *Mater. Sci. Eng. A* **2016**, *674*, 308–317. [CrossRef]

13. Xia, H.B.; Wang, S.G.; Ben, H.F. Microstructure and mechanical properties of Ti/Al explosive cladding. *Mater. Des.* **2014**, *56*, 1014–1019. [CrossRef]

14. Fronczek, D.M.; Wojewoda-Budka, J.; Chulist, R.; Sypien, A.; Korneva, A.; Szulc, Z.; Schell, N.; Zieba, P. Structural properties of Ti/Al clads manufactured by explosive welding and annealing. *Mater. Des.* **2016**, *91*, 80–89. [CrossRef]

15. Bazarnik, P.; Adamczyk-Cieslak, B.; Galka, A.; Plonka, B.; Sniezek, L.; Cantoni, M.; Lewandowska, M. Mechanical and microstructural characteristics of Ti6Al4V/AA2519 and Ti6Al4V/AA1050/AA2519 laminates manufactured by explosive welding. *Mater. Des.* **2016**, *111*, 146–157. [CrossRef]

16. Ege, E.S.; Inal, O.T.; Zimmerly, C.A. Response surface study on production of explosively-welded aluminum-titanium laminates. *J. Mater. Sci.* **1998**, *33*, 5327–5338. [CrossRef]

17. Inal, O.T.; Szecket, A.; Vigueras, D.J.; Pak, H. Explosive welding of Ti–6Al–4V to mild-steel substrates. *J. Vac. Sci. Technol. A Vac. Surf. Film* **1985**, *3*, 2605–2609. [CrossRef]

18. Ege, E.S.; Inal, O.T. Stability of interfaces in explosively-welded aluminum-titanium. *J. Mater. Sci. Lett.* **2000**, *19*, 1533–1535. [CrossRef]

19. Lazurenko, D.V.; Bataev, I.A.; Mali, V.I.; Bataev, A.A.; Maliutina, I.N.; Lozhkin, V.S.; Esikov, M.A.; Jorge, A.M.J. Explosively welded multilayer Ti-Al composites: Structure and transformation during heat treatment. *Mater. Des.* **2016**, *102*, 122–130. [CrossRef]

20. Pavliukova, D.V.; Bataev, I.A.; Bataev, A.A.; Smirnov, A.I. Formation of the intermetallic layers in Ti-Al multilayer composites. *Adv. Mater. Res.* **2011**, *311*, 236–239.

21. Bataev, I.A.; Bataev, A.A.; Pavliukova, D.V.; Yartsev, P.S.; Golovin, E.D. Nucleation and growth of titanium aluminide in an explosion welded laminate composite. *Phys. Met. Metallogr.* **2012**, *113*, 998–1007. [CrossRef]

22. Fan, M.Y.; Domblesky, J.; Jin, K.; Qin, L.; Cui, S.Q.; Guo, X.Z.; Kim, N.; Tao, J. Effect of original layer thicknesses on the interface bonding and mechanical properties of Ti-Al laminate composites. *Mater. Des.* **2016**, *99*, 535–542. [CrossRef]

23. Fan, M.Y.; Yu, W.W.; Wang, W.T.; Guo, X.Z.; Jin, K.; Miao, R.J.; Hou, W.Q.; Kim, N.; Tao, J. Microstructure and mechanical properties of thin-multilayer Ti/Al laminates prepared by one-step explosive bonding. *J. Mater. Eng. Perform.* **2017**, *26*, 277–284. [CrossRef]

24. Foadian, F.; Soltanieh, M.; Adeli, M.; Etinanbakhsh, M. A study on the formation of intermetallics during the heat treatment of explosively welded Al-Timultilayers. *Metall. Mater. Trans. A* **2014**, *45*, 1823–1832. [CrossRef]

25. Mousavi, A.A.A.; Al-Hassani, S.T.S. Numerical and experimental studies of the mechanism of the wavy interface formations in explosive/impact welding. *J. Mech. Phys. Solids* **2005**, *53*, 2501–2528.

26. Nassiri, A.; Kinsey, B. Numerical studies on high-velocity impact welding: smoothed particle hydrodynamics (SPH) and arbitrary Lagrangian–Eulerian (ALE). *J. Manuf. Process.* **2016**, *24*, 376–381. [CrossRef]

27. Wang, X.; Zheng, Y.Y.; Liu, H.X.; Shen, Z.B.; Hu, Y.; Li, W.; Gao, Y.Y.; Guo, C. Numerical study of the mechanism of explosive/impact welding using Smoothed Particle Hydrodynamics method. *Mater. Des.* **2012**, *35*, 210–219. [CrossRef]

28. Wang, Y.; Beom, H.G.; Sun, M.; Lin, S. Numerical simulation of explosive welding using the material point method. *Int. J. Impact Eng.* **2011**, *38*, 51–60. [CrossRef]

29. Saravanan, S.; Raghukandan, K.; Hokamoto, K. Improved microstructure and mechanical properties of dissimilar explosive cladding by means of interlayer technique. *Arch. Civ. Mech. Eng.* **2016**, *16*, 563–568. [CrossRef]

30. Tanaka, K. Numerical studies on the explosive welding by smoothed particle hydrodynamics (SPH). *Shock Compress. Condens. Matter.* **2007**, *955*, 1301–1304.

31. Brows, W.B.; Braithwaite, M. Analytical reprsentation of the adiabatic equation for detonation products based on statistical mechanics and intermolecular forces. *Phil. Trans. Roy. Soc. A* **1992**, *339*, 345–353.

32. Deribas, A.A. Acceleration of metal plates by a tangential detonation wave. *J. Appl. Mech. Tech. Phys.* **2000**, *41*, 824–830. [CrossRef]

33. Manikandan, P.; Hokamoto, K.; Fujita, M.; Raghukandan, K.; Tomoshige, R. Control of energetic conditions by employing interlayer of different thickness for explosive welding of titanium/304 stainless steel. *J. Mater. Process. Technol.* **2008**, *195*, 232–240. [CrossRef]

34. Cooper, P. *Explosives Engineering*; Wiley: New York, NY, USA, 1996.

35. Zukas, J.A.; Walters, W.P. *Explosive Effects and Applications*; Springer: New York, NY, USA, 1998.

36. Kamlet, M.J.; Jacobs, S.J. Chemistry of detonations. I. A simple method for calculating detonation properties of C-H-N-O explosives. *J. Chem. Phys.* **1968**, *48*, 23–35. [CrossRef]

37. Mousavi, S.A.A.A.; Al-Hassani, S.T.S. Finite element simulation of explosively-driven plate impact with application to explosive welding. *Mater. Des.* **2008**, *29*, 1–19. [CrossRef]

38. Crossland, B. *Explosive Welding of Metals and Its Application*; Oxford University Press: New York, NY, USA, 1982.

39. Lesuer, D. *Experiment Investigations of Material Models for Ti–6Al–4V Titanium and 2024-T3 Aluminum*; Final Report, DOT/FAA/AR-00/25; US Department of Transportation, Federal Aviation Administration: Washington, DC, USA, 2000.

40. Ye, L.; Zhu, X. Analysis of the effect of impact of near-wall acoustic bubble collapse micro-jet on Al 1060. *Ultrason. Sonochem.* **2017**, *36*, 507–516. [CrossRef]

41. Zhang, L.J.; Pei, Q.; Zhang, J.X.; Bi, Z.Y.; Li, P.C. Study on the microstructure and mechanical properties of explosive welded 2205/X65 bimetallic sheet. *Mater. Des.* **2014**, *64*, 462–476. [CrossRef]

42. Gloc, M.; Wachowski, M.; Plocinski, T.; Kurzydlowski, K.J. Microstructural and microanalysis investigations of bond titanium grade1/low alloy steel st52-3N obtained by explosive welding. *J. Alloys Compd.* **2016**, *671*, 446–451. [CrossRef]

43. Raoelison, R.N.; Racine, D.; Zhang, Z.; Buiron, N.; Marceau, D.; Rachik, M. Magnetic pulse welding: Interface of Al/Cu joint and investigation of intermetallic formation effect on the weld features. *J. Manuf. Process.* **2015**, *16*, 427–434. [CrossRef]

44. Su, Y.C.; Hua, X.M.; Wu, Y.X. Quantitative characterization of porosity in Fe-Al dissimilar materials lap joint made by gas metal arc welding with different current modes. *J. Mater. Process. Technol.* **2014**, *214*, 81–86. [CrossRef]

45. Tomashchuk, I.; Sallamand, P.; Cicala, E.; Peyre, P.; Grevey, D. Direct keyhole laser welding of aluminum alloy AA5754 to titanium alloy Ti6Al4V. *J. Mater. Process. Technol.* **2015**, *217*, 96–104. [CrossRef]

46. Murr, L.E.; Ferreyra, E.; Pappu, T.S.; Garcia, E.P.; Sanchez, J.C.; Huang, W.; Rivas, J.M.; Kennedy, C.; Ayala, A.; Niou, C.S. Novel deformation processes and microstructures involving ballistic penetrator formation and hypervelocity impact and penetration phenomena. *Mater. Charact.* **1996**, *37*, 245–276. [CrossRef]

47. Kacar, R.; Acarer, M. An investigation on the explosive cladding of 316L stainless steel-din-P355GH steel. *J. Mater. Process. Technol.* **2004**, *152*, 91–96. [CrossRef]

48. Prasanthi, T.N.; Kirana, R.; Saroja, S. Explosive cladding and post-weld heat treatment of mild steel and titanium. *Mater. Des.* **2016**, *93*, 180–193. [CrossRef]

49. Chulist, R.; Fronczek, D.M.; Szulc, Z.; Wojewoda-Budka, J. Texture transformations near the bonding zones of the three-layer Al/Ti/Al explosively welded clads. *Mater. Charact.* **2017**, *129*, 242–246. [CrossRef]

50. Szecket, A.; Inal, O.T.; Vigueras, D.J.; Rocco, J. A wavy versus straight interface in the explosive welding of aluminum to steel. *J. Vac. Sci. Technol. A* **1985**, *3*, 2588–2593. [CrossRef]

51. Wang, Z.J.; Wang, Z.; Li, M.X. Failure analysis of Al 1060 sheets under double-sided pressure deformation conditions. *Key Eng. Mater.* **2007**, *353*, 603–606. [CrossRef]

52. Kimura, M.; Iijima, T.; Kusaka, M.; Kaizu, K.; Fuji, A. Joining phenomena and tensile strength of friction welded joint between Ti–6Al–4V titanium alloy and low carbon steel. *J. Manuf. Process.* **2016**, *24*, 203–211. [CrossRef]

53. Boroński, D.; Kotyk, M.; Maćkowiak, P.; Śnieżek, L. Mechanical properties of explosively welded AA2519-AA1050-Ti6Al4V layered material at ambient and cryogenic conditions. *Mater. Des.* **2017**, *133*, 390–403. [CrossRef]

54. Holtzman, A.H.; Cowan, G.R. Bonding of metals with explosives. *Weld. Res. Counc.* **1965**, *104*, 1–40.

55. Cowan, G.R.; Holtzman, A.H. Flow configurations in colliding plates: explosive bonding. *J. Appl. Phys.* **1962**, *34*, 928–939. [CrossRef]

Article

Structural Properties of Interfacial Layers in Tantalum to Stainless Steel Clad with Copper Interlayer Produced by Explosive Welding

Henryk Paul *, Robert Chulist and Izabela Mania

Institute of Metallurgy and Materials Science, Polish Academy of Sciences, 25 Reymonta St.,
30-059 Krakow, Poland; r.chulist@imim.pl (R.C.); i.mania@imim.pl (I.M.)
* Correspondence: h.paul@imim.pl; Tel.: +48-12-2952833

Received: 12 June 2020; Accepted: 15 July 2020; Published: 17 July 2020

Abstract: A systematic study of explosively welded tantalum and 304 L stainless steel clad with M1E copper interlayer was carried out to characterize the microstructure and mechanical properties of interfacial layers. Microstructures were examined using transmission and scanning (SEM) electron microscopy, whereas mechanical properties were evaluated using microhardness measurements and a bending test. The macroscale analyses showed that both interfaces between joined sheets were deformed to a wave-shape with solidified melt zones located preferentially at the crest of the wave and in the wave vortexes. The microscopic analyses showed that the solidified melt zones are composed of nano-/micro-crystalline phases of different chemical composition, incorporating elements from the joined sheets. SEM/electron backscattered diffraction (EBSD) measurements revealed the microstructure of layers of parent sheets that undergo severe plastic deformation causing refinement of the initial grains. It has been established that severely deformed areas can undergo recovery and recrystallization already during clad processing. This leads to the formation of new stress-free grains. The microhardness of welded sheets increases significantly as the joining interface is approaching excluding the volumes directly adhering to large melted zones, where a noticeable drop of microhardness, due to recrystallization, is observed. On lateral bending the integrity of the all clad components is conserved.

Keywords: explosive welding; tantalum/copper/stainless steel clads; severe plastic deformation; SEM/EBSD; microhardness

---

1. Introduction

New strategies in the development of metallic materials for advanced structural applications involve the synthesis of bulk compounds that contain metallurgical bond. The bi- or multi-layered composites with built-in specific functionalities are an example of such materials. They offer an optimum balance between manufacturing and service costs, and the durability to perform under various conditions of usage. For materials used in the chemical industry, the proper combination of strength and high anticorrosive resistance, usually at high temperatures are especially important. The tantalum (Ta) and stainless steel composite is one of the industrially relevant bi-layered metallic material used in this field [1–4]; tantalum provides excellent corrosion resistance, while the stainless steel substrate is typically used as a load-bearing component. In most of the corrosion situations, it is capable to protect all installations exposed to highly oxidizing or caustic environments, where glass-lined equipment is subject to mechanical damage or thermal shock failures. Since the high cost of tantalum has traditionally been a major impediment in wide-scale industrial applications, such as large pressure vessels, therefore, it is advisable to use them rather as a coating on carbon, stainless or 'duplex'-type steels. Ta cladding is often used as an alternative to Ta coatings for fabricating coating parts out of solid Ta. Compound materials of this type (e.g., large vessels/tanks) are both structurally and cost-effective.

The Ta/stainless steel composites in the form of sheets/plates are difficult (or impossible) to produce via conventional methods of joining due to metallurgical incompatibility between joined components (high difference in the melting points of these metals, Ta at 3290 K and Fe at 1811 K). Therefore, explosive welding (EXW) is, at present, the only efficient way of surface joining of Ta and stainless steel sheets (Figure 1). However, further processing of bi-layered Ta/stainless or carbon steel composites is strongly restricted. This results from serious difficulties of a butt joint formation during 'conventional' welding due to limited heat transfer from the heat-affected zone. Moreover, if one attempts to use fusion welding to joint Ta to steel, the molten pool tends toward the eutectic composition (they are formed even well below the melting point of Ta) and then form brittle intermetallics upon solidification [1]. To solve problems associated with conventional welding of the Ta/stainless steel clads an intermediate layer, made of soft material and high thermal conductivity, such as copper (Cu), is used. Besides rapid heat dissipation, Cu has one more advantage, i.e., it does not react metallurgically with Ta.

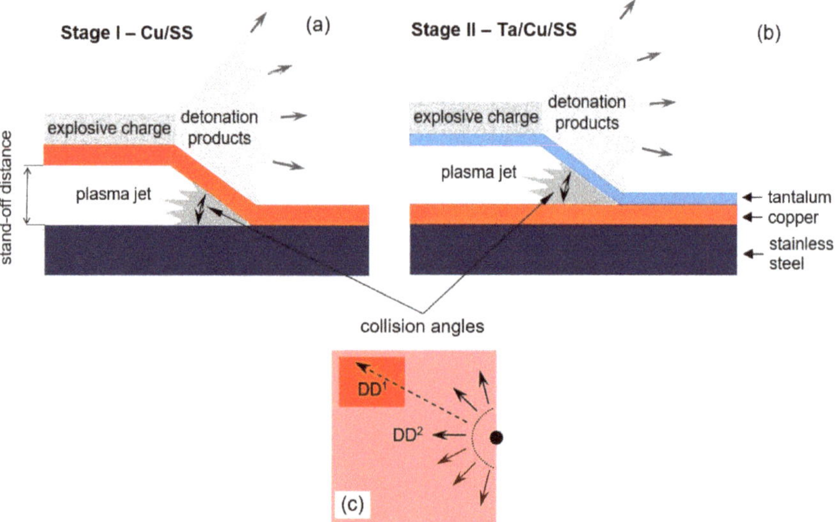

Figure 1. (**a**) Schematic representation of explosive welding in two steps: (**a**) stage I—formation of a Cu/stainless steel (SS) clad, (**b**) stage II—formation of Ta/Cu/SS clad, and (**c**) cutting of Cu/SS base plate before the second stage of joining.

In earlier works on different metal compositions, a lot of attention was put into an explanation of the correlation between the clad strength and the interface waviness and the quantity of solidified melt zones, e.g., [5,6]. However, it becomes increasingly apparent that not the details of the interface waviness but the complex microstructure of interfacial layers determines the mechanical and some physical properties of the clad [7–13]. In the light of such evidence, various microstructural transformations can be distinguished. On the one hand, the shear stresses, which occur due to oblique collision of the sheets are responsible for strain hardening and turbulent flow of the interfacial layers. This leads to the formation of wavy interfaces between the joined sheets. Since the interfacial layers are subjected to severe plastic deformation [14–18] they can easily undergo recovery and recrystallization. On the other hand, the processes of fast heating followed by fast cooling during clad preparation result in the formation of solidified melt zones of different structures, phase composition and mechanical properties [7–11,19,20].

Explosive welding of Ta and other metals have received much attention so far, referring to, e.g., Ta/Cu/steel [5,21], Cu/steel [22–24] and Cu/Ta [25–27]. The Cu–Ta system is characterized by nearly zero mutual solubility of the components in the solid-state [28,29] and high structural and mechanical stability at elevated temperatures. Greenberg et al. [30], Maliutina et al. [31] as well as Bataev et al. [11] have shown that explosively welded Ta and Cu sheets exhibit a heterophase mixture in the reaction region with the size of the dispersed Ta and Cu particles similar to those of colloids. As also documented by Bataev et al. [11] and Parchuri et al. [32], the Ta_xCu_{1-x} based intermetallics or decagonal quasicrystals were found to coexist along with pure Ta and Cu particles in the solidified melt zones at the Ta/Cu interface. The formation of metastable phases can also be expected due to rapid cooling during the solidification of melted volumes. In earlier works, the metastable phases in Ta–Cu system were investigated by Cullis et al. [33] who observed the metastable substitutional solid solution in the form of thin films. Furthermore, amorphous phases were observed by Natasi et al. [34] and Gong et al. [35], whereas the nano-crystalline phases by Purja Pun et al. [36] and Rajagopalan et al. [37]. On the other hand, a Cu/stainless steel interface contains solidified melt zones that are exclusively composed of intermetallic phases of different chemical compositions and various morphology of grains, e.g., [16,19]. Moreover, independently on metals composition, radical temperature changes [10,11] can lead to remarkable microstructural transformation in near-the-interface layers of the bonded sheets. In earlier works, a lot of attention was paid to the role of solidified melt zones with respect to strength properties, whilst significantly less effort has been directed towards characterizing their internal microstructure. To the best of the authors' knowledge, there is an absence of detailed studies of the Ta/Cu/stainless steel metals combination that specifically discussed the strain hardening, recovery, and recrystallization of interfacial layers.

Therefore, this work is intended to show interconnected phenomena that must be considered in the interfacial layers of joined sheets in the Ta/Cu (M1E)/(304L stainless steel) composite at the Ta/Cu (M1E) and Cu/304 L stainless steel interfaces. The microstructures and chemical composition changes are analyzed using scanning (SEM) and transmission (TEM) electron microscopes equipped with energy dispersive spectrometry (EDS) detectors. Since the interfacial layers are subjected to severe plastic deformation, which can undergo partial recrystallization, the high-resolution electron backscattered diffraction (EBSD) facility was used as a suitable tool to study the microstructural changes occurring in the parent sheets (areas of not mixed original sheets excluding melted zones). In order to support microstructural findings, the mechanical properties were evaluated using microhardness measurements, whereas the integrity of the joints via lateral bending test.

2. Experiment

EXW of Ta/Cu/stainless steel (SS) sheets was performed in two steps by High Energy Technologies Works 'Explomet' (Opole, Poland). In the first step (Figure 1a) the explosive welding of M1E Cu (flyer) to 304 L SS—base sheets with a size of 2400 mm × 2400 mm—was performed. The chemical composition of the joined sheets are presented in Table 1. After straitening the new plate was cut from the corner region of the Cu/SS plate (but still within the area of properly bonded sheets), as presented in Figure 1b. The dimension of this new, bimetallic sheet was 440 mm (length) × 205 mm (width). Then, in the second stage (Figure 1c), the Ta (flyer) sheet was clad onto the Cu/SS (base) plate. It is clear that the detonation direction during first (DD^1) and the second (DD^2) EXW steps are not parallel; the DD^2 is inclined at ~35° with respect to DD^1. The initial thicknesses of the sheets were: 1.8 mm (Ta), 3.0 mm (Cu) and 12 mm (SS). The contact surfaces of the joined sheets/plates were grounded, cleaned of solid particles and degreased. A detonator was located in the middle of the shorter edge of the flyer plate. The explosive was ammonium nitrate with fuel oil and a charge density of amount 0.75 g/cm^3. To manufacture high-quality clads, the detonation velocities during both steps of the EXW experiments were ranged between 2500–2600 m·s^{-1}. The welding conditions were tailored through the parallel geometry route with a 3 mm stand-off distance between the sheets (on each step).

Table 1. Chemical composition of joined components.

304 L Steel (Arcelor Mittal Certificate)										
Chemical element wt. %	C	Mn	P	S	Si	Cu	Ni	Cr	Mo	Co	Fe
	0.24–0.30	1.87–2.0	0.028–0.045	0.0017–0.015	0.323–0.75	0.257–0.750	8.037–10,5	18.035–19.5	0.238–0.75	0.129	balance

Tantalum (Hamilton Precision Metals®Certificate)									
Chemical element wt. %	C	O	N	H	Ni	Ti	W	Mo	Si	Ta
	0.01	0.015	0.01	0.0015	0.1	0.1	0.05	0.02	0.005	balance

M1E—Copper (Carl Schreiber GmbH Certificate)									
					ppm					wt.%
Chemical element ppm/wt.%	Ag	Ni	Fe	Sb	As	Sn	Zn	S	O	Cu
	12.0	3.0	2.0	2.0	1.7	1.7	1.7	5.0	30.0	99.95

Specimens for microstructural analyses were cut-off from the central part of the final clad in the as-welded state. The observation plane was perpendicular to the transverse direction (TD). This means that sample edges were parallel to the detonation (DD^2) and to the normal (ND) directions. The samples were mechanically ground up to 4000 SiC paper and polished in two steps with the use of the VibroMet-2 (Buehler, Lake Bluff, IL, USA) device and Al_2O_3 for 10 h and colloidal silica for 2 h. To study the microstructure evolution a high-resolution SEM (FEI Quanta 3D, Tokyo, Japan) equipped with EDS detector and high-speed Hikari EBSD camera by EDAX, were used. During SEM/EBSD measurements, the microscope control, pattern acquisition, and indexing were done using the Genesis TSL OIM Analysis 8 software (EDAX, Weiterstadt, Germany). The mappings were carried out in the beam-scanning mode. The applied step size ranged between 40 nm and 200 nm and with an accelerating voltage ranging between 15 and 30 kV. Supplementary analyses on the nanoscale were performed using TEM, FEI Technai Super Twin G^2 FEG (Tokyo, Japan) operating at 200 kV, equipped with an energy dispersive X-ray microanalysis system.

Vickers microhardness measurements were performed on the ND/DD^2 section to estimate the microhardness of intermetallic phases and the distribution in strain hardened layers across the interface. The tests were carried out on a finely-polished longitudinal section. The obtained microhardness values were the average of three indentation measurements. The average microhardness of the base materials in fully recrystallized states, i.e., before cladding, were 250 HV, 162 HV, and 110 HV for 304 L stainless steel, Ta and M1E-Cu, respectively. A three-point lateral bending test was employed to evaluate the resistance to delamination of the joints. The test was performed according to EN 13445-2:2014 (E) on samples cut along the DD^2 from the final clad. The specimens for bending testing with a size of 10 mm (width) × 13 mm (high) × 150 mm (length) were extracted from the central part of the clad in the plane parallel to the detonation direction.

3. Results

3.1. Macro-/Meso-Scale Interfaces Overview

Macroscale characterization of the interfacial layers includes light optical microscopy and low magnification SEM observations. The initial state of the sheets was characterized by a uniform, fully recrystallized microstructure. Both sections perpendicular to the rolling plane revealed structures of equiaxed grains with a diameter of ~80 μm, ~60 μm and ~100 μm for Ta, 304L (SS) and Cu, respectively (Figure 2). It confirmed the high quality of the Ta/Cu and Cu/SS interfaces, without voids and visible sheets delamination. The analysis made at the mesoscopic scale with low magnification SEM imaging shows wavy interfaces, however, with quite different wave parameters (Figure 3). Moreover, the character of waviness is different for both interfaces since the detonation direction during the first (DD^1) and the second (DD^2) EXW steps are not parallel. In the case of Ta/Cu interface the amplitude and the period of the wave, as observed in the ND/DD^2 section (Figure 1a), are close to ~100 μm and ~300 μm, respectively. In the case of Cu/SS sheets, a non-regular interface in this section is found to exist. On the contrary, the Cu/SS interface in the ND/DD^1 section shows a regular waviness with the wave amplitude and the wave period close to 450 μm and 950 μm, respectively (Figure 1c).

The wave formation coincides with the formation of solidified melt zones that are preferentially located at the wave crest and within the wave vortexes. The cracks within the solidified melt zones, commonly observed in other metal combinations, e.g., [4,7,9] were only occasionally detected in solidified melt zones formed at both interfaces of the Ta/Cu/SS clad. However, if observed, they were always limited to the zone of solidified melt and they propagated perpendicularly to the interface between solidified melt and pure metal. None of these cracks have shown any tendency to propagate across the base materials.

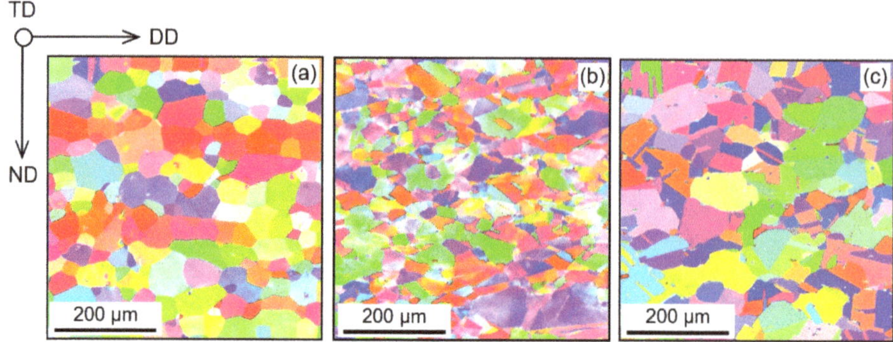

Figure 2. The initial microstructure of (**a**) tantalum, (**b**) copper and (**c**) stainless steel sheets taken in the normal/rolling direction (ND/RD) plane. Scanning electron microscopy/electron backscatter diffraction (SEM/EBSD) images of local orientation measurements with a step size of 200 nm.

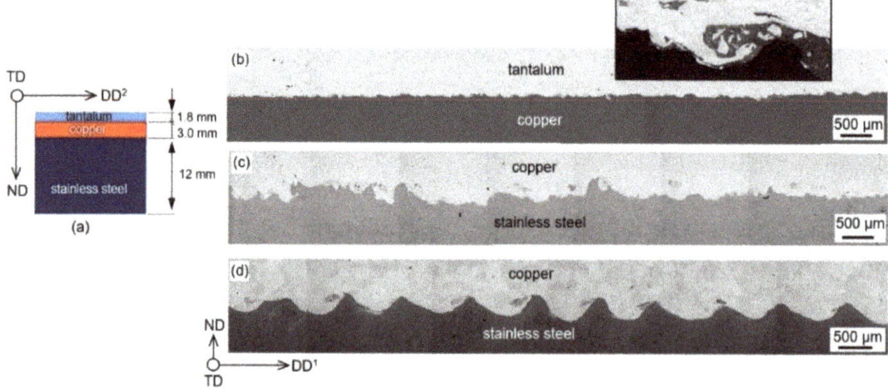

Figure 3. (**a**) Thickness of initial sheets, and the wavy interfaces, (**b**) regular wavy interface between Ta and Cu and (**c**) non-regular interface between Cu and SS observed in a final clad along ND/DD^2 section. (**d**) The regular wavy interface between Cu and SS observed along the ND/DD^1 section.

3.2. Microstructure of Severely Deformed Layers of Parent Sheets Near the Interface—SEM/EBSD Analysis

Figures 4 and 5 present the inverse pole figures (IPF) orientation maps combined with the image quality (IQ) factor. The IQ component emphasizes grain and interphase boundaries. The analysis is supported by direct SEM/EBSD/EDS chemical composition determination. The SEM/EBSD maps of the vortex region show a severely deformed microstructure of parent sheets composed of elongated cells/(sub)grains with a tangled network of dislocations (Figure 4a). The points inside the zone of solidified melt are mostly indexed as 'pure' metals with some quantity of not or mis-indexed pixels. Further chemical analysis reveals that the solidified melt zones contain both elements in the vortex region.

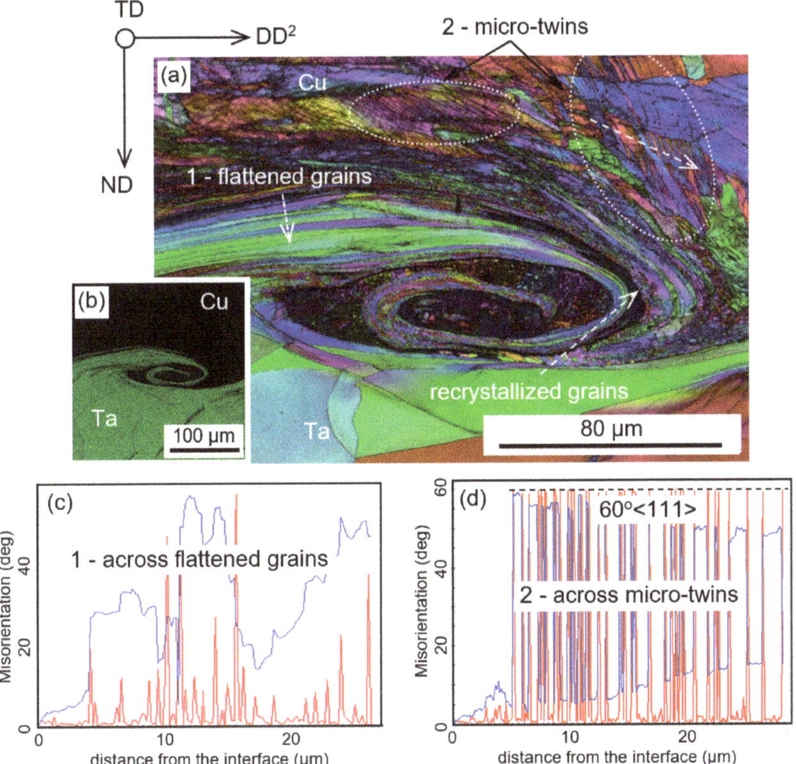

Figure 4. (**a**) Orientation map showing a typical vortex region close to the Cu/Ta interface, and corresponding (**b**) chemical composition map showing the distribution of Ta. Misorientation line scans across (**c**) flattened grains in Ta, and (**d**) clusters of twins in Cu. SEM/EBSD measurements with a step size of 100 nm along the ND/DD^2 section.

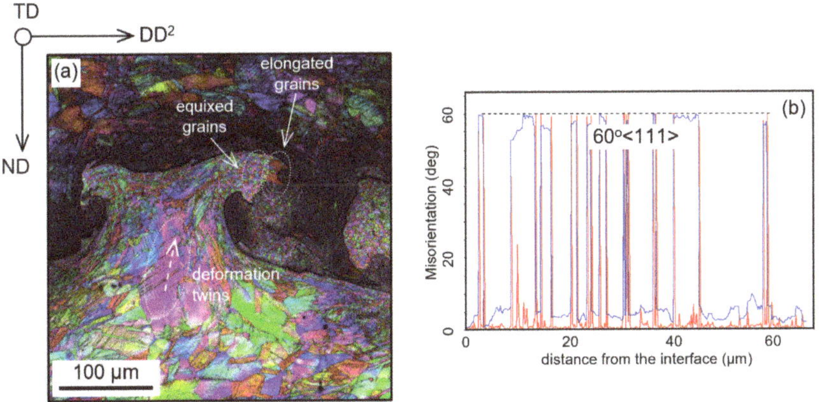

Figure 5. (**a**) Orientation map showing typical vortex near the SS/Cu interface along ND/DD^2 section. (**b**) Misorientation line scan across the cluster of twins. SEM/EBSD measurements with step size of 100 nm.

3.2.1. Ta/Cu Interface

The severely deformed layers of tantalum are relatively narrow and limited to the thickness of a few tens of microns. In contrast, in the copper sheet a significantly larger strain hardened layer is observed (a few hundreds of microns). For instance, Figure 4a shows the microstructure of the interfacial layer around and inside a wave vortex. The formation of vortexes leads to a characteristic interlock microstructure between tantalum and copper, as is clearly presented in Ta chemical composition map (Figure 4b). Interfacial layers of both parent sheets are composed of flattened grains with a thickness of 1–2 μm. Grain to grain misorientation plot shows that to a large extent the flattened grains are separated by high angle boundaries (Figure 4c). The orientations of neighboring, flattened grains are nearly symmetrical. Such a grain shape and misorientation relationship between grains refer to plane strain conditions and deformation banding mechanism dominating in medium-to-high stacking fault energy fcc or bcc metals [38]. However, in some places these flattened grains undergo intense recrystallization; this leads to the formation of layers composed of very small (<1 μm) equiaxed grains. Another characteristic feature of the as-deformed structure of copper (but not present in Ta sheet) is massive deformation twinning. This mechanism is supposed to accommodate the strains that emerged in the vortex formation. Figure 4d shows the misorientation angle distribution along the pathway marked by an arrow in Figure 4a. The misorientation angle between neighboring platelets is 60° with the misorientation axis corresponding to the <111> direction. This clearly indicates that the micro-twins formation is similar to the periodic twins that have been observed in pure copper deformed at extremely high strain rates, as observed earlier by Crossland and Williams [39] and by Lee et al. [17] in EXW copper to copper sheets. The line scan across the deformation twins can be contrasted with the misorientation vs. distance line scan presented in Figure 4c, where the opposite tendency of crystal lattice rotation in the neighboring layers (flattened grains) is observed. However, the misorientation angles between the layers always display values lower than 60°.

In the present work the most regions inside the solidified melt zones near the Ta/Cu interface, were indexed as 'pure' Ta or Cu, with some quantity of not or mis-indexed pixels [11,30,31].

3.2.2. SS/Cu Interface

Figure 5a shows an SEM/EBSD image with a vortex region along the SS/Cu interface. In ND/DD² section, unlike the typical wavy shape, the SS/Cu interface resembles a column capital. There are a molten and re-solidified regions near the vortices, marked by very low Kikuchi contrast (dark grey or black pixels). Poor indexing of those areas can be directly related to the formation of non-equilibrium ultra-fine grained (or even amorphous) phases based on Cu and elements present in stainless steel. Orientation maps show that interfacial layers of both parent sheets are severely deformed with a significantly refined microstructure. The interfacial layers of the SS sheet are composed of grains with the size ranged between 0.5 μm and 2 μm. The width of this layer is 50 μm–100 μm. However, at larger distances from the interface, much coarser and only slightly deformed grains can be observed. In the Cu sheet, the as-deformed grains are not exclusively observed near this interface. In areas situated near the wave crests and adhering to solidified melt zones, a more uniform structure of equiaxed grains (free of dislocations) with an average grain size of about 1–3 μm is observed. This strongly indicates intense recrystallization. In some places, this structure evolves into columnar grains indicating the occurrence of oriented grain growth along the preferred heat flow direction. The large columnar grains grow perpendicular to the melted zone/copper interface. The length of the larger axis of columnar grains is between 5 μm and 15 μm. However, in some of the Cu grains situated inside the vortex (pillars) region, the occurrence of dislocation slip is accompanied by intense deformation twinning. This is confirmed by a misorientation line scan in Figure 5b indicating compact clusters of deformation twins, similar to those observed in Cu near the Ta/Cu interface. The equilibrium phases corresponding with the Cu and Fe(Cr, Ni) binary diagrams were identified only accidentally near the Cu/SS interface. This observation confirms the early hypothesis [4,16] that the melted zones are mostly composed of a mixture of pure parent metals.

3.3. Dislocation Structure of Parent Sheets—TEM Analysis

The dislocation structures of interfacial layers of parent sheets were analyzed in areas of the wave valley, where large zones of solidified melt are not observed at the optical microscope or low magnification SEM scale. Generally, the structures were composed of fine equiaxed or elongated cells/(sub)grains with an increased density of dislocations inside them. Such microstructures are typical for strain hardened materials with a tangled network of dislocations, high vacancy concentration and possibly large numbers of microtwins (in the Cu and SS).

The most characteristic feature observed in the Ta sheet in layers near the Ta/Cu interface is the structure composed of relatively wide microbands (Figure 6a), whereas in the Cu sheet (at the same interface) the formation of extremely fine but equiaxed (sub)grains (Figure 6b) is observed. In layers near the Cu/SS interface the structure of Cu sheet is significantly less deformed as compared to the layers of Cu sheet situated near the Cu/Ta interface. Despite a huge shear strain, the structure of initial grains with recrystallization twins is still apparent (Figure 7a). The microstructure of SS plate in layers adjacent to the Cu sheet is composed of fine (sub)grains with a diameter of a few hundreds of nanometers (mostly <500 nm). A large number of dislocations accommodated in the cells/(sub)grains of both metals is a strong indication that the deformation processes are prevailing over the thermally activated softening ones, i.e., recovery and recrystallization, as suggested earlier for Al/Cu [9], Zr/(carbon steel) [10] and Al/Ti [14] clads.

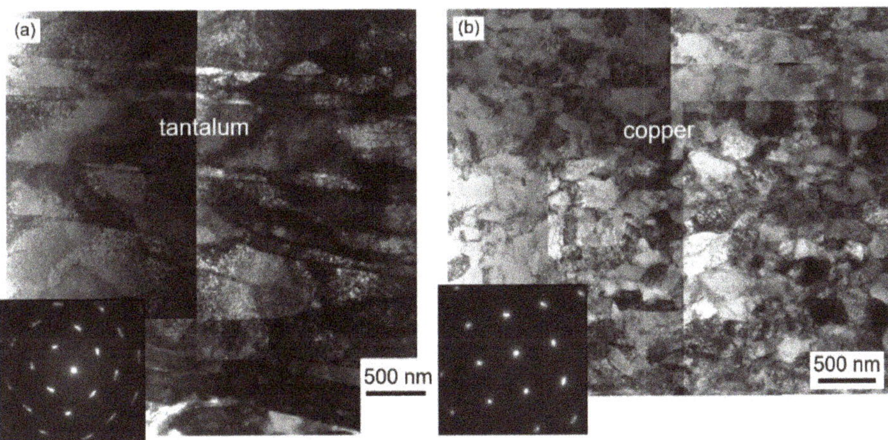

Figure 6. Transmission (TEM) bright-field images showing the structure composed of (**a**) elongated cells in Ta and (**b**) fine grains in Cu.

The wave valleys do not contain any macroscopically visible solidified melt zones (Figure 7a–c). However, a very thin layer of solidified melt, of few tens of nanometers is identified at large magnifications, as presented for Cu/SS interface in Figure 7d. This confirms the early thesis that the presence of a very thin reaction layer is one of the most important factors that guarantee a good bond between welded sheets and improves delamination resistance, as suggested in [4,8,10].

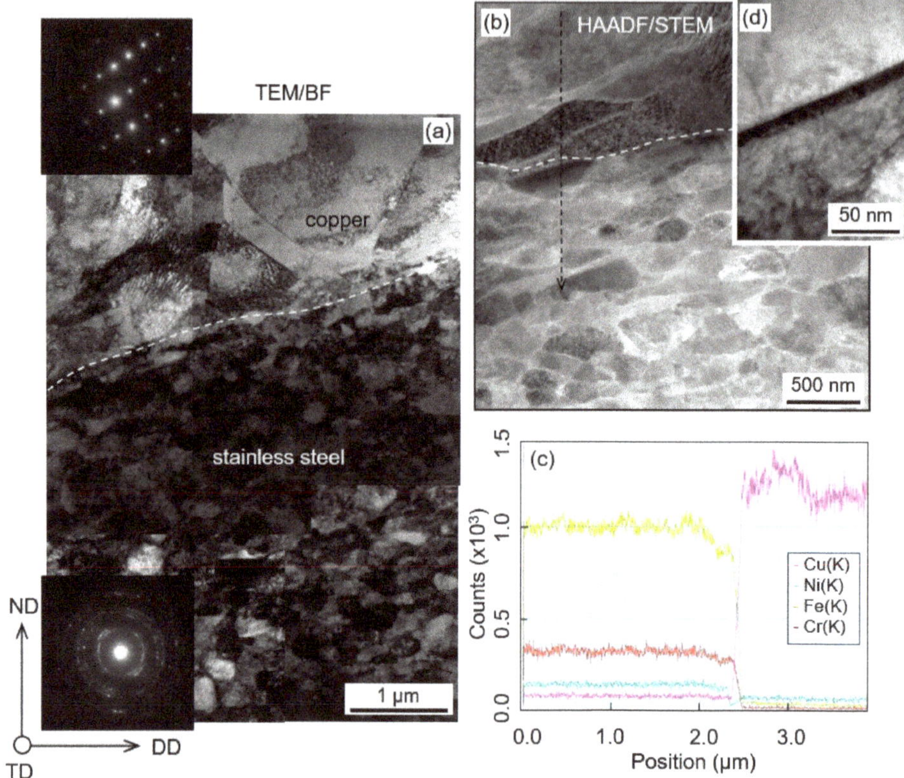

Figure 7. (**a**) TEM bright field and (**b**) TEM/High Angle Annular Dark Field (HAADF) images showing microstructure of copper and stainless steel near the Cu/SS interface. (**c**) Chemical composition line scan across the Cu/SS and (**d**) interfacial region observed at high magnification documenting the presence of a very thin layer of solidified melt.

3.4. Phase Constitution inside Solidified Melt Regions

The most spectacular microstructural changes are observed in the solidified melt zones. The melt zones may exhibit a different form—starting from large, nearly equiaxed ones, going through thin, and ending up with extremely thin layers. They are situated inside the wave vortexes (as inclusions inside the Ta or Cu sheet) and at the crest or on the bottom part of the wave (extremely thin layers). Large solidified melt zones situated inside the wave vortexes can be (i) entirely surrounded by Ta or Cu, without any contact with Cu or SS, respectively, or they can be (ii) still partly adjacent to the neighboring sheet (Cu or SS).

Some examples of solidified melt zones of the second type are presented in Figure 8a,b. These SEM images taken with a backscattered electron detector (SEM/BSE) show sharp interfaces between the zone of solidified melt and parent materials. This implies significant changes in chemical composition across the boundaries. Additionally, the areas of pure metals (mostly of a higher melting point) are detected inside the solidified melt.

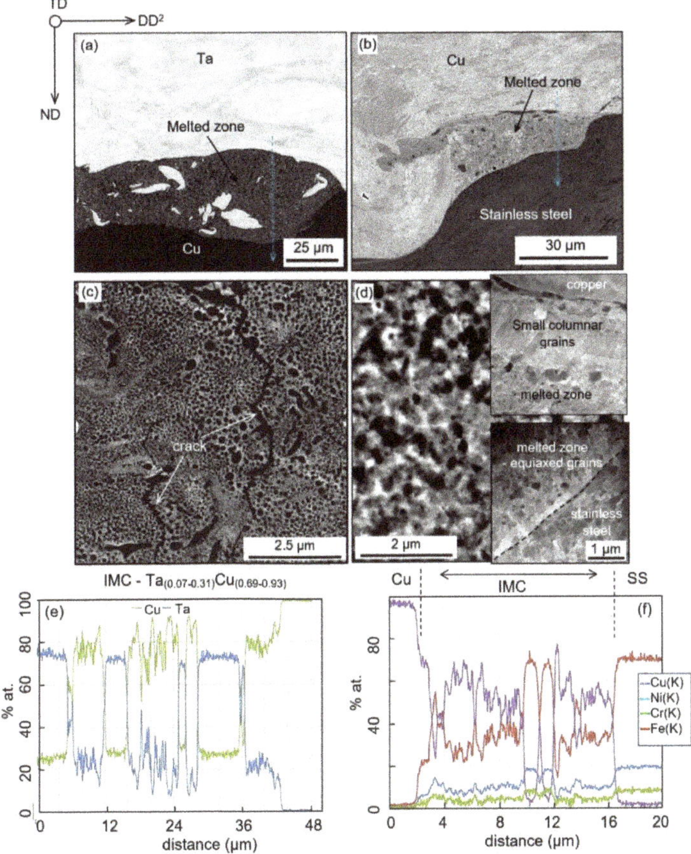

Figure 8. Microstructure of solidified melt zones formed between (**a**) Ta and Cu, and (**b**) Cu and SS, (**c,d**) details of internal microstructure. (**e,f**) Chemical composition line scan showing the distribution of main elements in the solidified melt region.

3.4.1. Chemical Composition of Solidified Melt Zones Formed between Tantalum and Copper

The internal microstructure of solidified melt zone shows small areas of alternating BSE contrast. For instance, Figure 8a depicts an enlarged part of the vortex between Ta and Cu sheets. In some cases, the pure metals reveal rotational character due to the vigorous stirring and mixing of the material in the liquid and/or semi-liquid states. A more detailed analysis shows that the internal structure of the solidified melt region can be defined as a mixture of fine Cu and Ta particles of different sizes. Figure 8c shows an SEM/BSE micrograph with a typical structure of the solidified melt. In most observed cases the size of spherical particles of Ta is ranged between 10 nm and 500 nm and they are homogeneously distributed in the Cu matrix. Nevertheless, there were experimentally discovered micro volumes in which the matrix material is Ta and Cu is in the form of compact spherical aggregates. However, as shown by Bataev et al. [14] and Parchuri et al. [32] they may coexist with Ta_xCu_{1-x} based intermetallics and decagonal quasicrystals at the Ta/Cu interface.

Despite the fact that the chemical composition inside the zone of solidified melt varies practically from 0% to 100% of a given element, a strong preference of chemical composition close to $Cu_{0.25}Ta_{0.75}$ and $Cu_{0.75}Ta_{0.25}$ is observed. This can be clearly seen in Figure 8e in the chemical composition line scan along the blue dashed arrow marked in Figure 8a.

3.4.2. Chemical Composition of Solidified Melt Zones Formed near the Stainless Steel and Copper Interface

In comparison to Ta/Cu interface, the grains inside the solidified melt zone near the Cu/SS interface show different morphology and chemical composition (Figure 8b,d,f). Most of the grains crystallize in the form of small dendrites. However, the grains that nucleate just near the Cu sheet form a characteristic sublayer composed of small columnar grains with longer axis perpendicular to boundary between solidified melt and copper. This indicates a more effective heat transfer across the interface towards Cu as compared to the other side of the melted zone, where only ultra-fine, but nearly equiaxed grains are formed (see insets in Figure 8d). Nevertheless, large volumes of the melted zone are occupied by ultra-fine-grained phases. Various chemical compositions are identified inside the zone. To a larger extent the phases are enriched in copper (ranging between 60% and 90%) (Figure 8f).

The solidified melt zone can also be enclosed inside the parent metal. Such an example is shown in Figure 9 where a solidified melt is surrounded by the Cu matrix. The corresponding chemical composition maps show the distribution of Cu and the main elements of stainless steel (Fe, Cr and Ni). The content of elements is quite similar to that observed inside the open zones and is close to $Cu_{(0.60-0.90)}Fe_{(0.20-0.72)}Cr_{(0.05-0.19)}Ni_{(0.02-0.08)}$. The maps also revealed large fragments of steel enclosed inside the solidified melt zone.

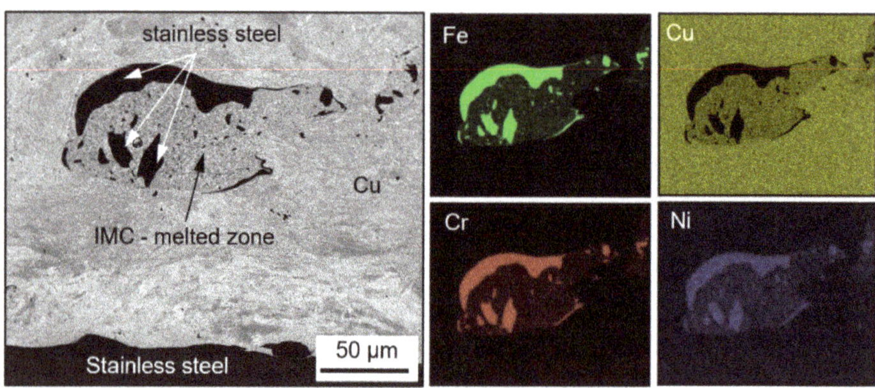

$$IMC - Cu_{(0.60-0.90)}Fe_{(0.20-0.72)}Cr_{(0.05-0.19)}Ni_{(0.02-0.08)}$$

Figure 9. SEM/backscattered electron detector (BSE) image showing solidified melt zone enclosed inside the deformed Cu and corresponding chemical composition maps documenting the distribution of Cu and main elements in stainless steel.

3.5. Microhardness across the Interface

In order to correlate the mechanical properties with the corresponding changes in microstructure the microhardness measurements were performed. The microhardness profiles along ND (in the ND/DD^2 section) in areas near the wave crests are presented in Figure 10a,b. They show the distribution of the strain hardening across the interface. The average microhardness values of the base materials in the fully recrystallized states, i.e., before cladding, are marked as blue horizontal dashed lines. For both interfaces, the microhardness of parent sheets strongly increases in layers near the interface, whereas at the center of the sheets does not differ significantly from those of the initial state (Figure 10a,b). It is apparent that the most radical increase of strain hardening is observed in the interfacial layers of SS sheet near the SS/Cu interfaces (Figure 10a,b). However, the microhardness values inside the solidified melt zones strongly scatter. For both interfaces, the microhardness of solidified melt zones reached values between 240 and 370 HV (Figure 10c) and between 210 and 260 HV (Figure 10d) for SS/Cu and Cu/Ta interfaces, respectively. These values significantly differ from those obtained for

SS/Ta metals composition, where the microhardness of solidified melt regions ranged between 720 HV and 1080 HV [14]. This is twice or even three times more than those observed in strongly refined (and strain-hardened) layers of SS sheets. However, in that case, a strong scattering of values was also observed.

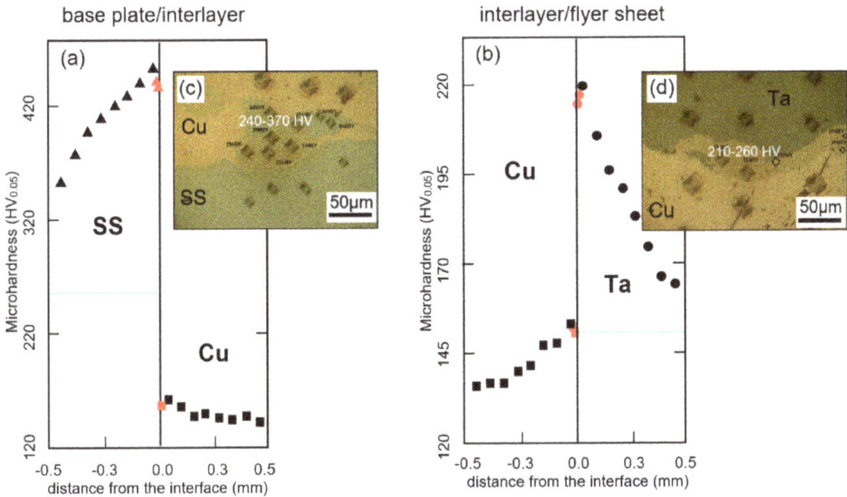

Figure 10. Vickers microhardness across the interfacial layers in (**a**) SS/Cu and (**b**) Cu/Ta clads, (**c**,**d**) corresponding optical micrographs showing values of microhardness of solidified melt zones. Red triangles (**a**,**b**) indicate the values of microhardness measured in the interfacial layers close to a large melted zone. Microhardness of copper in a fully recrystallized state is 110 HV.

As a next step, the lateral bending test was performed to evaluate the influence of the interfacial microstructures on the integrity of the joints. The tested specimens were deformed up to 90° bending angle (Figure 11). No tearing, fracture, or separation of the sheets was observed. The cracks, which were originally localized within the solidified melt region start to expand into the stainless steel substrate upon loading, whereas there was no propagation of the pre-existing cracks into the copper, stainless steel or tantalum sheets. This shows that Ta/Cu/SS welding is fully applicable in service even in a bent form.

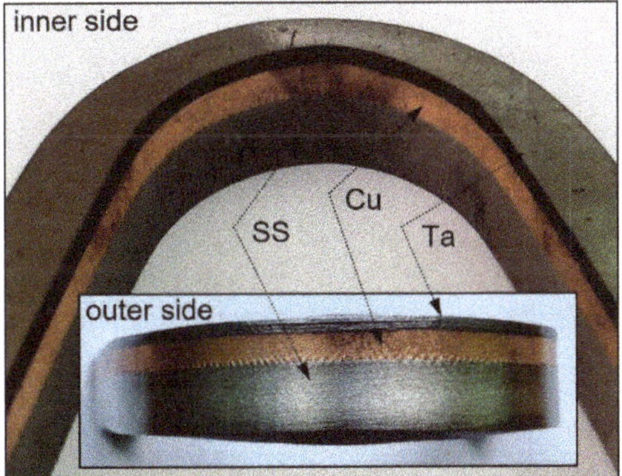

Figure 11. Optical micrographs showing macro-structure of the sample after (lateral) bending test. No delamination of the clad components and macro-cracks propagation is observed.

4. Discussion

This study describes several important aspects of our understanding of EXW and the evolution of microstructure in the process of joining. These changes are twofold. The first ones are due to strain hardening, recovery, and recrystallization of interfacial layers of the parent sheets. The latter ones are due to the phase transformations in the solidified melt zones due to fast heating followed by extremely fast cooling.

4.1. Microstructural Changes in the Parent Sheets

The strain hardening processes in the parent sheets prevail over the thermally activated softening ones in areas far from the large melted zones. As a consequence, the presence of strain-hardened structures can be regarded as a factor that increases the strength of the clad. However, in layers adhering to large melted zones the softening due to recrystallization is noticeable, as observed earlier for other metal compositions, e.g., [4,7,14,40]. The recrystallization starts to occur just during clad formation as a result of heat transfer between the large melted zones and severely deformed layers of parent sheets/plates. In areas near the interface, this process leads to the transformation of (sub) grains with randomly distributed dislocations into equiaxed fully recrystallized grains free of dislocations. Since the highest temperature is reached at the interface, it is often observed that these areas undergo intense recrystallization or even abnormal grain growth. This, in many cases leads to the situation where the average size of grains situated at the interface is larger than that of more distant from the interface, as also presented by Chu et al. [18] for Ti to carbon steel clad. As a consequence of the recovery and recrystallization processes, a pronounced drop of microhardness in layers of parent sheets directly adhering to the large melted zone can be detected.

4.2. Microstructural Changes due to Formation of the Solidified Melt Regions

The history of dramatic temperature changes during the impact process can be used to elucidate the microstructure evolution of the solidified melt region. During the cooling period (as the collision point is moved forward), the cooling rates calculated by various authors always exceed the 10^5–10^7 ks^{-1} range, e.g., [10–12,18]. These rates make the thermodynamic conditions very similar to that observed during spin melting, where the formation of extremely fine-grained or even amorphous phases is expected.

Although a low microhardness has been observed in the solidified melt zone (as compared to strain hardened layers of 304 L SS sheet), upon bending the strain is mostly accumulated by the fully recrystallized microstructure of Cu that surround the solidified melt region. It is clear that this behavior is quite different from those observed for other metal compositions, e.g., based on steels and reactive metals, where the formation of brittle solidified melt zones of high hardness is dominant, e.g., [4,7,10].

For the particular case of Ta/Cu metal composition, a possible mechanism of intermediate heterophase region formation was proposed by Maliutina et al. [31]. In this mechanism there is not necessary to exceed the melting point of Ta since the globular particles of Ta are plunged into solidified Cu. However, it is very difficult to state without any doubt, whether this mechanism is the only one or just dominant. The weak point of the mechanism proposed in [31] is that the matrix is not pure copper but is mostly a solid solution of Ta in the Cu (see Figure 8). However, aside from the limited information on the phase constitution in the Cu-Ta immiscible systems, the formation of metastable phases can be expected due to rapid cooling during solidification of melted volumes, as described for other processes in [33–37,41,42]. Due to the extremely high dynamic of the EXW process, there is no time for the diffusion in the solid state, hence the melting of Ta can be expected. However, to solve this problem further analyses with the use of TEM are needed.

4.3. Microstructure vs. Strength Properties

An important question then arises here is—how the processes occurring inside the deformed layers and melted regions can influence the strength of the clad? Although, the strain hardened layers of parent sheets exhibit high hardness, the recrystallized structure near the interface and cracks propagation inside the solidified melt zone can countervail the overall strengthening effect.

The presence of ultra-fine-grained phases inside the solidified melt zones improves the clad strength, but on the other hand, it may lead to an increase in the resistance of a material to plastic deformation. The cracks appear inside the solidified melt zones owing to the shrinkage during the solidification of molten materials which significantly decreases the clad strength. However, in Cu–Ta and Cu–SS solidified melt regions the cracks are observed only accidentally. During the bending test some newly generated small interfacial discontinuities appear and grow together with pre-existing ones only inside the Cu–Fe(Ni, Cr) solidified melt regions. However, the crack propagation towards parent materials was not observed. The strain hardening of the interfacial layers of parent sheets can be attributed to the increase of lattice defects, such as dislocations, vacancies, (sub)grain boundaries, and in more global scale intense deformation twinning. However, the recovery and recrystallization processes, initiated just during EXW can decrease the density of structural defects due to heat transfer from large melted zones towards severely deformed layers [43]. This, in turn, leads to a decrease of strength.

This study also contributes to the issue that cannot be completely solved by optical microscopy and low magnification of SEM analyses. We are referring here to the mechanisms responsible for bond formation. Generally, there are two interpretations strongly disputed in the literature. On the one hand, high-resolution TEM studies of similar [8] and dissimilar [4,10,12] metals show the formation of an extremely thin layer of solidified melt of a few tens of nanometers in thickness, between neighboring sheets. The formation of a very thin layer of solidified melt, as observed in Figure 8, was predicted by Carpenter [44] who suggested that the weld interface is in reality a thin melted layer. The open question is—if this is a requirement for all bonding conditions, or if it is a result of excessive explosive loading? However, in numerical modeling of EXW of Cu to Cu, Lee et al. [17] argue that an essential mechanism of bonding is the jetting and extremely high pressure which are responsible for squeezing two sides free of impurities to atomic contact. Despite this mechanism was used many times in the past by various authors, this study tends more to metallurgical bonding. This melting and mixing can give a continuous reaction layer between the matching pair.

This study also points to open questions and opportunities. The presence of a very thin (<300 nm) reaction layer along with the entire interface, free of cracks, can play a decisive role in enhanced resistance to delamination and may be responsible for proper bonding between the sheets. Although the appearance of a thin reaction layer is typical for most of the clads, it is difficult to state without any doubt if the presence of a thin intermetallic layer is the necessary condition for proper bonding or if it only supports (as suggested by some of the authors) the essential mechanism based on the jetting and squeezing two sides to atomic contact.

5. Summary

The present work describes the microstructure of the Ta/Cu and Cu/SS interfacial layers of explosively welded Ta/Cu/SS clad. It is shown that the hardening and softening processes are strongly related to the microstructure and clad strength evolution. The processes that increase the yield strength of the material are severe plastic deformation, the formation of thin layers of ultra-fine grains, and solidified melt zones, while dynamic recovery, recrystallization, and crack formation inside the solidified melt cause softening.

SEM/EBSD analyses revealed a complex microstructure of parent sheets, which consists of characteristic features such as interlock microstructures, elongated grains, or twins. Strain hardening processes predominate softening ones in the interfacial regions far from large solidified melt zone. It has been established that a high-rate shearing of interfacial layers leads to a strain-hardening due to intense slip and deformation twinning in copper sheets at both interfaces. In layers surrounding the melted zones, the as-deformed grains are replaced by new recrystallized (equiaxed or columnar) ones due to the heat transfer from the melt to the severely deformed metals. The microhardness of welded sheets increases significantly as the joining interface is approaching excluding the volumes directly adhering to large melted zones, where a noticeable drop of microhardness due to recrystallization is observed.

However, precise analysis of dynamic recrystallization and recovery is exceptionally complicated as both phenomena strongly depend on both temperature and strain distributions near the interface as well as the melting point of the joined metals. Another observation is a quite different morphology of the solidified melt zones at the Ta/Cu and Cu/SS interfaces. The melted zones close to the Ta/Cu interface consist mainly of a mixture of pure Cu and Ta particles of different sizes, whereas the melted zones near the Cu/SS interface are composed of nano-grained compounds based on elements of both neighboring joined sheets. The microhardness of solidified melt regions formed near both interfaces is significantly lower than that measured inside the strain hardened layers of stainless steel. Therefore, the integrity of all clad components is fully conserved on lateral bending.

Author Contributions: Conceptualization, H.P.; methodology, H.P. and R.C.; formal analysis, H.P.; investigation, R.C. and I.M.; writing—original draft preparation, H.P.; writing—review and editing, R.C.; visualization, H.P.; project administration, I.M; funding acquisition, H.P. All authors have read and agreed to the published version of the manuscript.

Funding: This research was funded by the National Centre for Research and Development of Poland under grant TECHMATSTRATEG2/412341/8/NCBR/2019 (EMuLiReMat).

Conflicts of Interest: The authors declare no conflict of interest.

References

1. Frey, D.; Banker, J. Recent successes in tantalum clad pressure vessels manufacture: A new generation of tantalum vessels. In *Proceedings of Corrosion Solutions Conference*; Paper no. 22; USA Wah Chang: Albany, OR, USA, 2003; pp. 163–169.

2. Danzig, I.F.; Dempsey, R.M.; la Conti, A.B. Characteristic of tantalided and hafnided samples in high corrosive electrolyte solutions. *Corrosion* **1971**, *27*, 55–62. [CrossRef]

3. Cardarelli, F.; Taxil, P.; Savall, A. Tantalum Protective Thin coating Techniques for the Chemical process Industry: Molten salts Electrocoating as a New Alternative. *Int. J. Refract. Hard Mater.* **1996**, *14*, 365–381. [CrossRef]

4. Paul, H.; Miszczyk, M.M.; Chulist, R.; Prażmowski, M.; Morgiel, J.; Gałka, A.; Faryna, M.; Brisset, F. Microstructure and phase constitution in the bonding zone of explosively welded tantalum and stainless steel sheets. *Mat. Des.* **2018**, *153*, 177–189. [CrossRef]

5. Blazynski, T.Z. *Explosive Welding, Forming and Compaction*; Applied Science Publishers LTD: New York, NY, USA, 1983. [CrossRef]

6. Kahraman, N.; Gülenç, B. Microstructural and mechanical properties of Cu-Ti plates bonded through explosive welding process. *J. Mater. Proc. Techn.* **2005**, *169*, 67–71. [CrossRef]

7. Song, J.; Kostka, A.; Veehmayer, M.; Raabe, D. Hierarhical microstructure of explosive joints: Example of titanium to steel cladding. *Mat. Sci. Eng.* **2011**, *528*, 2641–2647. [CrossRef]

8. Bataev, I.A.; Bataev, A.A.; Mali, V.I.; Pavliukova, D.V. Structural and mechanical properties of metallic-intermetallic laminate composites produced by explosive welding and annealing. *Mater. Des.* **2012**, *35*, 225–234. [CrossRef]

9. Paul, H.; Lityńska-Dobrzyńska, L.; Prażmowski, M. Microstructure and phase constitution near the interface of explosively welded aluminium/copper plates. *Metall. Mater. Trans. A* **2013**, *44*, 3836–3851. [CrossRef]

10. Paul, H.; Morgiel, J.; Faryna, M.; Prażmowski, M.; Miszczyk, M. Microstructure and interfacial reactions in the bonding zone of explosively welded Zr700 and carbon steel plates. *Int. J. Mater Res.* **2015**, *106*, 782–792. [CrossRef]

11. Bataev, I.A.; Lazurenko, D.V.; Tanaka, S.; Hokamoto, K.; Bataev, A.A.; Guo, Y.; Jorge, A.M., Jr. High cooling rates and metastable phases at the interfaces of explosively welded materials. *Acta Mater.* **2017**, *135*, 277–289. [CrossRef]

12. Bataev, I.A.; Tanaka, S.; Zhou, Q.; Lazurenko, D.V.; Junior, A.M.J.; Bataev, A.A.; Hokamoto, K.; Mori, A.; Chen, P. Towards better understanding of explosive welding by combination of numerical simulation and experimental study. *Mat. Des.* **2019**, *169*, 107649. [CrossRef]

13. Hammerschmidt, M.; Kreye, H. Microstructure and bonding mechanism in explosive welding. In *Shock Waves and High-Strain-Rate Phenomena in Metals*; Murr, M.A., Murr, L.E., Eds.; Plenum Press: New York, NY, USA, 1981. [CrossRef]

14. Paul, H.; Miszczyk, M.M.; Gałka, A.; Chulist, R.; Szulc, Z. Microstructural and chemical composition changes in the bonding zone of explosively welded sheets. *Arch. Metall. Mater.* **2019**, *64*, 683–694. [CrossRef]

15. Paul, H.; Miszczyk, M.; Prażmowski, M. Experimental investigation of texture gradients in aluminium/copper plates bonded through explosive welding process. *Mater. Sci. Forum* **2012**, *702–703*, 603–606. [CrossRef]

16. Paul, H.; Skuza, W.; Chulist, R.; Miszczyk, M.M.; Gałka, A.; Prażmowski, M.; Pstruś, J. The effect of interface morphology on the electro-mechanical properties of Ti/Cu clad composites produced by explosive welding. *Metall. Mater. Trans. A* **2020**, *51*, 750–766. [CrossRef]

17. Lee, T.; Nassiri, A.; Diettrich, T.; Vivek, A.; Daehn, G. Microstructure development in impact welding of a model system. *Scripta Mater.* **2020**, *178*, 203–206. [CrossRef]

18. Chu, Q.; Zhang, M.; Li, J.; Yan, C. Experimental and numerical investigation of microstructure and mechanical behaviour of titanium/steel interfaces prepared by explosive welding. *Mater. Sci. Eng. A* **2017**, *689*, 323–331. [CrossRef]

19. Mali, V.I.; Bataev, A.A.; Maliutina, I.N.; Kurguzov, V.D.; Bataev, I.A.; Esikov, M.A.; Lozhkin, V.S. Microstructure and mechanical properties of Ti/Ta/Cu/Ni alloy laminate composite materials produced by explosive welding. *Int. J Adv. Manuf. Technol.* **2017**, *93*, 4285–4294. [CrossRef]

20. Paul, H. Interfacial reactions during explosive bonding. *Mater. Sci. Forum* **2014**, *783–786*, 1476–1481. [CrossRef]

21. Bouckaert, G.P.; Hix, H.B.; Chelius, J. (Eds.) *(1974) Explosive-bonded tantalum-steel vessels. DECHEMA Monograph*; Applied Science Publishers LTD: New York, NY, USA, 1983; pp. 9–22.

22. Durgutlu, A.; Gülenc, B.; Findik, F. Examination of copper/stainless steel joints formed by explosive welding. *Mater. Des.* **2005**, *25*, 497–507. [CrossRef]

23. Zhang, H.; Jiao, K.X.; Zhang, J.L.; Liu, J. Experimental and numerical investigations of interface characteristics of copper/steel composite prepared by explosive welding. *Mater. Des.* **2018**, *154*, 140–152. [CrossRef]

24. Saravanan, S.; Raghukandan, K. Influence of Interlayer in Explosive Cladding of Dissimilar Metals. *Mater. Manuf. Process.* **2013**, *25*, 589–594. [CrossRef]

25. Pushkin, M.S.; Greenberg, B.A.; Ivanov, M.A.; Inozemtsev, A.V.; Patselov, A.M.; Besshaposhnikov, Y.P. Microstructure of joints Cu–Ta, Cu–Ti, Cu–Cu, produced by means of explosive welding: Fractal description of interface relief. *Compos. Interfaces* **2020**, 1–14. [CrossRef]

26. Maliutina, I.N.; Bataev, A.A.; Bataev, I.A.; Skorokhod, K.A.; Mali, V.I. Explosive welding of titanium with stainless steel using bronze—Tantalum as interlayer. In Proceedings of the Materials of the 9th International Forum on Strategic Technology (IFOST), Cox's Bazar, Bangladesh, 21–23 October 2014. [CrossRef]

27. Yang, M.; Ma, H.; Shen, Z.; Huang, Z.; Tian, Q.; Tian, J. Dissimilar material welding of tantalum foil and Q235 steel plate using improved explosive welding technique. *Mater. Des.* **2020**, *186*, 108348. [CrossRef]

28. Darling, K.A.; Roberts, A.J.; Mishin, Y.; Mathaudhu, S.N.; Kecskes, L.J. Grain size stabilization of nanocrystalline copper at high temperatures by alloying with tantalum. *J. Alloys Compd.* **2013**, *573*, 142–150. [CrossRef]

29. Frolov, T.; Darling, K.A.; Kecskes, L.J.; Mishin, Y. Stabilization and strengthening of nanocrystalline copper by alloying with tantalum. *Acta Mater.* **2012**, *60*, 2158–2168. [CrossRef]

30. Grenberg, B.A.; Ivanov, M.A.; Rybin, V.V.; Elkina, O.A.; Antonov, O.V.; Patselov, A.M.; Inozemtzev, A.V.; Plotnikov, A.V.; Volkova, A.Y.; Bessaposhnikov, Y.P. The problem of intermixing of metals possessing no mutual solubility upon explosion welding (Cu-Ta, Fe-Ag, Al-Ta). *Mater. Charact.* **2013**, *25*, 51–62. [CrossRef]

31. Maliutina, I.N.; Mali, V.I.; Bataev, I.A.; Bataev, A.A.; Esikov, M.A.; Smirnov, A.I.; Skorokhod, K.A. Structure and Microhardness of Cu-Ta Joints Produced by Explosive Welding. *Sci. World J.* **2013**, 56758. [CrossRef]

32. Parchuri, P.K.; Kotegawa, S.; Yamamoto, H.; Ito, K.; Mori, A.; Hokamoto, K. Benefits of intermediate-layer formation at the interface of Nb/Cu and Ta/Cu explosive clads. *Mater. Des.* **2019**, *166*, 107610. [CrossRef]

33. Cullis, A.G.; Orders, J.A.; Hirvonen, J.K.; Poate, J.M. Metastable alloy layers produced by implantation of Ag and Ta ions into Cu crystals. *Phil. Mag. B* **1978**, *37*, 615–630. [CrossRef]

34. Nastasi, M.; Saris, F.W.; Hung, L.S.; Mayer, J.W. Stability of amorphous Cu/Ta and Cu/W alloys. *J. Appl. Phys.* **1985**, *58*, 3052. [CrossRef]

35. Gong, H.R.; Kong, L.T.; Liu, B.X. Amorphous alloy formation in immiscible Cu-Ta and Cu-W systems by atomistic modelling and ion-beam mixing. *MRS Online Proc. Libr.* **2003**, *806*. [CrossRef]

36. Pun, G.P.P.; Darling, K.A.; Kecskes, L.J.; Mishin, Y. Angular-dependent interatomic potential for the Cu–Ta system and its application to structural stability of nano-crystalline alloys. *Acta Mat.* **2015**, *100*, 377–391. [CrossRef]

37. Rajagopalan, M.; Darling, K.; Turnage, S.; Koju, R.K.; Hornbuckle, B.; Mishin, Y.; Solanki, K.N. Microstructural evolution in a nanocrystalline Cu-Ta alloy: A combined in-situ TEM and atomistic study. *Mater. Des.* **2016**, *113*, 178–185. [CrossRef]

38. Lee, C.S.; Duggan, B.J. Deformation banding and copper-type rolling textures. *Acta Metallurgica et Materialia* **1993**, *41*, 2691–2699. [CrossRef]

39. Crossland, B.; Williams, J.D. Explosive welding. *Metall. Rev.* **1970**, *15*, 79–100. [CrossRef]

40. Liu, L.; Jia, Y.-F.; Xuan, F.-Z. Gradient effect in the waved interfacial layer of 304L/533B bimetallic plates induced by explosive welding. *Mater. Sci. Eng. A* **2017**, *704*, 493–502. [CrossRef]

41. Carvalho, G.H.S.F.L.; Galvão, I.; Mendes, R.; Leal, R.M.; Loureiro, A. Explosive welding of aluminium to stainless steel using carbon steel and niobium interlayers. *J. Mater. Process. Technol.* **2020**, *283*, 116707. [CrossRef]

42. Prażmowski, M.; Rozumek, D.; Paul, H. Static and fatigue tests of bimetal Zr-steel made by explosive welding. *Eng. Fail. Anal.* **2017**, *75*, 71–81. [CrossRef]

43. Paul, H.; Morawiec, A.; Baudin, T. Early Stages of Recrystallization in Equal-Channel Angular Pressing (ECAP)-Deformed AA3104 Alloy Investigated Using Scanning Electron Microscopy (SEM) and Transmission Electron Microscopy (TEM) Orientation Mappings. *Metall. Mater. Trans. A* **2012**, *43*, 4777–4793. [CrossRef]

44. Carpenter, S.H. Explosion welding: A Review. In *Shock Waves and High-Strain Rate Phenomena in Metals: Concepts and Applications*; Meyers, M.A., Murr, L.E., Eds.; Plenum Press: New York, NY, USA, 1981; Chapter 53; pp. 941–959.

Article

Aluminum-to-Steel Cladding by Explosive Welding

Gustavo H. S. F. L. Carvalho [1,*], Ivan Galvão [1,2], Ricardo Mendes [3], Rui M. Leal [1,4] and Altino Loureiro [1]

[1] Universidade de Coimbra, CEMMPRE, Departamento de Engenharia Mecânica, Rua Luís Reis Santos, 3030-788 Coimbra, Portugal; ivan.galvao@dem.uc.pt (I.G.); rui.leal@dem.uc.pt (R.M.L.); altino.loureiro@dem.uc.pt (A.L.)

[2] ISEL, Department of Mechanical Engineering, Polytechnic Institute of Lisbon, Rua Conselheiro Emídio Navarro 1, 1959-007 Lisboa, Portugal

[3] Universidade de Coimbra, ADAI/LEDAP, Departamento de Engenharia Mecânica, Rua Luís Reis Santos, 3030-788 Coimbra, Portugal; ricardo.mendes@dem.uc.pt

[4] LIDA-ESAD.CR, Polytechnic Institute of Leiria, Rua Isidoro Inácio Alves de Carvalho, 2500-321 Caldas da Rainha, Portugal

* Correspondence: gustavo.carvalho@dem.uc.pt; Tel.: +351-239-790-700

Received: 26 June 2020; Accepted: 3 August 2020; Published: 6 August 2020

Abstract: The production of aluminum-carbon steel and aluminum-stainless steel clads is challenging, and explosive welding is one of the most suitable processes to achieve them. The present work aims to investigate the coupled effect of two strategies for optimizing the production of these clads by explosive welding: the use of a low-density interlayer and the use of a low-density and low-detonation velocity explosive mixture. A broad range of techniques was used to characterize the microstructural and the mechanical properties of the welds, specifically, optical microscopy, scanning electron microscopy, energy dispersive spectroscopy, electron backscatter diffraction, microhardness and tensile-shear testing with digital image correlation analysis. Although aluminum-carbon steel and aluminum-stainless steel have different weldabilities, clads with sound microstructure and good mechanical behavior were achieved for both combinations. These results were associated with the low values of collision point and impact velocities provided by the tested explosive mixture, which made the weldability difference between these combinations less significant. The successful testing of this explosive mixture indicates that it is suitable to be used for welding very thin flyers and/or dissimilar materials that easily form intermetallic phases.

Keywords: explosive welding; interlayer; aluminum; carbon steel; stainless steel

---

1. Introduction

The successful production of hybrid welded structures is one of the main targets of the 21st century's industry. The development of solutions combining different materials is a great industrial challenge, which brings many technical, economic, and environmental advantages by enabling the achievement of highly efficient structures. However, manufacturing hybrid structures can be complex, especially when combining materials with significantly different physical properties. With the increase in the industrial relevance of hybrid components with unique characteristics, the complexity of welding increases and the use of conventional fusion welding technologies may not be possible. Some of the materials composing the hybrid structures tend to form very brittle intermetallic phases at high temperature, which easily results in cracking, and consequently, in a severe loss in the mechanical properties of the welded components. Thus, the solid-state welding technologies, such as the friction-based or the impact-based techniques, have a very high potential to join dissimilar materials.

The impact-based processes have the advantage of restricting the welding zone to a very narrow band at the interface of the materials and minimizing their interaction under high temperature and strain. As the impact is almost instantaneous, there is no time for heat dissipation towards the adjacent regions of the weld zone, avoiding the formation of an extensive heat affected zone and the consequent loss in mechanical properties often reported to occur in this region. In other words, the thermal cycle is short and narrow and hence causes minor microstructural changes, all very close to the welding interface. Among the impact-based technologies, the explosive welding has a prominent position since this process makes it possible to clad extensive areas, which is especially relevant for the naval, railway and automotive sectors. For these industries, aluminum and steel are widely used materials, and their welding has an especial interest by enabling the combination of the lightweight of the aluminum alloys with the low cost and the high mechanical strength of the carbon steel or with the corrosion resistance of the stainless steel.

Some research has been conducted in explosive welding of aluminum to carbon steel (Al-CS) and aluminum to stainless steel (Al-SS). Many authors have investigated the thermomechanical conditions experienced at the weld interface and their influence on the structure and on the mechanical properties of the welds [1–4]. The literature shows that these material combinations present notable differences in weldability, specifically, the range of welding parameters enabling the production of welds with good mechanical properties is wider for Al-CS welding [5]. The weldability range of the Al-SS couple is strongly conditioned by a considerable difference in the thermal conductivity of both materials [6]. However, despite the differences in Al-CS and Al-SS weldability, better welding conditions are usually achieved when the detonation and impact velocities used for joining both couples are not high [5]. High values of this parameter often lead to welds with poor mechanical properties or even to welding failure, i.e., the separation of the welded plates after the impact [7]. In dissimilar welding, this type of failure is usually associated with the formation of a thick and continuous molten layer with intermetallic composition at the interface of the welded plates [8,9].

Considering that high detonation and impact velocities may preclude or hinder the Al-CS and Al-SS welding, the optimization of the welding conditions for both combinations requires the testing of welding strategies focused on decreasing the values of these parameters. The most common strategies are the use of low-detonation velocity explosive mixtures, low explosive ratios and the use of interlayers. Among these, welding with interlayer has been the most tested. This strategy is reported to be effective for reducing the energy lost in the collision, and therefore, for preventing an extensive formation of brittle intermetallic phases. Regarding the Al-CS joining, the most reported material to be used as interlayer is commercially pure aluminum [5,10–13]. On the other hand, different materials have been tested as an interlayer in Al-SS welding, specifically, stainless steel [14,15], commercially pure aluminum [5], carbon steel [16], niobium [16], titanium [17], copper [17] and tantalum [17]. When materials from different families of both welded materials are tested as interlayers, the authors intend to explore the physical, chemical, and mechanical properties of these materials for achieving better welding conditions.

However, the use of interlayers adds a new material to the welded structure, increasing its weight and cost, especially when dense (for example, copper or carbon steel) and expensive materials (for example, niobium, tantalum, or titanium) are used with this purpose. It is already possible to weld plates with a thickness in the order of 1 mm by explosion welding. However, with the increase in design and engineering requirements, it becomes necessary to establish new strategies for welding even thinner plates. That said, it is essential to investigate not only the use of interlayers but also the development of energetic mixtures capable of providing low-detonation velocity together with lower explosive ratios. Thus, the present work is aimed to test two strategies for optimizing the production of Al-CS and Al-SS clads by explosive welding: the use of a low-density interlayer, and the development of a low-density and low-detonation velocity explosive mixture. This research analyzes the coupled effect of these two strategies on the microstructural and mechanical properties of the joints. The studied explosive mixture resulted from an intensive optimization work of the density of a previously tested explosive by controlling the volume of the sensitizer. An in-depth experimental characterization was conducted in the welds, using a broad range of techniques, such as optical microscopy, scanning electron microscopy (SEM), energy dispersive spectroscopy (EDS), electron backscatter diffraction (EBSD), microhardness and tensile-shear testing with digital image correlation (DIC).

2. Materials and Methods

Aluminum to carbon steel and aluminum to stainless steel welded clads were produced by explosive welding in parallel full overlap joint configuration, following the set-up presented in Figure 1. All welded plates had a length of 250 mm and a width of 70 mm. The flyer was a 3 mm-thick sheet of AA6082-T6 (112 HV0.2) for both weld series, whereas a 3 mm-thick sheet of EN10130 (DC06) carbon steel (100 HV0.2) or AISI 304 stainless steel (188 HV0.2) was used as the baseplate. All the welds were produced with a 1 mm-thick interlayer of AA1050 (38 HV0.2). The explosive mixture developed to produce the welds was a low-detonation velocity emulsion-based explosive. This mixture, which is based on standard emulsion explosives [18,19], resulted from an optimization work, that was developed in the Laboratory of Energetic and Detonics (LEDAP), focused on decreasing the density of an explosive mixture through the control of the volume of the sensitizer. The weld series are identified according to the alloys of the flyer and the base plate: the Al/CS and the Al/SS series concern to aluminum to carbon steel and aluminum to stainless steel welds, respectively. All the other conditions were kept constant, i.e. the interlayer alloy, the flyer/interlayer and the interlayer/baseplate stand-off distances (STD) and the explosive thickness and ratio. Table 1 summarizes the welding conditions.

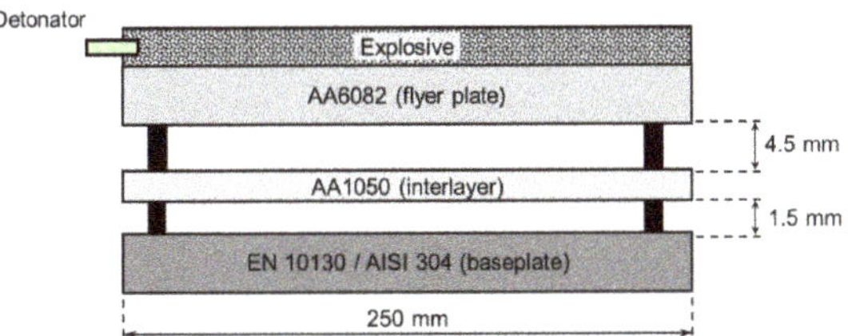

Figure 1. Schematic drawing of the welding set-up.

Table 1. Welding conditions.

Welding Conditions	Weld Series	
	Al/CS	Al/SS
Flyer plate alloy	AA6082	AA6082
Interlayer alloy	AA1050	AA1050
Baseplate alloy	EN10130	AISI 304
Flyer-interlayer STD	4.5 mm	4.5 mm
Interlayer-baseplate STD	1.5 mm	1.5 mm
Explosive Mixture	EE	EE
Explosive Mixture Density	485 kg.m^{-3}	485 kg.m^{-3}
Explosive Ratio	0.9	0.9

During welding, the detonation velocity (V_d), which has the same value as the collision point velocity (V_c) in the tested welding configuration, was measured according to Mendes et al. [20]. After the visual inspection of the welds, the samples were removed longitudinally to the welding direction and prepared for metallographic analysis according to ASTM E3-11. An optical microscope, Leica DM4000M LED (Wetzlar, Germany), was used to observe the samples, which were etched with Weck's etchant, 2% Nital and 10% oxalic acid for revealing the microstructure of the AA6082, EN10130 and AISI 304, respectively. The microstructural characterization of the welds was also conducted by SEM, using a Zeiss Merlin VP Compact microscope (Oberkochen, Germany), which was equipped with EDS. The semi-quantitative chemical composition of the welds was performed using this equipment. An accurate analysis of the grain structure of the weld interface was conducted by EBSD, using a FEI Quanta 400FEG SEM (Hillsboro, OR, USA) equipped with a TSL-EDAX EBSD unit.

The software TSL OIM Analysis 5.2 (EDAX Inc., Mahwah, NJ, USA) was used to analyze these results. The mechanical characterization of the welds was conducted by microhardness and tensile-shear testing. Microhardness profiles (HV0.2) were performed in longitudinal weld samples along the thickness direction, with a distance between indentations of 250 μm for the flyer and baseplate and 200 μm for the interlayer. Localized microhardness measurements (HV0.025) were performed at the weld interface. The microhardness tests were performed using an HMV-G Shimadzu tester (Kyoto, Japan). The tensile-shear tests were performed in quasi-static loading conditions (1 mm/min), using a 100 kN universal testing machine, Shimadzu AGS-X (Kyoto, Japan). Three specimens (removed longitudinally to the welding direction), whose design was similar to that reported by Carvalho et al. [5,16], were tested for each weld series. Figure 2 shows the geometry of the specimens, which have the original thickness of the plates. The local strain fields of the tested specimens were acquired by DIC using a GOM Aramis 5M system (Braunschweig, Germany). The procedures to prepare the specimens and to process/analyze the strain data are detailed in Leitão et al. [21]. After the tests, the fracture surface of the specimens was analyzed by SEM and the fracture mode fractions were computed by image processing.

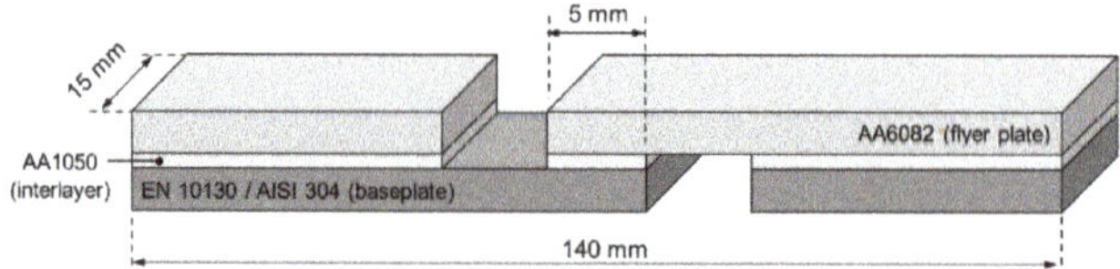

Figure 2. Details of the tensile-shear specimen.

3. Results and Discussion

3.1. Welding Results and Velocities

Table 2 shows the values measured for the detonation (V_d) and the collision point (V_c) velocities (the same values for parallel welding arrangement) and the values calculated for the impact velocity (V_p). Since the welds were produced with interlayer, two values of impact velocity were calculated: V_{pF} and V_{pFI}. V_{pF} corresponds to the velocity of the flyer plate at the instant of the first impact (the impact on the interlayer), which was computed using Gurney's equation for a one-dimensional problem in parallel configuration (Equation (1)) [22,23]. It must be noted that this equation, despite being widely accepted, presents some limitations. It ignores the acceleration of the flyer plate, and therefore, it represents only the terminal velocity [22,24]. That said, the proximity of the real value to the one calculated with Gurney's equation depends on the chosen STD.

$$V_{pF} = \sqrt{2E}\left(\frac{3R^2}{R^2 + 5R + 4}\right)^{\frac{1}{2}} \tag{1}$$

R is the explosive ratio (dimensionless), $\sqrt{2E}$ is the Gurney explosive's characteristic velocity (m·s^{-1}). An empirical correlation developed by Cooper [25] for ideal explosives, $\sqrt{2E} = V/2.97$, was used to estimate this parameter. The limitations of this approach were reported by Carvalho et al. [26].

V_{pFI} corresponds to the velocity of the set composed of the flyer plate and the interlayer at the instant of the second impact (the impact on the baseplate). This parameter was computed using an approximate method considering the perfectly inelastic collision theory and the momentum conservation (Equation (2)).

$$V_{pFI} = \frac{m_F \cdot V_{pF}}{m_F + m_I} \tag{2}$$

m_F is the mass of the flyer plate (kg), m_I is the mass of the interlayer (kg).

Table 2 shows that consistent welds were produced with low values of impact velocity (about 270 m·s^{-1}). In a previous study, Carvalho et al. [5] tested two explosive mixtures to produce the same type of joints: an emulsion explosive-based mixture ($V_d \approx 2800$ m·s^{-1}) and an ANFO-based mixture ($V_d \approx 2000$ m·s^{-1}). Compared to that work, the present mixture made it possible to weld with much lower values of impact velocity, which resulted from its lower detonation velocity (comparing to the emulsion explosive-based mixture) and lower density (comparing to the ANFO-based mixture). In fact, although the differences in detonation velocity between the present mixture and the ANFO-based mixture are not significant, a much lower explosive ratio was possible to be used with the present mixture (0.9 vs. 2.5), decreasing the impact velocity.

Table 2. Values of detonation/collision point and impact velocities and welding results.

Weld Series	V_d, V_c (m·s^{-1})	V_{pF} (m·s^{-1})	V_{pFI} (m·s^{-1})	Welding Results
Al/CS	2055	349	262	consistent
Al/SS	2055	357	268	consistent

3.2. Interface Morphology and Microstructure

Figure 3 presents the micrographs of the longitudinal interface of the welds. Figure 3a,b show that the AA6082/AA1050 interface (flyer/interlayer interface) was similar for both weld series and was composed of well-defined typical waves. On the other hand, significant differences in morphology were observed for the dissimilar interfaces (interlayer/baseplate interface). While small curled waves were formed at the interface of the Al/CS welds (Figure 3c), a flat interface was formed for the Al/SS welds (Figure 3d). In addition to this, there is also a morphological difference related to the formation of intermediate material. For the Al/CS welds, the intermediate material was mainly formed inside the curled waves, being totally encompassed by the ductile carbon steel (Figure 3c). On the other hand, a layer of intermediate material was intermittently formed at the interface of the Al/SS welds (Figure 3d). This layer is not discernible in some zones of the weld interface, in which a direct contact between the interlayer and the baseplate material exists.

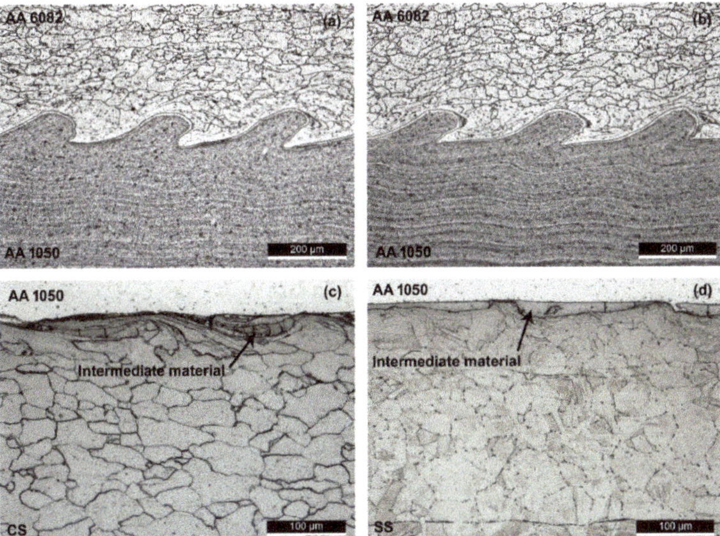

Figure 3. Micrographs of the weld interface: (**a**) Al/CS weld series—AA6082/AA1050 interface; (**b**) Al/SS weld series—AA6082/AA1050 interface; (**c**) Al/CS weld series—AA1050/CS interface; and (**d**) Al/SS weld series—AA1050/SS interface.

Comparing to the welds produced by Carvalho et al. [5], it is observed that the wave morphology of the present welds is much more similar to the morphology reported for welds produced with the emulsion explosive-based mixture than for welds produced with the ANFO-based mixture. This makes it possible to infer that the wave morphology is deeply affected by the nature of the explosive mixture and the explosive ratio, as claimed by Mendes et al. [20] and Plaksin et al. [27] for the SS-CS welding system. On the other hand, although the ANFO-based and the present mixtures have quite similar detonation velocities, the weld interfacial waves presented significant differences in amplitude and wavelength (higher in welds produced with ANFO).

Figure 4 presents the Vickers microhardness profiles of both weld series. It shows an increase in hardness compared to the base materials hardness. This is typical from the explosive welding process and is a consequence of the strong plastic deformation promoted by the impact. Both weld series presented a general increase in hardness throughout the thickness and a slightly more pronounced increase near the interface. The AISI 304 stainless steel presented the highest increase since work hardening is an effective hardening mechanism for this type of steel.

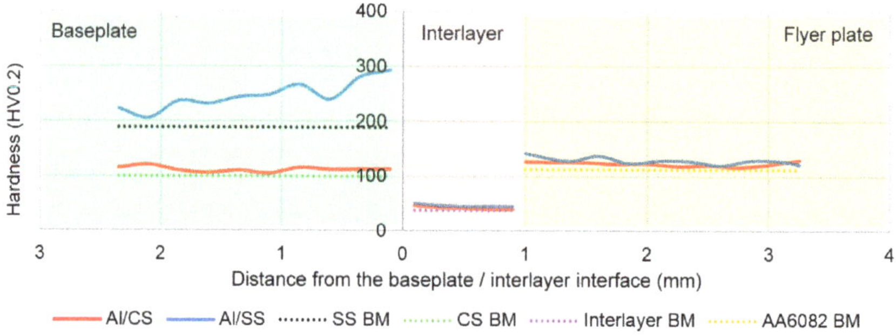

Figure 4. Microhardness (HV0.2) profiles.

Figure 5 shows the SEM micrographs of the dissimilar interface of the welds. From Figure 5a it can be observed that many cracks propagate along the intermediate material formed inside the curled waves of the Al/CS welds. The micrograph also shows that the propagation of the cracks is blocked by the wave structure, pointing to significant differences in ductility between the wave (CS) and the intermediate material. The results of the EDS analyses conducted in the regions indicated in Figure 5b (zones 1, 2 and 3) show that the intermediate material has a mixed chemical composition, being composed of Al and Fe (Table 3). The chemical composition of the intermediate region is fairly homogeneous and richer in Al. So, the formation of brittle Al-rich Fe_xAl_y intermetallic phases is expected to have occurred inside the curled waves. In good agreement with this, Carvalho et al. [7] reported the formation of Fe_4Al_{13} and Fe_2Al_5 at the interface of CS-Al explosive welds. These two intermetallic phases are the most reported in explosive welds between aluminum and steels. However, the heating and cooling conditions experienced at the interface of the explosive welds are far from equilibrium, and therefore, non-equilibrium phases may exist at the interface of the welds.

Figure 5c shows that fewer cracks propagate along the intermediate material of the Al/SS welds. The morphology of the intermediate material is also quite different for both weld series. From Figure 5d it can be observed that a much less homogenous intermediate region was formed at the interface of the Al/SS welds. Two zones are identified in this region, i.e., a lighter grey zone, which encompasses a larger area, and a darker grey zone. According to the results of the EDS analyses (Table 3), which were conducted in the regions indicated in Figure 5d, the lighter grey zone has a mixed Al-Fe chemical composition (zones 4 and 5). However, this region is richer in Al than the homogeneous intermediate material of the Al/CS welds. Besides Al and Fe, Ni and Cr, which are present in the stainless steel, were also detected. Regarding the darker grey zone, it is almost exclusively composed of Al. These results point to the formation of a heterogeneous intermediate region composed of both Al and

Al-Fe intermetallic phases. Considering the chemical composition of the stainless steel, a large range of intermetallic phases may be formed at the interface of the Al/SS welds, which makes it very difficult to indicate which phases were effectively formed.

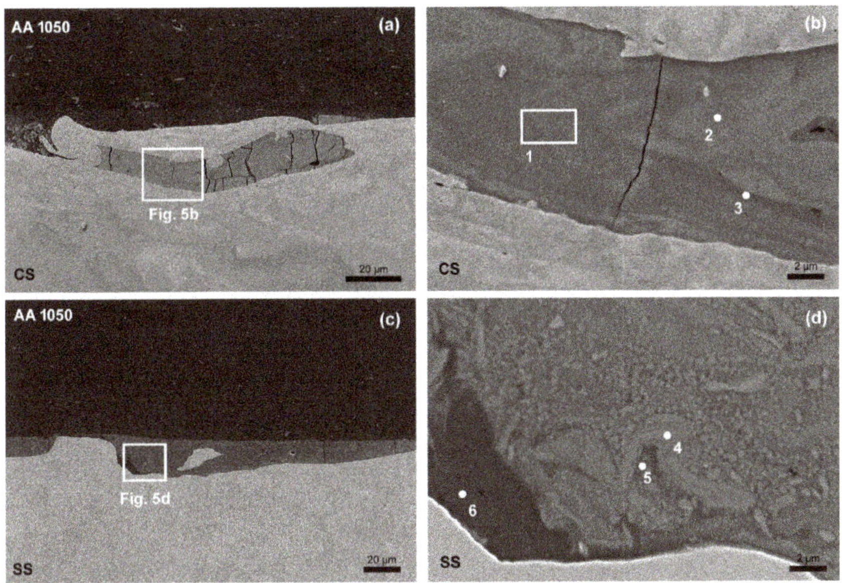

Figure 5. SEM micrographs of the dissimilar interface of the welds: (**a,b**) Al/CS weld series; (**c,d**) Al/SS weld series.

Table 3. Chemical composition (% at.) of the intermediate material.

Weld Series	Analysis Zone	Al	Fe	Cr	Ni	Average Microhardness (HV0.025)
Al/CS	1	67.0	33.0	——	——	702
	2	67.7	32.3	——	——	
	3	70.2	29.8	——	——	
Al/SS	4	82.6	13.1	3.1	1.2	414
	5	87.8	9.5	2.7	——	
	6	97.7	2.3	——	——	

Table 3 also presents the microhardness average values measured at the intermediate regions shown in Figure 5. The microhardness values are very high compared to those of the base materials. The mixed composition, high hardness, and presence of cracks (brittleness) agree well with the formation of intermetallic phases in the intermediate regions.

Regardless of the weld series, the intermediate regions formed at the weld interface have an Al-rich intermetallic composition, which agrees well with the significant differences in the melting temperature of the Al and the two Fe alloys. In fact, since the Al has a much lower melting temperature, there is a more substantial amount of this element in the interfacial molten volumes from which the intermetallic phases are generated. Moreover, although fewer cracks are observed in the intermediate regions of the Al/SS welds, they propagate throughout the intermetallic layer, i.e., from the SS until the Al. On the other hand, the cracks in the intermediate regions of the Al/CS welds are enclosed by the curled waves, which agrees with the results reported by Carvalho et al. [5]. The ductile waves act as a protection when brittle intermetallic phases are present.

Figure 6 presents the results of the EBSD analysis conducted at the dissimilar interface of the welds. For the Al/CS welds, Figure 6a shows that the AA1050 and the CS have an elongated grain structure, which agrees well with the plastic deformation experienced. However, the deformation experienced by the CS composing the wave structure is especially evident. Largely deformed grains with an impressive length to width ratio are observed in this zone. In turn, a much finer and equiaxed grain structure is observed inside the wave, pointing to the recrystallization of new grains. This region corresponds to the center of an interfacial vortex, where the most extreme strain and temperature values are usually achieved [28]. For the Al/SS welds, an elongated grain structure is also observed on the AA1050 side of the interface, as shown in Figure 6b. Regarding the SS side, this figure shows that fine equiaxed grains were formed in the nearest regions of the interface. The coupled effect of temperature and plastic deformation also promoted the recrystallization of the SS grains. Although the interaction of the welded materials is almost instantaneous in explosive welding, they experience a very strong plastic deformation and a high-temperature peak at the weld interface. The strong plastic deformation agrees with the increase in hardness next to the interface between the baseplate and the interlayer presented in Figure 4. The very high temperature and plastic deformation, as well as the occurrence of localized melting, boosted the interaction of the elements in this region, giving rise to the formation of intermetallic phases.

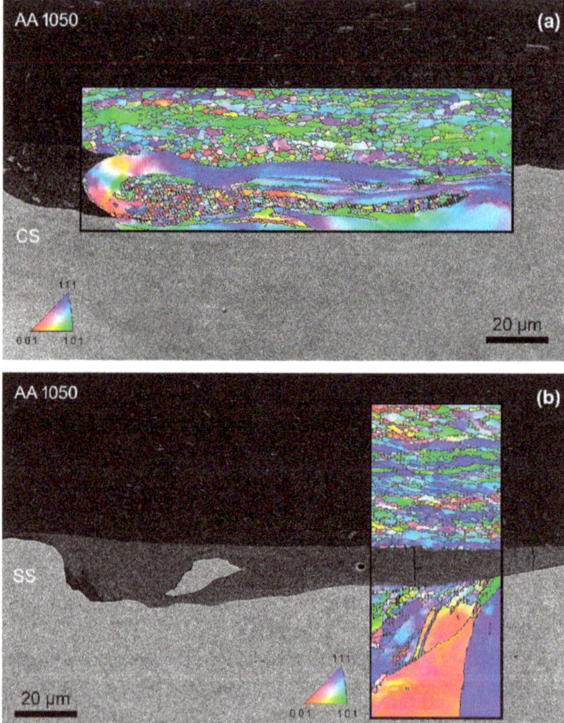

Figure 6. EBSD micrographs registered at the dissimilar interface of the welds: (**a**) Al/CS welds; (**b**) Al/SS welds.

3.3. Mechanical Properties

Table 4 displays the results of the tensile-shear tests. The table shows the maximum load value, the fracture region, and the fracture mode. For the maximum load, two values are presented for each weld series, which correspond to the lowest and the highest values obtained among all the tested specimens. Regarding the Al/CS series, it can be observed that the specimens had a very regular behavior, presenting all of them a 100% ductile fracture at the interlayer zone, with the maximum load value ranging between 4.8 kN and 5.1 kN. Figure 7a shows the Von Mises equivalent strain distribution map at the maximum load, which indicates that the strain was completely localized in the interlayer plate. In good agreement with this, Figure 8a and Table 5 indicate that the fracture surface of the specimens consisted of shear dimples and was exclusively composed of Al, which matches the chemical composition of the AA1050 interlayer. These results indicate that the strength of the similar and dissimilar weld interfaces was higher than that of the interlayer material. The mechanical behavior of the welds was conditioned by the strength of the AA1050 and not by poor interfacial bonding.

The maximum load value of the Al/SS specimens ranged between 4.5 kN and 5.0 kN (Table 4), which was approximately the same load range observed for the Al/CS weld series. Regarding the failure region, the specimens failed in the interlayer zone, which agrees well with the Von Mises equivalent strain distribution map at the maximum load shown in Figure 7b. However, some differences were observed in the fracture mode of these specimens. The specimens with a higher maximum load presented a 100% ductile fracture through the interlayer material. Figure 8b and Table 5 show that the fracture surface of these specimens consists of shear dimples and is exclusively composed of Al. On the other hand, the specimens with a lower maximum load presented a more heterogeneous fracture surface, because they fractured both through the interlayer and at the interlayer/SS interface. While the fracture through the interlayer was ductile, with the formation of Al shear dimples (Figure 8c and Table 5), the fracture at the interlayer/SS interface was brittle. Figure 8d and Table 5 indicate that the brittle fracture surface consists of cleavage patterns and is composed of both Al and Fe. The mixed chemical composition of the fracture surface agrees well with the chemical composition of the intermediate regions formed at the dissimilar interface of these welds, which indicates that the brittle intermetallic phases partially promoted the fracture. However, despite the presence of a brittle fracture, the maximum load was not very different. This is because the percentage of brittle fracture (17%) was much lower when compared to the percentage of ductile fracture (83%).

Table 4. Maximum load in tensile-shear testing, fracture region and fracture mode.

Weld Series	Maximum Load (kN)		Fracture Region	Fracture Mode
Al/CS	Lowest	4.8	Interlayer	Ductile (100%)
	Highest	5.1	Interlayer	Ductile (100%)
Al/SS	Lowest	4.5	Interlayer[1]	Ductile (83%) and Brittle (17%)
	Highest	5.0	Interlayer	Ductile (100%)

[1] The fracture occurred through the interlayer and through the interlayer/SS interface.

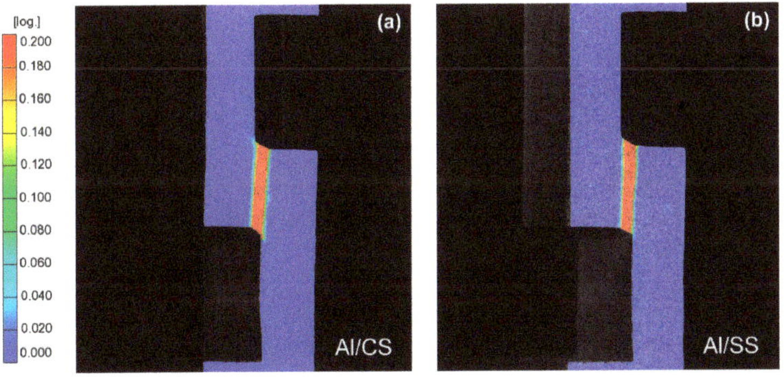

Figure 7. Von Mises equivalent strain distribution map: (**a**) Al/CS weld—ductile fracture; (**b**) Al/SS weld—ductile fracture.

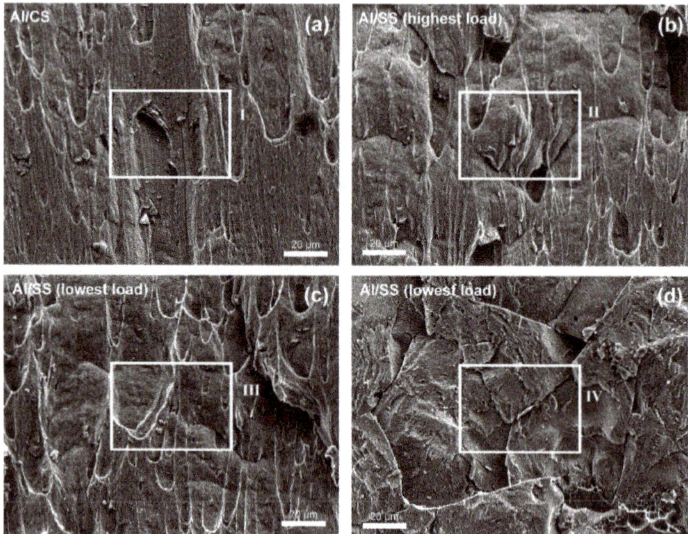

Figure 8. SEM micrographs of the fracture surface of the welds: (**a**) Al/CS welds; (**b**) Al/SS welds—highest maximum load; (**c**,**d**) Al/SS welds—lowest maximum load.

Table 5. Chemical composition (% at.) of the fracture surface of the welds.

Analysis Zone	Al	Fe	Cr	Ni	Si
I	100	——	——	——	——
II	99.8	——	——	——	0.2
III	100	——	——	——	——
IV	78.2	12.9	3.6	3.2	2.1

The differences between Al/CS and Al/SS welds in the tensile-shear tests occur mainly because of two reasons: the intermetallic formation and the interfacial microstructure. These differences lie mainly in the appearance of a brittle fracture percentage. The intermetallic formation of the Al/SS is more complex because of the presence of more alloying elements (Cr and Ni) that can form intermetallic phases with Al, Fe and complex phases combining more than these two elements. The interfacial microstructure also represents a critical factor on mechanical performance. The curled wavy interface of the Al/CS improved the mechanical performance, once the brittle intermetallic phases formed during the process were surrounded by ductile material (the waves). In other words, it avoids the intermetallic phases from being propitious regions to an uninterrupted propagation of a brittle fracture. Carvalho et al. [5] detailed this beneficial effect of the curled wave compared to flat interfaces.

The interfacial microstructure has a major role in explosive welding. The presence of waves (typical or curled wave) often leads to a better mechanical performance of the joint. Some recent works make it possible to better understand and predict the weld interface. While Carvalho et al. [26] study the prediction of a wavy interface in general, Carvalho et al. [16] study the prediction of curled waves specifically. These studies support the fact that the interfacial microstructure of the Al/CS welds significantly contributed to the best mechanical performance.

In a previous work [5], the Al-CS and Al-SS pairs were welded with a slightly lower collision point velocity and higher explosive ratio than the present work. The collision point velocity was not significantly different, but the higher explosive ratio led to higher impact velocity values. This means that the impact pressure was also more intense, which led to more substantial plastic deformation and strain hardening. It is important to note that two aspects should be balanced for the selection of the most suitable impact velocity in dissimilar welding of materials with easy formation of brittle intermetallic phases, specifically, the volume of intermetallic phases formed and the plastic deformation experienced by the materials at the interface. Until a threshold in impact velocity, higher values provide welds with better mechanical behavior by increasing the interfacial plastic deformation. After reaching this threshold, the increase in impact velocity only leads to the increase in the volume of brittle phases at the interface, and consequently, to the weakening of the welds. However, this threshold strongly depends on the microstructure of the weld interface, since, when curled waves are formed, the interface presents a higher ability to accommodate the volume of brittle phases generated during welding.

When higher collision point velocities were used on the same previous work [5], the results were rather irregular, considering that some welds failed during specimens preparation due to the poor bonding strength. The present work is between the two situations, offering an alternative of parameters that can be used especially in cases of flyers with lower thickness and density, in which mixtures of low velocity and high ratio, as used in previous works [5], may not be suitable. It is a significant advance for the welding of very thin plates of low-density materials, such as the aluminum alloys.

3.4. Energetic Mixture Analysis

In explosive welding, the challenge of the process is not only to define suitable welding parameters. After defining the best parameters, it is necessary to find energetic materials that can provide the parameters needed, such as detonation and impact velocities. One of the usual problems is that there are not too many energetic materials with adequate parameters, especially when low detonation velocities are preferred. In order to detonate, many of the low-detonation velocity explosive mixtures need a higher thickness of material than the high-detonation velocity mixtures. This fact leads to an issue, i.e., despite a lower detonation velocity, once the thickness of explosive mixture is high, the explosive ratio will be higher, and consequently, the impact velocity will increase too. In other words, choosing a low-detonation velocity explosive does not mean that the energy of the collision will be significantly lower due to the increase in the explosive ratio [5,6,9,20].

One of the topics of the present work is the study of a novel explosive mixture that has a low detonation velocity, low density and that does not need a high thickness to detonate, i.e. it can be used with a low explosive ratio. Figure 9 shows a sample of results from different works relating the detonation velocity with the explosive ratio used. The figure illustrates the abovementioned fact that, except for the mixture used in the present research, the explosive mixtures of lower detonation velocities tend to be used with higher explosive ratios. Since low-detonation velocity explosives usually need higher thickness, most explosive mixtures depart from the graph's origin. This proves the importance of developing mixtures that provide low detonation velocities and can detonate with thin layers. As referred above, the mixture developed in LEDAP and used in the present research has low detonation velocity, low density and does not need a high thickness to detonate. This is an important issue, especially for the welding of low-thickness and low-density flyers.

Deepening the analysis beyond the low detonation velocity and ratio, the density of the energetic material deserves attention due to its influence on the ratio. The achievement of low ratio explosive welds is facilitated by the use of low-density explosive mixtures. That said, an important novelty of the present research is the achievement of sound welds and proper detonation using a very low-density energetic mixture. Figure 10 compares 58 works from the literature relating the density (ρ_{exp}) and the detonation velocity (V_d) of explosive mixtures used for planar explosive welding. A point with multiple references indicates that all the references listed for that point have the same values of density and detonation velocity and are therefore located in the same position on the graph. For works in which the detonation velocity of a specific mixture was informed as a range (instead of a single measured value); in order to represent any possibility of detonation velocity within the range, two points concerning its maximum and minimum values were plotted. The figure shows that the mixture developed in the present work belongs to a small group of mixtures located closer to the origin of the graph. It should be noted that the detonation velocity of an explosive mixture may vary according to other properties beyond the density, such as the thickness of the mixture [15]. However, the data on the properties of the tested explosive mixtures are very limited in the literature.

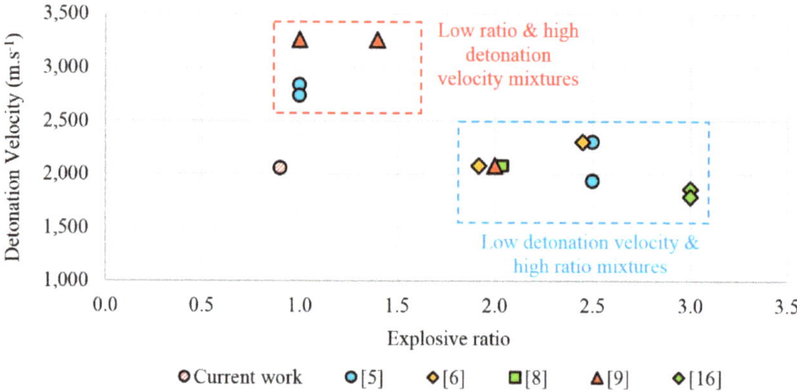

Figure 9. Graph relating the detonation velocity and the explosive ratio of tested explosive mixtures.

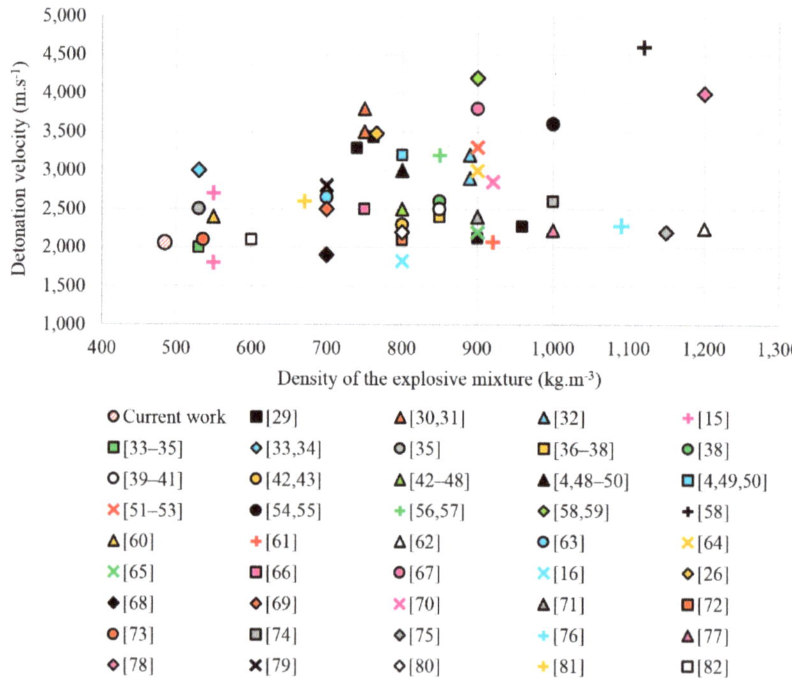

Figure 10. Graph relating the detonation velocity and the density of tested explosive mixtures. Data from [4,15,16,26,29–82].

In order to examine the detonation velocity and the density together, they were multiplied by one another. In this way, the two parameters were ordered on a single axis and are shown in Figure 11. The mixture tested in the present work presents one of the lowest values of $V_d \cdot \rho_{exp}$ among all the 58 works analyzed. Figures 9–11 show that the mixture is outside the groups of mixtures most tested in explosive welding literature. One of the limitations of the explosive welding process is precisely the difficulty of welding very thin plates (mainly the flyer plate). It happens due to the complexity of detonating thin layers of energetic mixtures that can provide the parameters needed, and because of the damage that the detonation causes on thin plates. This new mixture of energetic material enables the possibility of a mixture with low-density capable of providing low-detonation velocity, low explosive ratio and low-impact velocity. This is particularly important because it facilitates the welding of thinner materials. Beyond this fact, the tested mixture also has the advantage of reducing the energy conditions at the weld interface, which is significant for material combinations that tend to form intermetallic phases. On the one hand, if these combinations do not form favorable interface morphologies, such as curled waves, the large intermetallic volumes are very detrimental to the mechanical properties of the welds. On the other hand, even when favorable interfaces are formed, it is already established that large volumes of intermetallic phases may affect the physical phenomena at the weld interface, specifically, the solidification time of the interfacial molten material, conditioning the bonding conditions [7].

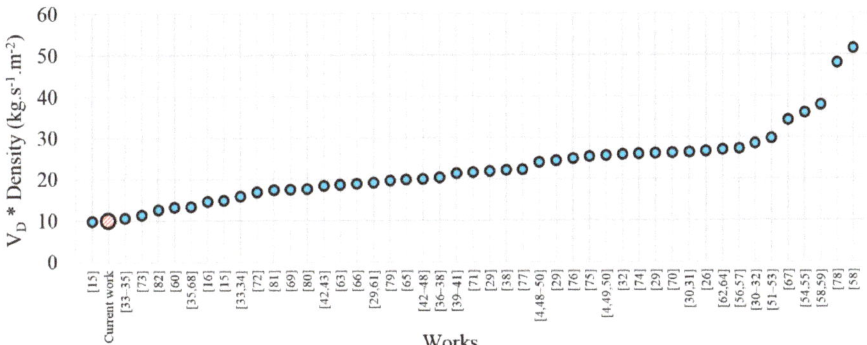

Figure 11. Graph relating on a single axis the detonation velocity and the density of the explosive mixture.

4. Conclusions

The present work has investigated the coupled effect of two strategies for optimizing the production of aluminum-carbon steel and aluminum-stainless steel clads by explosive welding: the use of a low-density interlayer, and the development of a low-density and low-detonation velocity explosive mixture. The following conclusions can be drawn:

- The coupled use of an interlayer and a low-density and low-detonation velocity explosive mixture is an effective strategy for producing aluminum-to-carbon steel and aluminum-to-stainless steel clads with sound microstructure and good mechanical behavior;
- The difference in weldability of aluminum-carbon steel and aluminum-stainless steel couples are less significant when welding under low energetic conditions;
- The tested low-density explosive mixture detonated with low detonation velocity, using a low explosive ratio, which resulted in welding with low values of both collision point velocity and impact velocity;
- Given to its properties of low-detonation velocity, low-density and the ability to detonate in small explosive thickness, the tested mixture is suitable to be used for welding very thin flyers and for welding dissimilar materials that tend to form intermetallic phases.

Author Contributions: Conceptualization, G.H.S.F.L.C. and I.G.; methodology, G.H.S.F.L.C., I.G. and R.M.L.; investigation, G.H.S.F.L.C., I.G. and R.M.L.; resources, R.M. and A.L.; writing—original draft preparation and edition, G.H.S.F.L.C. and I.G.; writing—review, R.M.L., R.M. and A.L.; supervision, R.M. and A.L.; project administration, A.L.; funding acquisition, A.L. All authors have read and agreed to the published version of the manuscript.

Funding: This research was funded by FEDER and FCT, as indicated in the acknowledgements.

Acknowledgments: This research is sponsored by FEDER funds through the program COMPETE—Programa Operacional Factores de Competitividade—and by national funds through FCT—Fundação para a Ciência e a Tecnologia, under the project UIDB/00285/2020.

Conflicts of Interest: The authors declare no conflict of interest.

References

1. Li, Y.; Hashimoto, H.; Sukedai, E.; Zhang, Y.; Zhang, Z. Morphology and structure of various phases at the bonding interface of Al/steel formed by explosive welding. *J. Electron Microsc.* **2000**, *49*, 5–16. [CrossRef] [PubMed]

2. Li, X.; Ma, H.; Shen, Z. Research on explosive welding of aluminum alloy to steel with dovetail grooves. *Mater. Des.* **2015**, *87*, 815–824. [CrossRef]

3. Guo, X.; Fan, M.; Wang, L.; Ma, F. Bonding Interface and Bending Deformation of Al/316LSS Clad Metal Prepared by Explosive Welding. *J. Mater. Eng. Perform.* **2016**, *25*, 2157–2163. [CrossRef]

4. Kaya, Y. Microstructural, Mechanical and Corrosion Investigations of Ship Steel-Aluminum Bimetal Composites Produced by Explosive Welding. *Metals* **2018**, *8*, 544. [CrossRef]

5. Carvalho, G.H.S.F.L.; Galvão, I.; Mendes, R.; Leal, R.M.; Loureiro, A. Microstructure and mechanical behaviour of aluminium-carbon steel and aluminium-stainless steel clads produced with an aluminium interlayer. *Mater. Charact.* **2019**, *155*, 109819. [CrossRef]

6. Carvalho, G.H.S.F.L.; Galvão, I.; Mendes, R.; Leal, R.M.; Loureiro, A. Explosive welding of aluminium to stainless steel. *J. Mater. Process. Technol.* **2018**, *262*, 340–349. [CrossRef]

7. Carvalho, G.H.S.F.L.; Galvão, I.; Mendes, R.; Leal, R.M.; Loureiro, A. Formation of intermetallic structures at the interface of steel-to-aluminium explosive welds. *Mater. Charact.* **2018**, *142*, 432–442. [CrossRef]

8. Carvalho, G.H.S.F.L.; Galvão, I.; Mendes, R.; Leal, R.M.; Loureiro, A. Influence of base material properties on copper and aluminium–copper explosive welds. *Sci. Technol. Weld. Join.* **2018**, *23*, 501–507. [CrossRef]

9. Carvalho, G.H.S.F.L.; Galvão, I.; Mendes, R.; Leal, R.M.; Loureiro, A. Weldability of aluminium-copper in explosive welding. *Int. J. Adv. Manuf. Technol.* **2019**, *103*, 3211–3221. [CrossRef]

10. Han, J.H.; Ahn, J.P.; Shin, M.C. Effect of interlayer thickness on shear deformation behavior of AA5083 aluminum alloy/SS41 steel plates manufactured by explosive welding. *J. Mater. Sci.* **2003**, *38*, 13–18. [CrossRef]

11. Tricarico, L.; Spina, R.; Sorgente, D.; Brandizzi, M. Effects of heat treatments on mechanical properties of Fe/Al explosion-welded structural transition joints. *Mater. Des.* **2009**, *30*, 2693–2700. [CrossRef]

12. Costanza, G.; Crupi, V.; Guglielmino, E.; Sili, A.; Tata, M.E. Metallurgical characterization of an explosion welded aluminum/steel joint. *Metall. Ital.* **2016**, *108*, 17–22.

13. Corigliano, P.; Crupi, V.; Guglielmino, E.; Mariano Sili, A. Full-field analysis of AL/FE explosive welded joints for shipbuilding applications. *Mar. Struct.* **2018**, *57*, 207–218. [CrossRef]

14. Izuma, T.; Hokamoto, K.; Fujita, M.; Aoyagi, M. Single-shot explosive welding of hard-to-weld A5083/SUS304 clad using SUS304 intermediate plate. *Weld. Int.* **1992**, *6*, 941–946. [CrossRef]

15. Hokamoto, K.; Izuma, T.; Fujita, M. New explosive welding technique to weld aluminum alloy and stainless steel plates using a stainless steel intermediate plate. *Metall. Trans. A* **1993**, *24*, 2289–2297. [CrossRef]

16. Carvalho, G.H.S.F.L.; Galvão, I.; Mendes, R.; Leal, R.M.; Loureiro, A. Explosive welding of aluminium to stainless steel using carbon steel and niobium interlayers. *J. Mater. Process. Technol.* **2020**, *283*, 116707. [CrossRef]

17. Aceves, S.M.; Espinosa-Loza, F.; Elmer, J.W.; Huber, R. Comparison of Cu, Ti and Ta interlayer explosively fabricated aluminum to stainless steel transition joints for cryogenic pressurized hydrogen storage. *Int. J. Hydrogen Energy* **2015**, *40*, 1490–1503. [CrossRef]

18. Mendes, R.; Ribeiro, J.; Plaksin, I.; Campos, J.; Tavares, B. Differences between the detonation behavior of emulsion explosives sensitized with glass or with polymeric micro-balloons. *J. Phys. Conf. Ser.* **2014**, *500*, 1–6. [CrossRef]

19. Mendes, R.; Ribeiro, J.B.; Plaksin, I.; Campos, J. Non Ideal Detonation of Emulsion Explosives Mixed with Metal Particles. *AIP Conf. Proc.* **2012**, *1426*, 267–270. [CrossRef]

20. Mendes, R.; Ribeiro, J.B.; Loureiro, A. Effect of explosive characteristics on the explosive welding of stainless steel to carbon steel in cylindrical configuration. *Mater. Des.* **2013**, *51*, 182–192. [CrossRef]

21. Leitão, C.; Galvão, I.; Leal, R.M.; Rodrigues, D.M. Determination of local constitutive properties of aluminium friction stir welds using digital image correlation. *Mater. Des.* **2012**, *33*, 69–74. [CrossRef]

22. Kennedy, J.E. *Gurney Energy of Explosives: Estimation of the Velocity and Impulse Imparted to Driven Metal*; Sandia Laboratories: Albuquerque, NM, USA, 1970; ISBN SC-RR-70-790.

23. El-Sobky, H. Mechanics of Explosive Welding. In *Explosive Welding, Forming and Compaction*; Blazynski, T.Z., Ed.; Springer: Dordrecht, The Netherlands, 1983; pp. 189–217.

24. Patterson, R.A. Fundamentals of Explosion Welding. In *ASM Handbook—Volume 6: Welding, Brazing and Soldering*; ASM International: Materials Park, OH, USA, 1993; pp. 160–164.

25. Cooper, P.W. *Explosive Engineering*; Wiley-VCH: New York, NY, USA, 1996; ISBN 0-471-18636-8.

26. Carvalho, G.H.S.F.L.; Mendes, R.; Leal, R.M.; Galvão, I.; Loureiro, A. Effect of the flyer material on the interface phenomena in aluminium and copper explosive welds. *Mater. Des.* **2017**, *122*, 172–183. [CrossRef]

27. Plaksin, I.; Campos, J.; Ribeiro, J.; Mendes, R.; Direito, J.; Braga, D.; Pruemmer, R. Novelties in physics of explosive welding and powder compaction. *J. Phys. IV* **2003**, *110*, 797–802. [CrossRef]

28. Bataev, I.A.; Bataev, A.A.; Mali, V.I.; Bataev, V.A.; Balaganskii, I.A. Structural Changes of Surface Layers of Steel Plates in the Process of Explosive Welding. *Met. Sci. Heat Treat.* **2014**, *55*, 509–513. [CrossRef]

29. Loureiro, A.; Mendes, R.; Ribeiro, J.B.; Leal, R.M.; Galvão, I. Effect of explosive mixture on quality of explosive welds of copper to aluminium. *Mater. Des.* **2016**, *95*, 256–267. [CrossRef]

30. Durgutlu, A.; Gülenç, B.; Findik, F. Examination of copper/stainless steel joints formed by explosive welding. *Mater. Des.* **2005**, *26*, 497–507. [CrossRef]

31. Kahraman, N.; Gülenç, B.; Findik, F. Joining of titanium/stainless steel by explosive welding and effect on interface. *J. Mater. Process. Technol.* **2005**, *169*, 127–133. [CrossRef]

32. Zhao, H.; Li, P.; Zhou, Y.; Huang, Z.; Wang, H. Study on the Technology of Explosive Welding Incoloy800-SS304. *J. Mater. Eng. Perform.* **2011**, *20*, 911–917. [CrossRef]

33. Manikandan, P.; Hokamoto, K.; Raghukandan, K.; Chiba, A.; Deribas, A.A. The effect of experimental parameters on the explosive welding of Ti and stainless steel. *Sci. Technol. Energ. Mater.* **2005**, *66*, 370–374.

34. Manikandan, P.; Hokamoto, K.; Fujita, M.; Raghukandan, K.; Tomoshige, R. Control of energetic conditions by employing interlayer of different thickness for explosive welding of titanium/304 stainless steel. *J. Mater. Process. Technol.* **2008**, *195*, 232–240. [CrossRef]

35. Inao, D.; Mori, A.; Tanaka, S.; Hokamoto, K. Explosive Welding of Thin Aluminum Plate onto Magnesium Alloy Plate Using a Gelatin Layer as a Pressure-Transmitting Medium. *Metals* **2020**, *10*, 106. [CrossRef]

36. Mousavi, S.A.A.A.; Sartangi, P.F. Effect of post-weld heat treatment on the interface microstructure of explosively welded titanium–stainless steel composite. *Mater. Sci. Eng. A* **2008**, *494*, 329–336. [CrossRef]

37. Mousavi, S.A.A.A.; Sartangi, P.F. Experimental investigation of explosive welding of cp-titanium/AISI 304 stainless steel. *Mater. Des.* **2009**, *30*, 459–468. [CrossRef]

38. Duan, M.; Wang, Y.; Ran, H.; Ma, R.; Wei, L. Study on Inconel 625 Hollow Structure Manufactured by Explosive Welding. *Mater. Manuf. Process.* **2014**, *29*, 1011–1016. [CrossRef]

39. Zhang, H.; Jiao, K.X.; Zhang, J.L.; Liu, J. Experimental and numerical investigations of interface characteristics of copper/steel composite prepared by explosive welding. *Mater. Des.* **2018**, *154*, 140–152. [CrossRef]

40. Zhang, H.; Jiao, K.X.; Zhang, J.L.; Liu, J. Microstructure and mechanical properties investigations of copper-steel composite fabricated by explosive welding. *Mater. Sci. Eng. A* **2018**, *731*, 278–287. [CrossRef]

41. Yang, M.; Ma, H.; Shen, Z. Study on explosive welding of Ta2 titanium to Q235 steel using colloid water as a covering for explosives. *J. Mater. Res. Technol.* **2019**, *8*, 5572–5580. [CrossRef]

42. Ma, R.; Wang, Y.; Wu, J.; Duan, M. Explosive welding method for manufacturing ITER-grade 316L(N)/CuCrZr hollow structural member. *Fusion Eng. Des.* **2014**, *89*, 3117–3124. [CrossRef]

43. Ma, R.; Wang, Y.; Wu, J.; Duan, M. Investigation of microstructure and mechanical properties of explosively welded ITER-grade 316L(N)/CuCrZr hollow structural member. *Fusion Eng. Des.* **2015**, *93*, 43–50. [CrossRef]

44. Liu, Y.; Li, C.; Hu, X.; Yin, C.; Liu, T. Explosive Welding of Copper to High Nitrogen Austenitic Stainless Steel. *Metals* **2019**, *9*, 339. [CrossRef]

45. Honarpisheh, M.; Asemabadi, M.; Sedighi, M. Investigation of annealing treatment on the interfacial properties of explosive-welded Al/Cu/Al multilayer. *Mater. Des.* **2012**, *37*, 122–127. [CrossRef]

46. Sedighi, M.; Honarpisheh, M. Experimental study of through-depth residual stress in explosive welded Al–Cu–Al multilayer. *Mater. Des.* **2012**, *37*, 577–581. [CrossRef]

47. Asemabadi, M.; Sedighi, M.; Honarpisheh, M. Investigation of cold rolling influence on the mechanical properties of explosive-welded Al/Cu bimetal. *Mater. Sci. Eng. A* **2012**, *558*, 144–149. [CrossRef]

48. Gong, S.; Li, Z.; Xiao, Z.; Zheng, F. Microstructure and property of the composite laminate cladded by explosive welding of CuAlMn shape memory alloy and QBe2 alloy. *Mater. Des.* **2009**, *30*, 1404–1408. [CrossRef]
49. Kaçar, R.; Acarer, M. Microstructure–property relationship in explosively welded duplex stainless steel–steel. *Mater. Sci. Eng. A* **2003**, *363*, 290–296. [CrossRef]
50. Liu, L.; Jia, Y.-F.; Xuan, F.-Z. Gradient effect in the waved interfacial layer of 304L/533B bimetallic plates induced by explosive welding. *Mater. Sci. Eng. A* **2017**, *704*, 493–502. [CrossRef]
51. Rajani, H.R.Z.; Mousavi, S.A.A.A.; Madani Sani, F. Comparison of corrosion behavior between fusion cladded and explosive cladded Inconel 625/plain carbon steel bimetal plates. *Mater. Des.* **2013**, *43*, 467–474. [CrossRef]
52. Rajani, H.R.Z.; Mousavi, S.A.A.A. The effect of explosive welding parameters on metallurgical and mechanical interfacial features of Inconel 625/plain carbon steel bimetal plate. *Mater. Sci. Eng. A* **2012**, *556*, 454–464. [CrossRef]
53. Rajani, H.R.Z.; Mousavi, S.A.A.A. The Role of Impact Energy in Failure of Explosive Cladding of Inconel 625 and Steel. *J. Fail. Anal. Prev.* **2012**, *12*, 646–653. [CrossRef]
54. Lazurenko, D.V.; Bataev, I.A.; Mali, V.I.; Lozhkina, E.A.; Esikov, M.A.; Bataev, V.A. Structural Transformations Occurring upon Explosive Welding of Alloy Steel and High-Strength Titanium. *Phys. Met. Metallogr.* **2018**, *119*, 469–476. [CrossRef]
55. Lazurenko, D.V.; Bataev, I.A.; Mali, V.I.; Esikov, M.A.; Bataev, A.A. Effect of Hardening Heat Treatment on the Structure and Properties of a Three-Layer Composite of Type 'VT23-08ps-45KhNM' Obtained by Explosion Welding. *Met. Sci. Heat Treat.* **2019**, *60*, 651–658. [CrossRef]
56. Chen, P.; Feng, J.; Zhou, Q.; An, E.; Li, J.; Yuan, Y.; Ou, S. Investigation on the Explosive Welding of 1100 Aluminum Alloy and AZ31 Magnesium Alloy. *J. Mater. Eng. Perform.* **2016**, *25*, 2635–2641. [CrossRef]
57. Feng, J.; Chen, P.; Zhou, Q. Investigation on Explosive Welding of Zr53Cu35Al12 Bulk Metallic Glass with Crystalline Copper. *J. Mater. Eng. Perform.* **2018**, *27*, 2932–2937. [CrossRef]
58. Bataev, I.A.; Bataev, A.A.; Mali, V.I.; Pavliukova, D.V. Structural and mechanical properties of metallic–intermetallic laminate composites produced by explosive welding and annealing. *Mater. Des.* **2012**, *35*, 225–234. [CrossRef]
59. Lazurenko, D.V.; Bataev, I.A.; Mali, V.I.; Bataev, A.A.; Maliutina, I.N.; Lozhkin, V.S.; Esikov, M.A.; Jorge, A.M.J. Explosively welded multilayer Ti-Al composites: Structure and transformation during heat treatment. *Mater. Des.* **2016**, *102*, 122–130. [CrossRef]
60. Hokamoto, K.; Chiba, A.; Fujita, M.; Izuma, T. Single-shot explosive welding technique for the fabrication of multilayered metal base composites: Effect of welding parameters leading to optimum bonding condition. *Compos. Eng.* **1995**, *5*, 1069–1079. [CrossRef]
61. Chu, Q.L.; Zhang, M.; Li, J.H.; Jin, Q.; Fan, Q.Y.; Xie, W.W.; Luo, H.; Bi, Z.Y. Experimental investigation of explosion-welded CP-Ti/Q345 bimetallic sheet filled with Cu/V based flux-cored wire. *Mater. Des.* **2015**, *67*, 606–614. [CrossRef]
62. Wang, P.; Chen, J.; Li, Q.; Liu, D.; Huang, P.; Jin, F.; Zhou, Y.; Yang, B. Study on the microstructure and properties evolution of CuCrZr/316LN-IG explosion bonding for ITER first wall components. *Fusion Eng. Des.* **2017**, *124*, 1135–1139. [CrossRef]
63. Wang, Y.; Li, X.; Wang, X.; Yan, H. Fabrication of a thick copper-stainless steel clad plate for nuclear fusion equipment by explosive welding. *Fusion Eng. Des.* **2018**, *137*, 91–96. [CrossRef]
64. Yang, M.; Ma, H.; Shen, Z.; Sun, Y. Study on explosive welding for manufacturing meshing bonding interface of CuCrZr to 316L stainless steel. *Fusion Eng. Des.* **2019**, *143*, 106–114. [CrossRef]
65. Wang, T.; Zhang, F.; Li, X.; Jiang, S.; Feng, J. Interfacial evolution of explosively welded titanium/steel joint under subsequent EBW process. *J. Mater. Process. Technol.* **2018**, *261*, 24–30. [CrossRef]
66. Yang, M.; Ma, H.; Shen, Z.; Chen, D.; Deng, Y. Microstructure and mechanical properties of Al-Fe meshing bonding interfaces manufactured by explosive welding. *Trans. Nonferr. Met. Soc. China* **2019**, *29*, 680–691. [CrossRef]
67. Bataev, I.A.; Lazurenko, D.V.; Tanaka, S.; Hokamoto, K.; Bataev, A.A.; Guo, Y.; Jorge, A.M. High cooling rates and metastable phases at the interfaces of explosively welded materials. *Acta Mater.* **2017**, *135*, 277–289. [CrossRef]
68. Andreevskikh, L.A.; Drozdov, A.A.; Mikhailov, A.L.; Samarokov, Y.M.; Skachkov, O.A.; Deribas, A.A. Producing bimetallic steel-copper composites by explosive welding. *Steel Transl.* **2015**, *45*, 84–87. [CrossRef]
69. Blatter, A.; Peguiron, D.A. Explosive joining of precious metals. *Gold Bull.* **1998**, *31*, 93–98. [CrossRef]

70. Chu, Q.; Tong, X.; Xu, S.; Zhang, M.; Li, J.; Yan, F.; Yan, C. Interfacial Investigation of Explosion-Welded Titanium/Steel Bimetallic Plates. *J. Mater. Eng. Perform.* **2020**, *29*, 78–86. [CrossRef]
71. Greenberg, B.A.; Ivanov, M.A.; Inozemtsev, A.V.; Pushkin, M.S.; Patselov, A.M.; Besshaposhnikov, Y.R. Comparative characterisation of interfaces for two- and multi-layered Cu-Ta explosively welded composites. *Compos. Interfaces* **2020**, *27*, 705–715. [CrossRef]
72. Sun, Z.; Shi, C.; Wu, X.; Shi, H. Comprehensive investigation of effect of the charge thickness and stand-off gap on interface characteristics of explosively welded TA2 and Q235B. *Compos. Interfaces* **2020**, 1–17. [CrossRef]
73. Guo, X.; Ma, Y.; Jin, K.; Wang, H.; Tao, J.; Fan, M. Effect of Stand-Off Distance on the Microstructure and Mechanical Properties of Ni/Al/Ni Laminates Prepared by Explosive Bonding. *J. Mater. Eng. Perform.* **2017**, *26*, 4235–4244. [CrossRef]
74. Zhou, Q.; Feng, J.; Chen, P. Numerical and Experimental Studies on the Explosive Welding of Tungsten Foil to Copper. *Materials* **2017**, *10*, 984. [CrossRef]
75. Tao, C.; Li, J.; Lu, M.; Yang, X.; Zhao, H.; Wang, W.; Zhu, W. Microstructure and mechanical properties of Cu/CuCrZr composite plates fabricated by explosive welding. *Compos. Interfaces* **2020**, 1–12. [CrossRef]
76. Sahul, M.; Sahul, M.; Lokaj, J.; Čaplovič, L.; Nesvadba, P. Influence of Annealing on the Properties of Explosively Welded Titanium Grade 1—AW7075 Aluminum Alloy Bimetals. *J. Mater. Eng. Perform.* **2018**, *27*, 5665–5674. [CrossRef]
77. Sahul, M.; Sahul, M.; Lokaj, J.; Čaplovič, Ľ.; Nesvadba, P.; Odokienová, B. The Effect of Annealing on the Properties of AW5754 Aluminum Alloy-AZ31B Magnesium Alloy Explosively Welded Bimetals. *J. Mater. Eng. Perform.* **2019**, *28*, 6192–6208. [CrossRef]
78. Saravanan, S.; Raghukandan, K.; Kumar, P. Effect of wire mesh interlayer in explosive cladding of dissimilar grade aluminum plates. *J. Cent. South Univ.* **2019**, *26*, 604–611. [CrossRef]
79. Yazdani, M.; Toroghinejad, M.R.; Hashemi, S.M. Investigation of Microstructure and Mechanical Properties of St37 Steel-Ck60 Steel Joints by Explosive Cladding. *J. Mater. Eng. Perform.* **2015**, *24*, 4032–4043. [CrossRef]
80. Fang, Z.; Shi, C.; Shi, H.; Sun, Z. Influence of Explosive Ratio on Morphological and Structural Properties of Ti/Al Clads. *Metals* **2019**, *9*, 119. [CrossRef]
81. Mahmood, Y.; Dai, K.; Chen, P.; Zhou, Q.; Bhatti, A.A.; Arab, A. Experimental and Numerical Study on Microstructure and Mechanical Properties of Ti-6Al-4V/Al-1060 Explosive Welding. *Metals* **2019**, *9*, 1189. [CrossRef]
82. Cui, Y.; Liu, D.; Zhang, Y.; Deng, G.; Fan, M.; Chen, D.; Sun, L.; Zhang, Z. The Microstructure and Mechanical Properties of TA1-Low Alloy Steel Composite Plate Manufactured by Explosive Welding. *Metals* **2020**, *10*, 663. [CrossRef]

Article

Explosive Welding of Thin Aluminum Plate onto Magnesium Alloy Plate Using a Gelatin Layer as a Pressure-Transmitting Medium

Daisuke Inao [1,*], Akihisa Mori [2], Shigeru Tanaka [3] and Kazuyuki Hokamoto [3]

[1] Faculty of Engineering, Kumamoto University, Kumamoto 860-8555, Japan
[2] Department of Mechanical Engineering, Sojo University, Kumamoto 860-0082, Japan;
    makihisa@mec.sojo-u.ac.jp
[3] Institute of Pulsed Power Science, Kumamoto University, Kumamoto 860-8555, Japan;
    tanaka@mech.kumamoto-u.ac.jp (S.T.); hokamoto@mech.kumamoto-u.ac.jp (K.H.)
*   Correspondence: inao@tech.eng.kumamoto-u.ac.jp; Tel.: +81-96-342-3901

Received: 7 November 2019; Accepted: 7 January 2020; Published: 9 January 2020

Abstract: Mg alloys are extensively used in various automotive, aerospace, and industrial applications. Their limited corrosion resistance can be enhanced by welding a thin Al plate onto the alloy surface. In this study, we perform the explosive welding of a thin Al plate, accelerated by the detonation of an explosive through a gelatin layer as a pressure-transmitting medium, onto two Mg alloy samples: $Mg_{96}Zn_2Y_2$ alloy containing a long-period stacking ordered phase in an α-Mg matrix and commercial AZ31. The bonding interface is characterized using optical microscopy, scanning electron microscopy, X-ray diffraction, and electron probe microanalysis. Under moderate experimental conditions, the thin Al plates are successfully welded onto the Mg alloys, showing typical wavy interfaces without intermediate layers. Due to the decreased energetic condition corresponding to the use of a thin flyer plate and gelatin medium, the resulting bonding quality is better than that obtained using a regular explosive welding technique. Further, based on the well-known window for explosive welding, we estimate that the experimental conditions for successful bonding are close to the lower welding limit for a thin Al plate with the two Mg alloys considered. These findings may contribute to improving the quality of materials welded with explosive welding.

Keywords: explosive welding; gelatin; thin aluminum plate; magnesium alloys; LPSO phase

---

1. Introduction

Mg is the lightest among metals that have industrial uses and, thus, Mg-based alloys are extensively used in automotive, aerospace, and other applications [1]. However, Mg alloys are known to be flammable, and their mechanical properties, particularly ductility, are inferior to those of Al and its alloys. Therefore, researchers recently developed a heat-resistant Mg alloy with increased mechanical strength at elevated temperatures. The alloy structure can be described as Mg-M-RE (M = Co, Ni, Cu, Zn, or Al; RE = Y, Gd, Tb, Dy, Ho, Er, or Tm), wherein the α-Mg phase exhibits a long-period stacking order (LPSO) [2–8]. However, as with conventional Mg alloys, inadequate corrosion resistance limits its further application. Therefore, Mg alloys covered with a thin Al layer can yield improved corrosion resistance, and such composites may afford improved balance between strength and ductility [9].

Thus far, researchers have developed various techniques for producing Al/Mg alloy composite plates, such as laser welding [10], diffusion welding [11], friction stir welding [12], and ultrasonic welding [13–15]. However, because the quality of the weld joint is affected by the bonding temperature and holding time, Al/Mg tends to form intermetallic compounds at the bonding interface with heating. The formation of these intermetallic compounds can degrade the mechanical properties of the alloy [16].

In this regard, to the best of our knowledge, there is no report on the bonding of Al and $Mg_{96}Zn_2Y_2$ alloy with ultrasonic welding.

The technique of explosive welding (EXW) can be used to solve such problems due to heating. EXW is a solid-state bonding technique that can be used to obtain metallurgical welds of two or more dissimilar metals via the acceleration of a flyer plate by the detonation of an explosive [17]. EXW is useful in solving the problem of intermetallic formation because of its short processing time; the time available is insufficient for the diffusion through welds obtained under moderate conditions near the lower limit of the weldability window [18,19]. In addition, we note that it is preferable to reduce the energy deposited at the interface because of the high reactivity of Mg; such energy reduction can suppress the formation of intermetallic compounds. The technique of underwater EXW has been developed for this purpose, and the method is effective for accelerating thin metal plates [20–23]. The formation of intermetallic compounds (or mixing zones) is associated with a kinetic energy loss ΔKE [24]. Since flyer plates used in the underwater method (including gelatin) are targeted for thicknesses (especially below 1 mm), the value of ΔKE is inevitably small. In other words, in the case of EXW through a medium, the kinetic energy loss at the collision point is smaller than that in the normal method, leading to a decrease in intermetallic compounds (or mixing zones). However, because the use of water complicates the assembly setup, we are currently developing a modified method that uses a gelatin layer instead of water as the pressure-transmitting medium [25–27]. With this approach, it is possible to simplify the assembly. Further, it must be noted that the weld responses under shockwaves in water and gelatin are quite similar, implying that the experimental conditions can be easily designed based on stock data of the parameters utilized for underwater EXW.

The wave shape formed at the interface is similar to those formed during EXW in air and is clearly different from that formed during underwater EXW [28]. Low collision point velocities in the horizontal direction (EXW through gelatin medium) result in a wavy or flat interface without an intermediate layer (IL). Conversely, high collision point velocities in the horizontal direction and low collision angles (underwater EXW) result in the formation of continuous or discontinuous ILs due to the rapid solidification in EXW [29]. As the formation of a continuous IL containing a brittle intermetallic compound at the bonding interface may lead to a decrease in bonding quality, the method employing a gelatin medium can be expected to suppress the formation of ILs.

In this study, a thin Al plate was bonded to extruded $Mg_{96}Zn_2Y_2$ alloy with the use of gelatin as a pressure-transmitting medium [25,26]. Further, the bonding of an Al plate with AZ31, which is a commercially available Mg alloy, was performed for comparison. The microstructures of the recovered samples were characterized using optical microscopy (OM), scanning electron microscopy (SEM), and microfocus X-ray diffraction (XRD) analysis, and electron probe microanalysis (EPMA). Furthermore, the experimental parameters, calculated based on numerical simulations, were estimated based on the welding window and found to lie close to the lower limit of welding.

2. Materials and Methods

The experimental conditions and a schematic of the experimental assembly are presented in Table 1 and Figure 1, respectively, and the chemical composition of $Mg_{96}Zn_2Y_2$ is listed in Table 2. In this study, we employed a parallel layer arrangement with 20 wt.% gelatin as the pressure-transmitting medium for the EXW process.

The explosive used for welding was ANFO-A (explosion velocity: 2–2.5 km/s, density: 530 kg/m^3, explosive characteristics were the same as described in [30]), which is mainly composed of ammonium nitrate and fuel oil. The explosive was detonated with the use of SEP (detonation velocity: 7 km/s, density: 1300 kg/m^3, 3 g) as a booster, which was ignited by a No. 6 electric detonator (ED). The flyer plate was composed of industrially pure Al (>99%) pasted under a cover plate (0.2-mm-thick 304 stainless steel), which was used for surface protection and for decreasing momentum. The parent plate was composed of AZ31 or $Mg_{96}Zn_2Y_2$ (5-mm thickness). The stand-off distance (SOD) between the flyer and parent plate was set at 0.5 mm. Other Mg alloy blocks (AZ31) were placed around the parent plate to act as momentum traps to reduce the cracks induced by the reflected tensile waves [22]. To stabilize the detonation velocity of the ANFO-A explosive, the welding plates were set to have a sufficient horizontal distance of >80 mm.

We estimated the experimental conditions via numerical analysis using AUTODYN-2D (Century Dynamics, a subsidiary of ANSYS, Inc., TK, USA) [20,23,31] to study the influence of certain parameters based on the welding window proposed by Wittman and Deribas [17,32].

Table 1. Experimental conditions for the explosive welding of metal and Mg alloy plates.

No.	Parent Plate ($l \times w \times t$, mm^3)		Flyer Plate (t, mm)	Cover Plate (t, mm)	Explosive Thickness (T, mm)	Weld
1	AZ31	$50 \times 50 \times 5$	Al (0.2)	JIS-SUS304 (0.2)	18	Yes
2	AZ31	$50 \times 50 \times 5$	Al (0.2)	JIS-SUS304 (0.2)	23	Yes
3	AZ31	$50 \times 50 \times 5$	Al (0.2)	JIS-SUS304 (0.2)	29	Yes
4	$Mg_{96}Zn_2Y_2$	$50 \times 50 \times 5$	Al (0.2)	JIS-SUS304 (0.2)	18	No
5	$Mg_{96}Zn_2Y_2$	$50 \times 50 \times 5$	Al (0.2)	JIS-SUS304 (0.3)	18	No
6	$Mg_{96}Zn_2Y_2$	$50 \times 50 \times 5$	Al (0.2)	JIS-SUS304 (0.2)	23	Yes
7	$Mg_{96}Zn_2Y_2$	$95 \times 50 \times 5$	Al (0.2)	JIS-SUS304 (0.2)	23	Yes

Table 2. Chemical composition of Mg alloy (wt.%).

Material	Mg	Zn	Y	Al
AZ31	bal.	0.7–0.13	-	2.5–3.5
$Mg_{96}Zn_2Y_2$	bal.	4.9	6.36	0.25

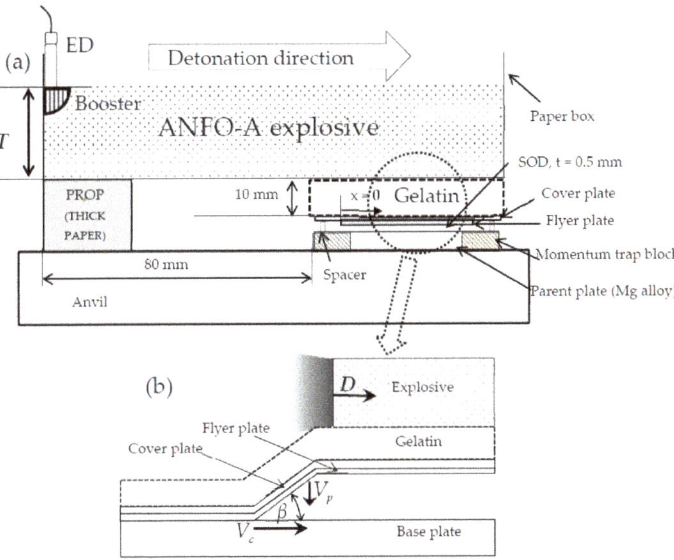

Figure 1. Schematic of (**a**) the explosive welding of Al/Mg alloy plates and (**b**) the process of explosive welding.

3. Results and Discussion

3.1. Microstructure of the Weld Interface

The recovered Al/Mg$_{96}$Zn$_2$Y$_2$ composite plate for experiment No. 6 (top view) is shown in Figure 2. We note that the Al thin plate is successfully welded onto Mg alloy by EXW through gelatin, and the upper surface appears fairly uniform and "smooth" after the EXW process. The resulting clear surface of the recovered sample has already been reported for EXW using underwater shockwaves, and a similar positive effect with the use of gelatin is confirmed by our experiments.

Figure 2. Welded Al/Mg$_{96}$Zn$_2$Y$_2$ plate (No. 6) sample.

From Table 1, we note that two experiments (No. 4 and No. 5) were unsuccessful because of the lack of energy for welding. The welding was successful in all other cases. The welding of Mg$_{96}$Zn$_2$Y$_2$ is more difficult than that of AZ31 because of the lower strength of the former; the relevant details are discussed later in the context of the welding window.

As shown in Figure 2, a portion near the end of the Mg plate appears to have undergone shear fracture. Even with the use of momentum trap blocks, it is difficult to eliminate such shear cracks in the case of Mg$_{96}$Zn$_2$Y$_2$ alloy, which is not ductile [33]. No such shear cracks are observed in the case of AZ31.

Figure 3 shows the microstructure of the Al/Mg alloy interface obtained with an optical microscope. The bonded interface exhibits a typical wavy structure, which is considered as evidence for successful welding based on the mechanism of explosive welding. Further, the wavelength and amplitude increase with increasing explosive thickness (Table 3). These results are similar to those obtained with regular EXW in air [34]. In addition, the wavelength and amplitude of the "waves" of explosively-welded Al/Mg$_{96}$Zn$_2$Y$_2$ and Al/AZ31 obtained under the same explosive thickness (No. 2 and No. 6) are nearly identical (Table 3). Upon comparison with a previous result obtained for regular EXW in air by Ghaderi et al. [34] for Al/AZ31, we note that the wavelength of the wavy structure in our case is greater than 1 mm.

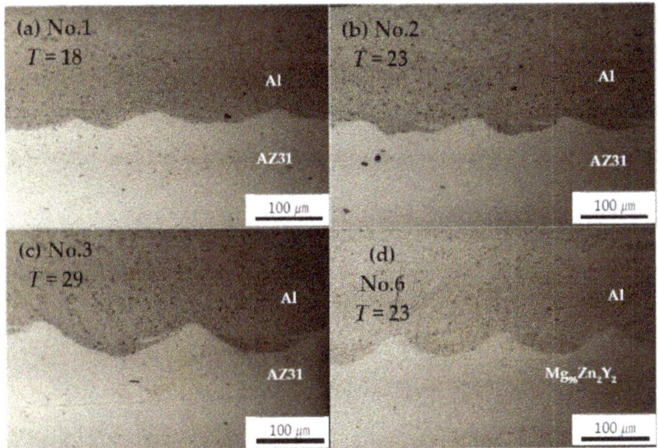

Figure 3. Optical micrograph of the welded interface for Al/Mg alloys at $x = 25$ mm: (**a**) No. 1, (**b**) No. 2, (**c**) No. 3, and (**d**) No. 6.

Table 3. Effect of explosive thickness on the formation of the wavy interface.

No.	Thickness of Explosive (*T*)	Combination	Wavelength (Average)	Amplitude (Average)
No. 1	$T = 18$ mm	Al/AZ31	102 µm	20 µm
No. 2	$T = 23$ mm	Al/AZ31	167 µm	27 µm
No. 3	$T = 29$ mm	Al/AZ31	216 µm	51 µm
No. 6	$T = 23$ mm	Al/Mg$_{96}$Zn$_2$Y$_2$	161 µm	35 µm

The use of Al thin plates and the gelatin medium in the present study causes a decrease in the energy dissipated upon collision. Therefore, the waves formed are smaller than those formed during regular EXW.

Additionally, it has been reported that an IL, which is associated with vortices, is formed during the EXW of Al/AZ31 in air and underwater [34,35]. During EXW in air, an IL is formed in a vortex confined locally at the crest of a wave, and during underwater EXW, a continuous IL tends to form along the interface. Intermetallic compounds confined in inconspicuous vortices do not affect bond strength [34,36]. In our study, IL formation was not observed clearly in any of the trials.

3.2. Quality of Bonding Across Interface

Figures 4 and 5 show the results of microfocus X-ray diffraction analyses of Al/AZ31 and Al/Mg$_{96}$Zn$_2$Y$_2$, respectively. The XRD measurements were performed using a collimator with a diameter of 50 µm. The figures show the XRD measurements (a) on the Al side 50 µm away from the interface, (b) at the center of the bonding interface, and (c) on the Mg-alloy side. As can be inferred from the figures, our attempts to measure the diffraction at the interface, including the interface region, yielded only the peak value of the metal considered for the measurement, and no peaks corresponding to the formation of intermetallic compounds were observed. This result suggests that no (or little) intermetallic compound was formed at the interface.

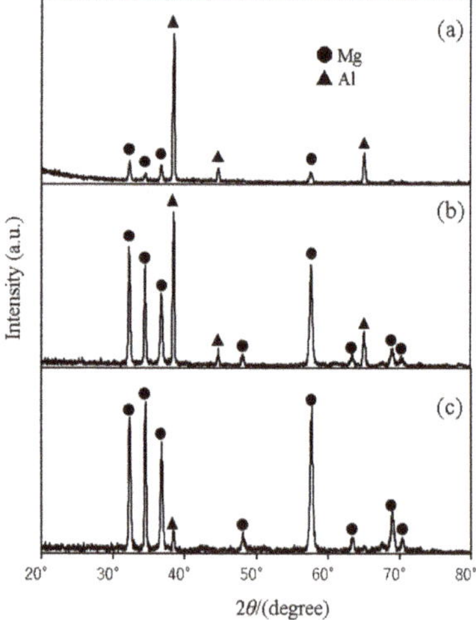

Figure 4. Al/AZ31 (No. 2) characterized by microfocus XRD analysis: (**a**) Al side, (**b**) interface, and (**c**) AZ31 side.

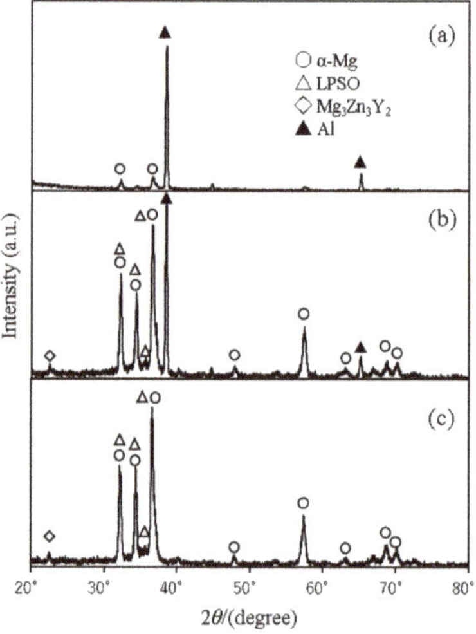

Figure 5. Al/Mg$_{96}$Zn$_2$Y$_2$ (No. 6) characterized by microfocus XRD analysis: (**a**) Al side, (**b**) interface, and (**c**) Mg$_{96}$Zn$_2$Y$_2$ side.

The weld interface was also characterized via SEM and EPMA. The SEM images and elemental mapping of the Al/Mg alloys are shown in Figure 6 (Al/AZ31) and Figure 7 (Al/Mg$_{96}$Zn$_2$Y$_2$). The SEM images illustrate that the composite plates show no significant diffusion or IL formation. The mapping images confirm a small melting zone, but this melting zone does not affect the bonding strength because the melting occurs only at the wave peak [34,36]. In addition, there exists a ripple-like zone close to the top of the "waves" (indicated by arrows), with a height of 10 μm or less, that appears deformed and stretched at the Mg alloy region. In a previous study [34] on the regular EXW of Al/AZ31, the presence of Al$_2$Mg was detected in a ripple of size 200–300 μm via microfocus XRD analysis.

In this regard, Higashi et al. [15] reported the formation of a "band region," which is considered to arise from recrystallization, as evidence of a temperature increase; however, our results did not yield such fine grains. This indicates that the processing time was fairly short, which corroborates with the lack of evidence of heating at the bonding interface.

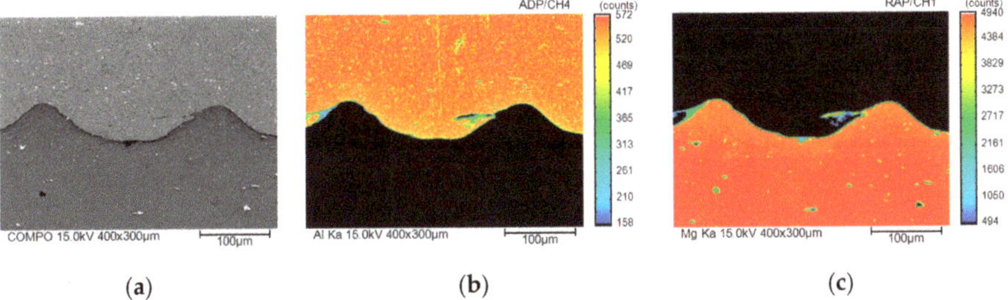

Figure 6. Scanning electron microscope (SEM) image of the welded interface and elemental mapping of Al/AZ31 (No. 2) across the interface, which was acquired using electron probe microanalysis (EPMA): (**a**) SEM image, (**b**) Al mapping, and (**c**) Mg mapping.

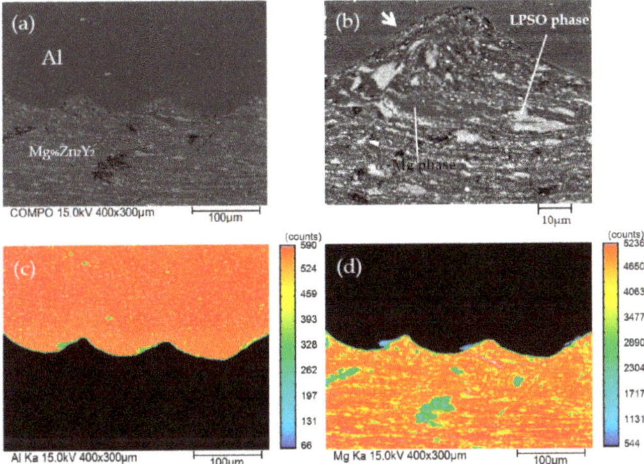

Figure 7. Scanning electron microscope (SEM) image of the welded interface and elemental mapping of Al/$Mg_{96}Zn_2Y_2$ (No. 6) across the interface, which was acquired using electron probe microanalysis (EPMA): (**a**) SEM image, (**b**) magnified view of the SEM image, (**c**) Al mapping, and (**d**) Mg mapping.

We further measured the Vickers hardness of certain samples (Figure 8). Figure 8a,b shows the measurements under a load of 10 gf (98 mN), whereas Figure 8c shows measurements around the consolidated melt observed by mapping analysis under a load of 5 gf (49 mN). Figure 9 shows the hardness profile across the bonding interface. The Vickers hardness close to the bonded interface exhibits high values due to work hardening, and a gradual decrease in the hardness is observed with increasing distance from the interface; however, the hardness is still greater than the Vickers hardness of the as-received metals. In this regard, for the regular EXW of Al/AZ31, Ghaderi et al. [34] reported a high hardness ranging from 170 HV to 280 HV due to the existence of a large IL. In addition, similar hardness values were obtained at the Al and Mg alloy areas.

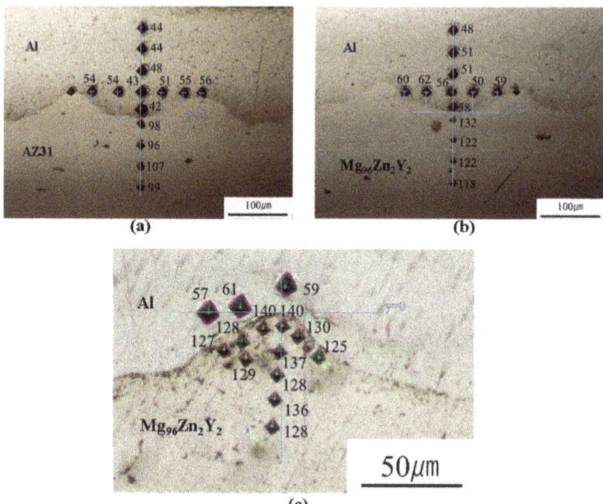

Figure 8. Hardness distribution across the welded Al/Mg alloy interface: (**a**) Al/AZ31, (**b**) Al/Mg$_{96}$Zn$_2$Y$_2$, and (**c**) Al/Mg$_{96}$Zn$_2$Y$_2$ interface around the consolidated melt.

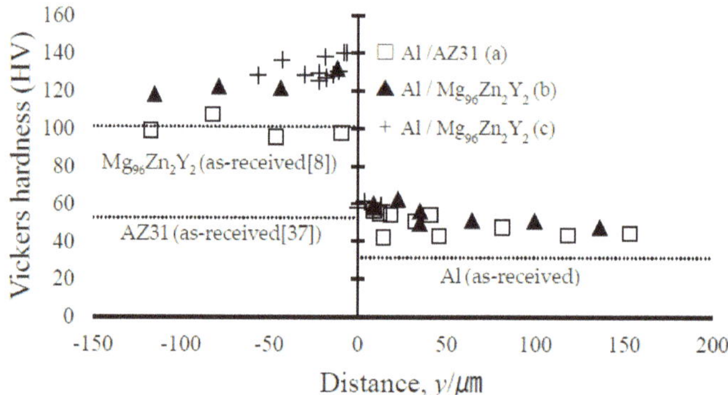

Figure 9. Microhardness distribution near the solidified melt area. The as-received hardness of Mg alloy is referenced from existing literature [8,37], and the as-received hardness value of Al is the average measured value.

Next, to prepare a large test piece, we performed another EXW using gelatin to obtain Al/Mg$_{96}$Zn$_2$Y$_2$ (No. 7). The recovered sample had a crack-free length of 80 mm (Figure 10). As in the case of experiment No. 6, the sheared crack at the end side could not be eliminated. The thickness of the bonded Al was only 0.2 mm, and it was difficult to measure the bonding strength directly. Because there is no general testing method for measuring the quality of welding for such composites, we performed a three-point bending test to qualitatively determine the bonding strength, which is identical to the method employed by Habib et al. [19]. Figure 11 shows a schematic of the method and the results of the three-point bending test for the Al/Mg$_{96}$Zn$_2$Y$_2$ composite. We note that after the bending test, the composite is fractured into two pieces without showing ductile behavior; however, there is no evidence of separation at the interface (as can be inferred from Figure 11b), which implies that the bonding strength is suitably high for practical applications.

Figure 10. Al/Mg$_{96}$Zn$_2$Y$_2$ composite long plate (No. 7).

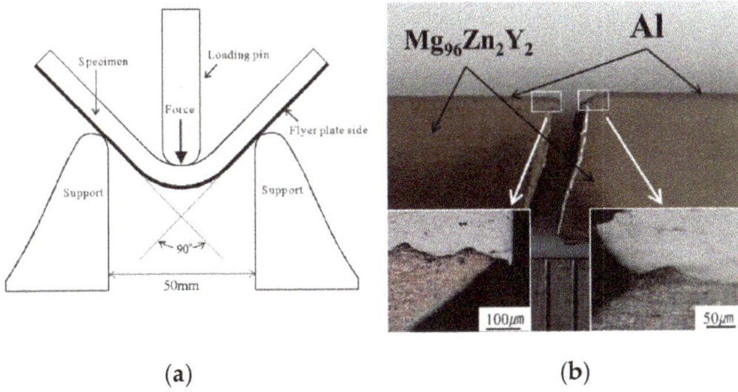

(**a**)   (**b**)

Figure 11. (**a**) Schematic of the three-point bending test and (**b**) photograph of Al/Mg$_{96}$Zn$_2$Y$_2$ composite long plate after the bending test.

3.3. Welding Conditions

The experimental parameters of interest for EXW are the horizontal collision point velocity, Vc, dynamic bending angle at collision, β, and flyer plate velocity, Vp [38,39]. Here, we discuss the welding conditions based on these parameters [17,32]. For the parallel assembly shown in Figure 1a, Vc is equal to the detonation velocity [39], and the following relationship holds true [38,39]:

$$Vp = 2Vc \times \sin(\beta/2) \tag{1}$$

Normally, the welding window is plotted based on the relationship between Vc and β. In addition, Vp is an important parameter that is directly related to the kinetic energy due to collision. We first determine the flyer plate velocity Vp For regular EXW, the equations reported by Gurney [40] and Chadwick et al. [41] as well as other equations are utilized to estimate the flyer plate velocity Vp; however, in our case, we cannot assume that the flyer mass is equal to the mass of the gelatin + cover and flyer plate, and it is more appropriate to consider the gelatin as the pressure-transmitting medium. In such a case, it is necessary to use numerical simulations, similar to case with the use of water, as reported previously [20,22,23]. Therefore, we calculated the flyer plate velocity with gelatin as the medium by using the AUTODYN-2D code.

The analytical model was developed with the shell solver for the cover and flyer plates, and the Euler solver was applied for the ANFO-A explosive and 20 wt.% gelatin. The Euler-Lagrange interaction was applied to the boundary between gelatin and the cover plate. The upper side of the interaction line interacts with the Euler boundary, whereas the lower side is calculated only for Lagrangian conditions. In other words, the shock waves propagated in gelatin act only on the surface of the cover plate. Points for the measurement of values such as velocity were set at the center position of the flyer plate. Table 4 lists the parameters [26] applied to the Mie-Gruneisen-type shock Hugoniot equation of state (EOS) for gelatin.

Table 4. Parameters of the Mie–Gruneisen form of the shock Hugoniot equation of state [26].

Pressure Medium	Reference Density $(\rho_0/\text{kg·m}^{-3})$	Gruneisen Coefficient (Γ)	Sound Velocity $(c_0/\text{m·s}^{-1})$	Material Constant (s)
20 wt.% gelatin	1060	0.00	1570	1.77

Here, we note that the ANFO-A explosive can be treated as a high-pressure gas (ideal gas), on AUTODYN-2D code, and consequently, the EOS can be expressed as follows [42]:

$$P_0 = (\gamma - 1) \times \rho_0 \times e_0 \tag{2}$$

Here, P represents the pressure, γ the specific heat ratio, e the internal energy, and ρ the density. The subscript 0 refers to the initial condition. The initial conditions P_0, ρ_0, and γ are calculated using Hino's Equation (3) [43] and express as follows: The detonation velocity V_d of the explosive was calculated using an approximation of the experimental value given by the data reported by Hokamoto et al. [30], and the velocity change with the thickness of the explosive was considered.

$$P_0 = 416 \times U^2 \times \frac{\rho_f}{10^3}\left(1 - \frac{0.543 \times \rho_f}{10^3} + \frac{0.193 \times \rho_f^2}{10^6}\right) \tag{3}$$

$$P_0 = \rho_f \times U \times u \tag{4}$$

$$\rho_f U = \rho_0 \times (U - u) \tag{5}$$

$$U = c + u \tag{6}$$

$$n = \frac{c^2}{\left(\frac{P_0}{\rho_0}\right)} \tag{7}$$

Here, ρ_f represents the packing density of the explosive (kg/m^3), U the propagation velocity of the shock wave (equal to detonation velocity, m/s), u the particle velocity (m/s), c the sound velocity (m/s) of the detonation gas, and n represents the polytropic index (equal to γ). Further, the packing density is $\rho_f = 530$ kg/m^3.

Figure 12 shows the results of our numerical analysis of the velocity change in the flyer plate Vp as a function of vertical distance y at the center position of the sample along the welding direction ($x = 25$ mm). The figure indicates that the flyer plate is immediately accelerated within a very small SOD, and the velocity is > 350 m/s at $y = 0.5$ mm. This velocity Vp is sufficient to achieve EXW [29]. We next discuss the relevant conditions based on the welding window [17,32].

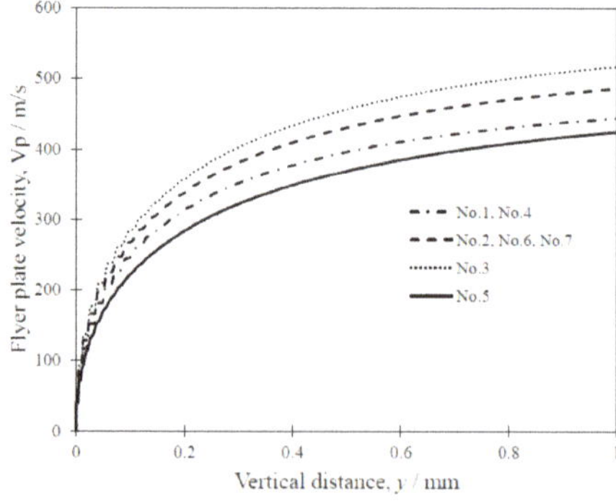

Figure 12. Flyer plate velocity as a function of vertical distance.

Figure 13 shows the welding window for Al/Mg alloys, including the plots for the present experiments. In the present research, we do not focus on the upper limit, because the welding conditions lie closer to the lower limit, as can be observed from the figure. The lower limit can be expressed by the following formula [17,32]:

$$\sin\left(\frac{\beta}{2}\right) = k_1 \sqrt{\frac{H_v}{\rho V_c^2}}$$

(8)

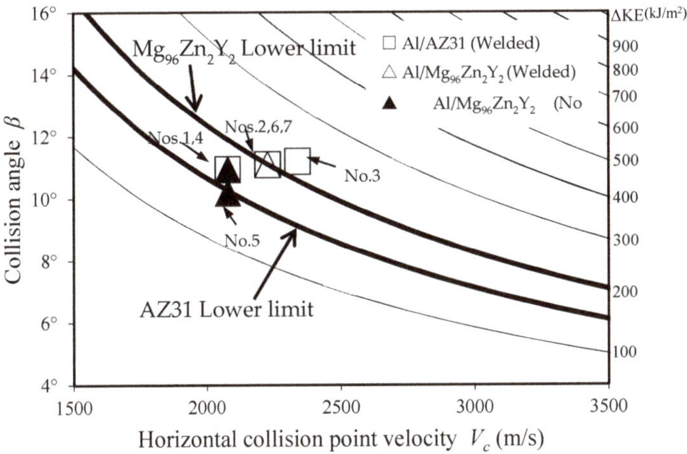

Figure 13. Welding window for Al/Mg alloys calculated using the AUTODYN-2D code.

Here, H_v (MPa) denotes the Vickers hardness of the hard component and k_1 a constant. The hardness values of each parent plate were obtained from the literature (AZ31: $H_v = 70$, $\rho = 1780$ kg/m³ [37]) and ($Mg_{96}Zn_2Y_2$: $H_v = 101$, $\rho = 1890$ kg/m³ [8]), and k_1 was set to 0.6 [19]. The contour lines corresponding to the kinetic energy lost due to collision are plotted in Figure 13, and we note that the lines are parallel to the lower-limit lines indicated as thick lines for welding. The kinetic energy lost due to collision, ΔKE, can be expressed by the following equation:

$$\Delta KE = \frac{(M_c + M_f)V_p^2}{2}$$

(9)

Here, ΔKE denotes the kinetic energy lost in the collision (kJ/m²), and M_c and M_f denote the masses of the cover and flyer plates per unit area (kg/m²), respectively. The calculated values of the kinetic energy loss (ΔKE) for $T = 18$, 23, and 29 mm are summarized in Table 5.

Table 5. Effect of explosive thickness on kinetic energy loss.

Thickness of Explosive (*T*)	Kinetic Energy Lost (ΔKE) kJ/m²
T = 18 mm (Cover plate thickness 0.2 mm)	168
T = 18 mm (Cover plate thickness 0.3 mm)	200
T = 23 mm	199
T = 29 mm	224

Based on the welding window shown in Figure 13, we observe that our experimental results correspond to the window. For example, the conditions for trials No. 4 and No. 5 (Al/$Mg_{96}Zn_2Y_2$) lie below the lower limit, suggesting that no welding occurred. In this case, $Mg_{96}Zn_2Y_2$ alloy is not sufficiently deformed to induce fluidization, which may cause jetting and the formation of the wavy interface. The welding of Al/AZ31 is considerably easier owing to the lower hardness and strength of the Mg alloy. Furthermore, the results in this study corresponding to the weldability window and the observed interface morphology are similar to those reported by Hoseini [36] et al., which were also calculated using the Wittman and Deribas [17,32] lower-limit equations, confirming the validity of the present calculations.

4. Conclusions

In this study, we proposed and demonstrated a modified method of explosive welding using a gelatin layer as the pressure-transmitting medium to weld a thin Al plate onto AZ31 and $Mg_{96}Zn_2Y_2$ Mg alloys. The resulting microstructure was characterized via OM, SEM, XRD, and EPMA analysis, and we confirmed that there was no significant reaction (formation of an intermediate layer) at the interface. The bonding strength at the interface was also found to be satisfactory as per the results of three-point bending tests. The experimental results were investigated based on the welding window obtained from numerical simulations, and the conditions corresponding to our experiments were found to lie close to the lower limit of welding; further, the welding conditions suitably corresponded with the experimental results. In summary, we believe that our findings can significantly contribute to advancements in explosive welding based on the use of gelatin as the pressure-transmitting medium.

Author Contributions: Conceptualization, data curation, investigation, writing—original draft: D.I.; methodology: A.M.; performed the experiments: S.T.; funding acquisition, supervision: K.H. All authors have read and agreed to the published version of the manuscript.

Funding: This research was supported by an MRC research grant program from Magnesium Research Center, Kumamoto University.

References

1. Mordike, B.L.; Ebert, T. Magnesium: Properties—Applications—Potential. *Mater. Sci. Eng. A* **2001**, *302*, 37–45. [CrossRef]
2. Yamasaki, M.; Hashimoto, K.; Hagihara, K.; Kawamura, Y. Effect of multimodal microstructure evolution on mechanical properties of Mg–Zn–Y extruded alloy. *Acta Mater.* **2011**, *59*, 3646–3658. [CrossRef]
3. Noda, M.; Mayama, T.; Kawamura, Y. Evolution of mechanical properties and microstructure in extruded Mg 96Zn2Y2 alloys by annealing. *Mater. Trans.* **2009**, *50*, 2526–2531. [CrossRef]
4. Hagihara, K.; Kinoshita, A.; Sugino, Y.; Yamasaki, M.; Kawamura, Y.; Yasuda, H.Y.; Umakoshi, Y. Effect of long-period stacking ordered phase on mechanical properties of Mg97Zn1Y2 extruded alloy. *Acta Mater.* **2010**, *58*, 6282–6293. [CrossRef]
5. Abe, E.; Ono, A.; Itoi, T.; Yamasaki, M.; Kawamura, Y. Polytypes of long-period stacking structures synchronized with chemical order in a dilute Mg–Zn–Y alloy. *Philos. Mag. Lett.* **2011**, *91*, 690–696. [CrossRef]
6. Hagihara, K.; Yokotani, N.; Umakoshi, Y. Plastic deformation behavior of Mg12YZn with 18R long-period stacking ordered structure. *Intermetallics* **2010**, *18*, 267–276. [CrossRef]
7. Kawamura, Y.; Yamasaki, M. Formation and Mechanical Properties of Mg97Zn1RE2 Alloys with Long-Period Stacking Ordered Structure. *Mater. Trans.* **2007**, *48*, 2986–2992. [CrossRef]
8. Noda, M.; Kawamura, Y.; Mayama, T.; Funami, K. Thermal Stability and Mechanical Properties of Extruded Mg-Zn-Y Alloys with a Long-Period Stacking Order Phase and Plastic Deformation. In *New Features on Magnesium Alloys*; InTechOpen: London, UK, 2012.
9. Liu, F.; Liang, W.; Li, X.; Zhao, X.; Zhang, Y.; Wang, H. Improvement of corrosion resistance of pure magnesium via vacuum pack treatment. *J. Alloys Compd.* **2008**, *461*, 399–403. [CrossRef]
10. Borrisutthekul, R.; Miyashita, Y.; Mutoh, Y. Dissimilar material laser welding between magnesium alloy AZ31B and aluminum alloy A5052-O. *Sci. Technol. Adv. Mater.* **2005**, *6*, 199–204. [CrossRef]
11. Mahendran, G.; Balasubramanian, V.; Senthilvelan, T. Developing diffusion bonding windows for joining AZ31B magnesium–AA2024 aluminium alloys. *Mater. Des.* **2009**, *30*, 1240–1244. [CrossRef]

12. Çam, G.; Mistikoglu, S. Recent Developments in Friction Stir Welding of Al-alloys. *J. Mater. Eng. Perform.* **2014**, *23*, 1936–1953. [CrossRef]

13. Kong, C.Y.; Soar, R.C.; Dickens, P.M. Optimum process parameters for ultrasonic consolidation of 3003 aluminium. *J. Mater. Process. Technol.* **2004**, *146*, 181–187. [CrossRef]

14. Mariani, E.; Ghassemieh, E. Microstructure evolution of 6061 O Al alloy during ultrasonic consolidation: An insight from electron backscatter diffraction. *Acta Mater.* **2010**, *58*, 2492–2503. [CrossRef]

15. Higashi, Y.; Iwamoto, C.; Kawamura, Y. Microstructure evolution and mechanical properties of extruded Mg96Zn2Y2 alloy joints with ultrasonic spot welding. *Mater. Sci. Eng. A* **2016**, *651*, 925–934. [CrossRef]

16. Luo, C.; Liang, W.; Chen, Z.; Zhang, J.; Chi, C.; Yang, F. Effect of high temperature annealing and subsequent hot rolling on microstructural evolution at the bond-interface of Al/Mg/Al alloy laminated composites. *Mater. Charact.* **2013**, *84*, 34–40. [CrossRef]

17. Carpenter, S.H.; Wittman, R.H. Explosion Welding. *Annu. Rev. Mater. Sci.* **1975**, *5*, 177–199. [CrossRef]

18. Blazynski, T.Z. *Explosive Welding, Forming, and Compaction*; Applied Science: London, UK, 1983; ISBN 9401197512.

19. Habib, M.A.; Keno, H.; Uchida, R.; Mori, A.; Hokamoto, K. Cladding of titanium and magnesium alloy plates using energy-controlled underwater three layer explosive welding. *J. Mater. Process. Tech.* **2015**, *217*, 310–316. [CrossRef]

20. Mori, A.; Tamaru, K.; Hokamoto, K.; Fujita, M. Underwater Explosive Welding, Discussion Based on Weldable Window. *AIP Conf. Proc.* **2006**, *845*, 1543–1546.

21. Mori, A.; Hokamoto, K.; Fujita, M. Characteristics of the New Explosive Welding Technique Using Underwater Shock Wave-Based on Numerical Analysis. *Mater. Sci. Forum* **2004**, *465*, 307–312. [CrossRef]

22. Hokamoto, K.; Fujita, M.; Shimokawa, H.; Okugawa, H. A new method for explosive welding of Al/ZrO$_2$ joint using regulated underwater shock wave. *J. Mater. Process. Technol.* **1999**, *85*, 175–179. [CrossRef]

23. Hokamoto, K.; Nakata, K.; Mori, A.; Tsuda, S.; Tsumura, T.; Inoue, A. Dissimilar material welding of rapidly solidified foil and stainless steel plate using underwater explosive welding technique. *J. Alloys Compd.* **2009**, *472*, 507–511. [CrossRef]

24. Mori, A.; Nishi, M.; Hokamoto, K. Underwater shock wave weldability window for Sn-Cu plates. *J. Mater. Process. Technol.* **2019**, *267*, 152–158.

25. Nagayama, K.; Mori, Y.; Motegi, Y.; Nakahara, M. Shock hugoniot for biological materials. *Shock Waves* **2006**, *15*, 267–275. [CrossRef]

26. Shepherd, C.J.; Appleby-Thomas, G.J.; Hazell, P.J.; Allsop, D.F. The dynamic behaviour of ballistic gelatin. In Proceedings of the AIP Conference Proceedings, Nashville, TN, USA, 28 June–3 July 2009; Volume 1195, pp. 1399–1402.

27. Inao, D.; Tanaka, S.; Yamashita, T.; Hokamoto, K. Visualization of shockwave behavior in water and gelatin. *Measurement* **2019**, *148*, 106929. [CrossRef]

28. Parchuri, P.; Kotegawa, S.; Yamamoto, H.; Ito, K.; Mori, A.; Hokamoto, K. Benefits of intermediate-layer formation at the interface of Nb/Cu and Ta/Cu explosive clads. *Mater. Des.* **2019**, *166*, 107610. [CrossRef]

29. Crossland, B. *Explosive Welding of Metals and Its Application*; Clarendon Press: Oxford, UK; New York, NY, USA, 1982; ISBN 9780198591191.

30. Hokamoto, K.; Izuma, T.; Fujita, M. New explosive welding technique to weld aluminum alloy and stainless steel plates using a stainless steel intermediate plate. *Metall. Mater. Trans. A* **1993**, *24*, 2289–2297. [CrossRef]

31. Akbari Mousavi, A.A.; Al-Hassani, S.T.S. Numerical and experimental studies of the mechanism of the wavy interface formations in explosive/impact welding. *J. Mech. Phys. Solids* **2005**, *53*, 2501–2528. [CrossRef]

32. Deribas, A. Physics of explosive hardening and welding. *Nauk. Novosib.* **1980**, *220*.

33. Yoshimoto, S.; Yamasaki, M.; Kawamura, Y. Microstructure and Mechanical Properties of Extruded Mg-Zn-Y Alloys with 14H Long Period Ordered Structure. *Mater. Trans.* **2006**, *47*, 959–965. [CrossRef]

34. Ghaderi, S.H.; Mori, A.; Hokamoto, K. Analysis of Explosively Welded Aluminum-AZ31 Magnesium Alloy Joints. *Mater. Trans.* **2008**, *49*, 1142–1147. [CrossRef]

35. Bataev, I.A.; Lazurenko, D.V.; Tanaka, S.; Hokamoto, K.; Bataev, A.A.; Guo, Y.; Jorge, A.M. High cooling rates and metastable phases at the interfaces of explosively welded materials. *Acta Mater.* **2017**, *135*, 277–289. [CrossRef]

36. Hoseini Athar, M.M.; Tolaminejad, B. Weldability window and the effect of interface morphology on the properties of Al/Cu/Al laminated composites fabricated by explosive welding. *Mater. Des.* **2015**, *86*, 516–525. [CrossRef]

37. Ghaderi, S.H.; Mori, A.; Hokamoto, K. Explosion Joining of Magnesium Alloy AZ31 and Aluminum. *Mater. Sci. Forum* **2008**, *566*, 291–296. [CrossRef]

38. Kennedy, J. Explosive Output for Driving Metal. *Behav. Util. Explos. Eng. Des.* **1972**, 109–124.

39. Cowan, G.R.; Bergmann, O.R.; Holtzman, A.H. Mechanism of Bond Zone Wave Formation in Explosion-Clad Metals. *Metall. Mater. Trans. B* **1971**, *2*, 3145–3155. [CrossRef]

40. Gurney, R.W. The Initial Velocities of Fragments from Bombs, Shell, and Grenades. *Army Ballist. Res. Lab Aberd. Proving Ground Md.* **1943**, 450.

41. Chadwick, M.D.; Howd, D.; Wildsmith, G.; Cairns, J.H. Explosive welding of tubes and tube plates. *Br. Weld. J.* **1968**, *15*, 480–492.

42. Liu, R.Q.; Nie, J.X.; Jiao, Q.J. Study on Lee-Tarver Model Parameters of CL-20 Explosive Ink. In Proceedings of the Smart Innovation, Systems and Technologies, Jilin, China, 18–20 July 2019; Springer: Singapore, 2019; Volume 157, pp. 205–215.

43. Hino, K. Fragmentation of rock through blasting. *J. Ind. Explos. Soc. Jpn.* **1956**, *17*, 2–11.

Article

Fabrication of Composite Unidirectional Cellular Metals by Using Explosive Compaction

Masatoshi Nishi [1], Shigeru Tanaka [2], Matej Vesenjak [3], Zoran Ren [3,4] and Kazuyuki Hokamoto [2,*

[1]   National Institute of Technology, Kumamoto College, Kumamoto 866-8501, Japan;
      nishima@kumamoto-nct.ac.jp
[2]   Institute of Pulsed Power Science, Kumamoto University, Kumamoto 860-8555, Japan;
      tanaka@mech.kumamoto-u.ac.jp
[3]   Faculty of Mechanical Engineering, University of Maribor, 2000, Maribor, Slovenia;
      matej.vesenjak@um.si (M.V.); zoran.ren@um.si (Z.R.)
[4]   International Research Organization for Advanced Science and Technology, Kumamoto University,
      860-8555 Kumamoto, Slovenia
*   Correspondence: hokamoto@mech.kumamoto-u.ac.jp; Tel.: +81-96-342-3740

Received: 27 December 2019; Accepted: 23 January 2020; Published: 29 January 2020

Abstract: Development of a small and highly efficient heat exchanger is an important issue for energy saving. In this study, the fabrication method of unidirectional (UniPore) composite cellular structure with long and uniform unidirectional cells was investigated to be applied as a heat exchanger. The composite UniPore structure was achieved by the unique fabrication method based on the explosive compaction of a particular arrangement of thin copper and stainless steel pipes. Slightly smaller thin stainless steel pipes filled with paraffin are inserted into small thin copper pipes, which are then arranged inside bigger and thicker outer copper pipes. Such an arrangement of pipes is placed centrally into a cylindrical explosion container and surrounded with explosive. Upon explosive detonation, the pipes are compacted and welded together, which results in a UniPore structure with a stainless steel covered inner surface of unidirectional pores to improve the corrosion resistance and high temperature resistance performance. Two different composite UniPore structures arrangements were studied. The microstructure of the new composite UniPore structure was investigated to confirm good bonding between the components (pipes).

Keywords: cellular metal; composite structure; unidirectional cellular metal; explosive welding; explosive compaction; high-velocity impact welding; high-energy-rate forming

---

1. Introduction

Cellular metals with countless small pores have various applicable characteristics such as low density, efficient damping, high grade of deformation, high-energy absorption capability, durability in dynamic loadings, and high thermal and acoustic isolation [1,2]. They can be used in a wide range of applications, since various (multi) functions can be obtained by a proper combination of the pore shape/size/distribution and base metal. Recently, cellular metals have been quite successfully applied as small and efficient heat exchangers [3–6] to improve energy saving from the viewpoint of environmental challenges and requirements.

Sato Y. et al. demonstrated that a heat exchanger could be downsized to one-tenth of its usual size by applying the unidirectional (UniPore) copper structure as the inner pipe of a double-pipe heat exchanger [7]. Hokamoto et al. proposed the fabrication method of the UniPore copper structure with an outer copper pipe completely filled with smaller inner copper pipes [8–11]. This proposed fabrication method is based on the explosive compaction of cylindrical copper pipes assembly [12–16]. It is possible to fabricate specimens with a constant cross-section in the order of several meters by using an explosive compaction technique. There is a lotus-type metal that is known to be similar to the UniPore material and is fabricated by unidirectional solidification in a pressurized gas atmosphere, as described by Nakajima [17,18]. However, the shape of its pores is mostly non-uniform and the length of the final products is expected to be limited [17].

Several studies have been conducted on the fabrication method and mechanical properties of the single metal UniPore structure. For instance, Vesenjak et al. [9] conducted extensive research on the microstructural and mechanical analyses of UniPore copper and confirmed that the fabricated UniPore specimens had good compressive properties with high-energy absorption capability under quasi-static compression. On the other hand, there have been studies regarding the composite cellular structure [19–21], since it is possible to combine the benefits of each material. Sun et al. [19] proposed a metal–foam-composite hybrid tubular sandwich structures, which combine low-cost metallic materials and high-strength composites with low-density cellular materials. In the other case, Gunji Co. Ltd. (Osaka, Japan) [20] fabricated a grooved double-tube heat exchanger with corrosion resistance and high temperature resistance by using stainless steel for one inner pipe. This grooved double tube heat exchanger was fabricated by the drawing process.

In this study, we conducted experiments to fabricate two types of copper & stainless steel composite UniPore structures as heat exchangers to combine the benefit of copper with high thermal conductivity and stainless steel with improved corrosion and high temperature resistance. These structures had a stainless steel cover layer for all of the inner surfaces of the copper pipes. Microstructure analysis of the fabricated samples demonstrated good interface bonding between the pipe walls.

2. Experimental Investigation

2.1. Sample Preparation

Two types of metal composite samples were produced in this study. Sample A, shown in Figure 1a, consists of an arrangement of inner pipes completely filling the inside of the outer pipe while Sample B, shown in Figure 1b, comprises a concentric arrangement of pipes and rods positioned between the outer and central pipes. Figure 1 shows the composite UniPore structures: the inner pipes in Figure 1a consist of a copper pipe and a steel pipe from the outside, and there is a steel inner pipe and copper inner solid bar in Figure 1b.

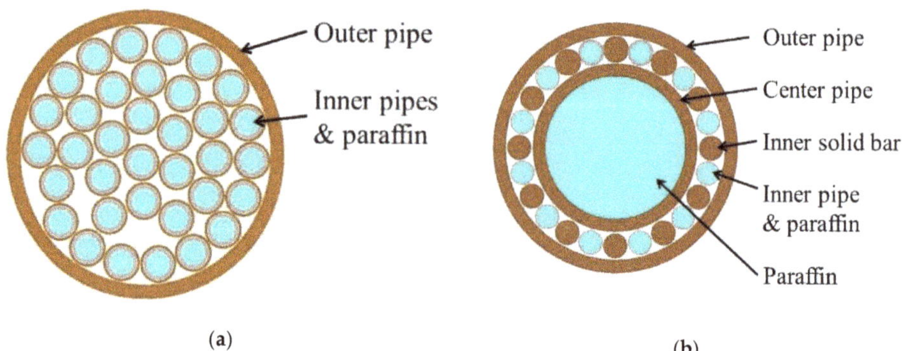

Figure 1. Schematic illustration of samples. (**a**) Sample A; (**b**) Sample B.

The original UniPore copper structures [8] consist of the outer copper pipe being completely filled with a number of inner copper pipes [9–11]. Sample A represents the upgraded UniPore copper structure, where slightly smaller and corrosion resistant stainless steel pipes were inserted into the inner copper pipes. The inner pipes in both cases were filled with paraffin to prevent their collapse during explosive compaction. The dimensions and number of used pipes to fabricate Sample A are given in Table 1, while the arrangement of pipes before the explosive compaction is shown in Figure 2.

Table 1. Dimensions and composition of Sample A.

Component	Material	Outer Diameter (mm)	Inner Diameter (mm)	Number	Length (mm)
Outer pipe	Cu (JIS-C1220)	30	27	1	210
Inner pipe	Cu (JIS-C1220)	4.0	3.4	35	260
Inner pipe	Stainless Steel 304	3.3	2.3	35	260

Figure 2. The pipe assembly of Sample A before explosive compaction (outer copper pipe, inner copper and steel pipes with paraffin).

Sample B consists of a concentric structure of the outer and center copper pipe. Solid copper bars and stainless steel pipes with paraffin were placed in the space between the outer and the central pipe. Table 2 provides the dimension and composition of Sample B, while Figure 3 shows its assembly before explosive compaction.

Table 2. Dimensions and composition of Sample B.

Component	Material	Outer Diameter (mm)	Inner Diameter (mm)	Number	Length (mm)
Outer pipe	Cu (JIS-C1220)	30	27	1	210
Center pipe	Cu (JIS-C1220)	20	17	1	200
Inner solid bar	Cu (JIS-C1220)	3.0	-	12	200
Inner pipe	Stainless Steel 304	3.0	2.4	12	200

For reference, the mechanical characteristics of copper, stainless steel, and paraffin are shown in Table 3.

Table 3. Value of mechanical characteristics.

Material	Density (kg/m^3)	Ultimate Tensile Strength (MPa, Lower Limit) (Source: JIS)	Elongation at Break (Lower Limit) (Source: JIS)
JIS-C1220	8940	315	-
Stainless Steel 304	8000	520	35
Paraffin	918	-	-

Figure 3. The pipe assembly of Sample B before explosive compaction (concentric outer and center copper pipe, inner solid copper bars, and steel pipes with paraffin).

2.2. Fabrication Method

The UniPore structure is produced by the cylindrical explosion welding method. Figure 4 shows the schematic setup of the fabrication method based on the principle of the radial explosion welding method [8]. A vinyl chloride pipe (PVC tube with outer diameter 89 mm, inner diameter 83 mm and length 270 mm, Figure 4) was used as the explosion container. The sample was aligned with the central axis of the PVC tube by acrylic support plates (PMMA). The space between the PVC tube and the UniPore specimen was filled with the primary explosive (ANFO-A: initial density 764 kg/m^3 and initial internal energy 1.254 MJ/kg, ratio of the specific heat 1.98), as shown in Figure 4. The electric detonator was used for the ignition of the high-performance explosive (SEP: density 1310 kg/m^3, Chapman-Jouguet detonation velocity 6.7 km/s, Chapman-Jouguet detonation pressure 15.9 GPa, source [22]), acting as a booster (to prevent any unexploded primary explosive), which was mounted at the center top of the PVC tube. A cylindrical plaster was placed on the top of the sample to absorb and reduce the impact from the top.

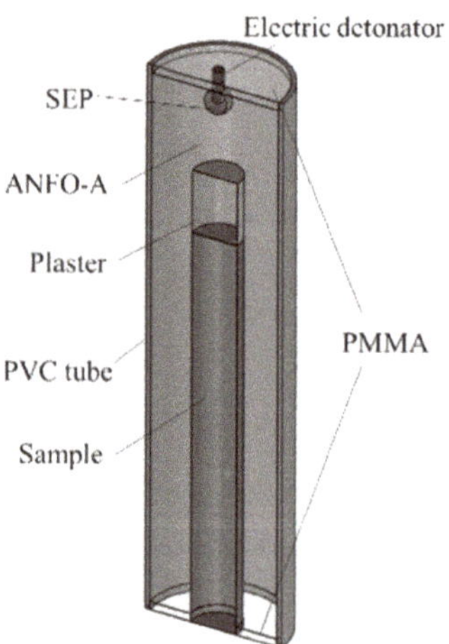

Figure 4. Schematic illustration of the experimental explosive compaction method (cross-section).

The pipes filled with paraffin were welded together by high radial pressure acting toward the central axis of the assembly upon explosive detonation (Figure 5), resulting in the UniPore cellular structure [8]. The length and porosity of the proposed cellular metals can be easily controlled by changing the diameter, thickness and number of the outer and inner pipes.

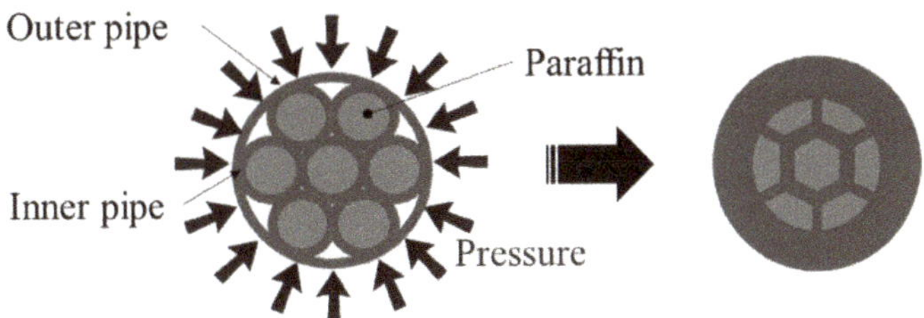

Figure 5. Schematic illustration of the top view of the fabrication process.

3. Results and Discussion—Sample A

The recovered Sample A is shown in Figure 6. No cracks or wrinkles or any other damage were observed after the fabrication process. Figure 7a shows the transversal cross-section of the sample of Sample A after cutting off both ends of the recovered structure, and the outer and inner diameter after forming are shown in Figure 7b. The remained paraffin in the sample was easily removed by melting just by heating the sample up to 373 K. All cells were empty and completely separated between each other, thus a fluid could flow through them without physical interaction. However, a few unevenly shaped cells could be noticed, since some inner pipes in the outer layer deformed more during the fabrication process. This can be attributed to the slightly irregular arrangement of the inner pipes in this region before explosive compaction [7], as can be seen in Figure 2. However, the shape of the cells can be improved with respect to uniformity by setting up a regular arrangement of the inner pipes.

Figure 6. The recovered Sample A (longitudinal direction).

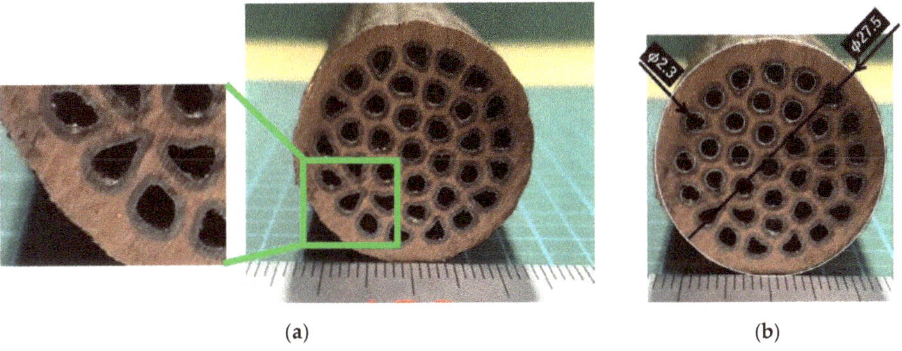

(**a**)   (**b**)

Figure 7. The transversal cross-section of Sample A: (**a**) final state form and enlarged view and (**b**) the approximate diameter of pipes after the experiment.

The magnified transversal cross-section of the sample at the welded pipes junction is shown in Figure 8. Here, the cross-section of the sample was observed by polishing in the etching process with ammonia solution and a small amount of hydrogen peroxide. It was confirmed that the copper pipes were tightly joined together with only a few points, where welding between the copper pipe and stainless-steel pipe was not achieved. This can be attributed to the insufficient velocity to achieve explosive welding since the copper pipe could sustain only a limited acceleration due to insufficient gap between the pipes. Wavy interfaces commonly encountered in impact welding were not observed in Figure 8, and the cause was also related to the low collision velocity between the pipes. Since it is difficult to obtain the collision by experiment, the information during compression such as the collision velocity, the impact pressure, etc. shall be analyzed by computational simulation in future. The Vickers hardness test was used to determine the bonding strength at the pipe junctions, Figure 9. The original hardness of the phosphorous deoxidized copper and stainless steel before explosive compaction was approximately HV 120 and HV 280, respectively. A significant decrease in hardness was observed in the melted area at the triple collision point located at the center of the joint boundary, while the hardness of the pipes generally increased elsewhere due to the work-hardening process. In addition, the interfacial bonding between the pipes was analyzed by scanning electron microscopy (SEM; JEOL Ltd., Tokyo, Japan, JSM-6390LV). From the result of the SEM measurements shown in Figure 10, strong bonding was achieved since the bonding interface could not be observed.

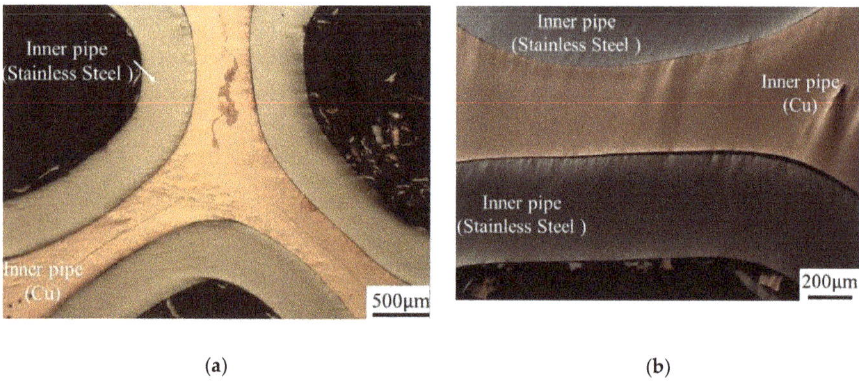

(a) (b)

Figure 8. The magnified transversal cross-section of Sample A: (**a**) the copper pipes were tightly joined together and (**b**) welding between the copper pipe and stainless-steel pipe was not achieved.

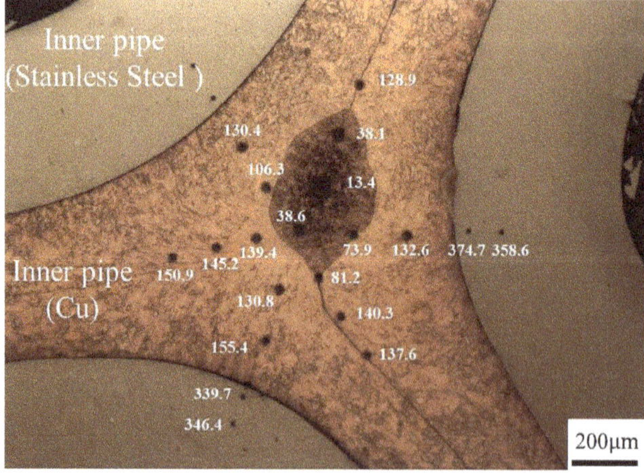

Figure 9. The results of the Vickers hardness test in the vicinity of the triple collision point (transversal cross-section of Sample A).

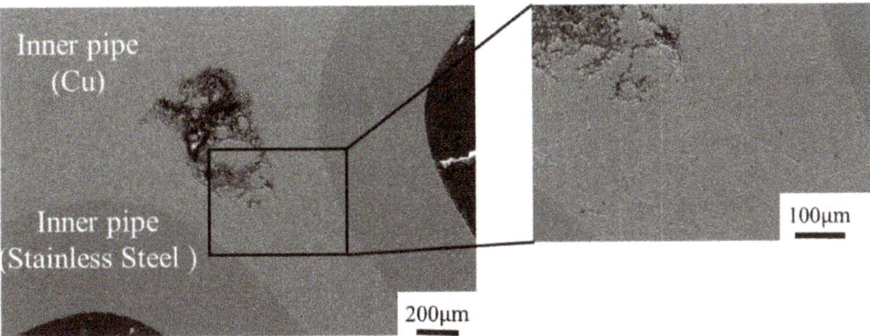

Figure 10. The results of scanning electron microscopy (SEM) in the vicinity of the triple collision point (transversal cross-section of Sample A).

4. Results and Discussion—Sample B

The longitudinal and transversal views of Sample B are shown in Figures 11 and 12, respectively. It can be seen that during the fabrication process, the outer pipe did not crack. All inner pipes with an initially circular cross-section were formed into a rectangular shape by explosive compaction [8]. It was confirmed that the inner pipes were well-bonded and that the transversal cross-section remained generally uniform through the length of the specimen.

Figure 11. The recovered Sample B (longitudinal direction).

Figure 12. The transversal cross-section of Sample B.

Figure 13 shows the transversal cross-section of one cell of Sample B observed with an optical microscope. The inner pipes and inner solid bars were well joined together without any gaps. The connectivity of the interfaces confirmed good bonding conditions during the fabrication process. In Figure 13a, the melted parts generated during the compaction process were observed and confirmed by optical microscopy.

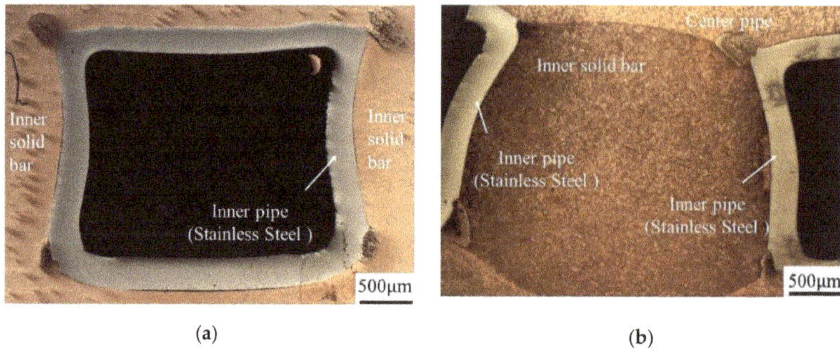

(a) (b)

Figure 13. The transversal cross-section of Sample B. (**a**) Inner pipe and (**b**) inner solid bar.

The Vickers hardness test was performed on the transversal cross-section of Sample B to check and confirm the bonding strength. In Figure 14, the melted area at the triple collision point between the outer pipe, inner pipe, and inner solid bar is shown. The hardness of the stainless steel was remarkably increased due to work hardening by plastic deformation of the materials. However, the hardness of the outer copper pipe and inner copper bar decreased in the melted area. The reason for this is presumed to be the thermal effect during the compaction process. Strong bonding was achieved between the outer copper pipes and copper bar, which can be deduced from the high hardness in the welding area, Figure 15.

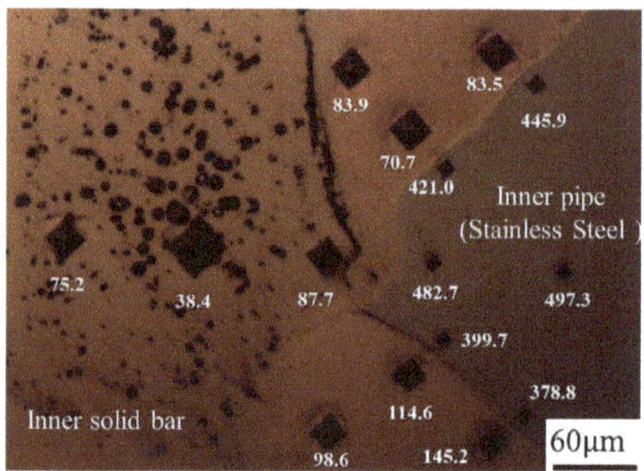

Figure 14. The results of the Vickers hardness test around the melted area in the triple collision point (Enlarged view of Figure 13a).

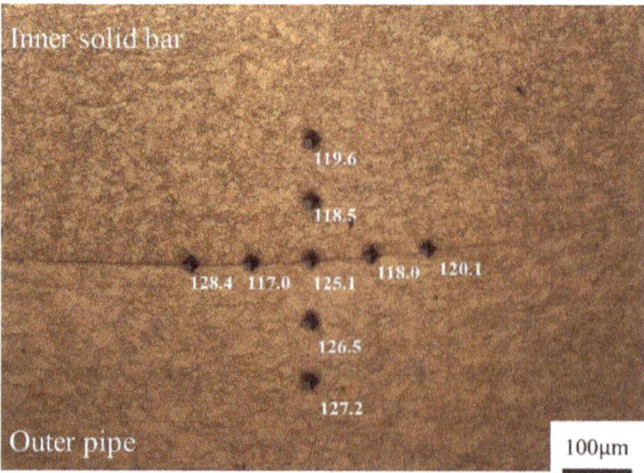

Figure 15. The results of the Vickers hardness test at the welding point between the outer copper pipe and copper bar.

5. Conclusions

The fabrication method of a novel composite metal UniPore structure with longitudinal pores was investigated in this study. For the first time, the surfaces of the longitudinal copper cellular structure were covered by a second metal, a thin layer of corrosion resistant stainless steel. Two assembly arrangements of the copper UniPore structure with stainless steel pipes layered at the inner surface of the inner copper pipes were successfully fabricated. Microstructure analysis of the fabricated samples demonstrated good interface bonding between the component walls as well as providing efficient and mechanical properties in the case of structural applications. The authors believe that the proposed composite UniPore structures have corrosion resistance, high temperature resistance, and high efficiency when used as heat exchangers or heat sinks, which will be proven in further study.

Author Contributions: Conceptualization, K.H.; Methodology, K.H.; Validation, M.N., S.T., and M.V.; Investigation, M.N. and M.V.; Resources, S.T.; Data evaluation, M.N. and S.T.; Writing—original draft preparation, M.N.; Writing—review and editing, M.V., Z.R., and K.H.; Supervision, Z.R. and K.H.; Project administration, Z.R. and K.H. All authors have read and agreed to the published version of the manuscript.

Funding: The authors acknowledge the financial support from the Slovenian Research Agency—ARRS (research core funding No. P2-0063).

Conflicts of Interest: There are no conflicts of interest.

References

1. Banhart, J. Manufacture, characterization and application of cellular metals and metal foams. *Prog. Mater. Sci.* **2001**, *46*, 559–632. [CrossRef]
2. Meyers, M.A.; Lin, A.Y.M.; Seki, Y.; Chen, P.Y.; Kad, B.; Bodde, S. Structural biological composites: An overview. *J. Min. Met. Mat. Soc.* **2006**, *7*, 35–41. [CrossRef]
3. Ashby, M.F.; Evans, A.; Fleck, N.A.; Gibson, L.J.; Hutchinson, J.W.; Wadley, H.N.G.; Gibson, L.J. *Metal Foams: A Design Guide*; Elsevier Science: Amsterdam, Netherlands, 2000.
4. Fiedler, T.; White, N.; Dahari, M.; Hooman, K. On the electrical and thermal contact resistance of metal foam. *Int. J. Heat Mass Transf.* **2014**, *72*, 565–571. [CrossRef]
5. Lefebvre, L.P.; Banhart, J.; Dunand, D.C. Porous metals and metallic foams: Current status and recent developments. *Adv. Eng. Mater.* **2008**, *10*, 775–787. [CrossRef]
6. Fiedler, T.I.; Belova, V.I.; Murch, G.E. On the thermal properties of expanded perlite–Metallic syntactic foam. *Int. J. Heat Mass Transf.* **2015**, *90*, 1009–1014. [CrossRef]
7. Sato, Y.; Yuki, K.; Abe, Y.; Kibushi, R.; Unno, N.; Hokamoto, K.; Tanaka, S.; Tomimura, T. Heat transfer characteristics of a gas flow in uni-directional porous copper pipes. *Int. Heat Trans. Conf.* **2018**, *16*, 8189–8193.
8. Hokamoto, K.; Vesenjak, M.; Ren, Z. Fabrication of cylindrical uni-directional porous metal with explosive compaction. *Mater. Lett.* **2014**, *137*, 323–327. [CrossRef]
9. Vesenjak, M.; Hokamoto, K.; Sakamoto, M.; Nishi, T.; Krstulović-Opara, L.; Ren, Z. Mechanical and microstructural analysis of unidirectional porous (UniPore) copper. *Mater. Des.* **2016**, *90*, 867–880. [CrossRef]
10. Vesenjak, M.; Hokamoto, K.; Matsumoto, S.; Marumo, Y.; Ren, Z. Uni-directional porous metal fabricated by rolling of copper sheet and explosive compaction. *Mater. Lett.* **2016**, *170*, 39–43. [CrossRef]
11. Hokamoto, K.; Shimomiya, K.; Nishi, M.; Krstulovic-Opara, L.; Vesenjak, M.; Ren, Z. Fabrication of unidirectional porous-structured aluminum through explosive compaction using cylindrical geometry. *J. Mater. Process. Tech.* **2018**, *251*, 262–266. [CrossRef]
12. Crossland, B. *Explosive Welding of Metals and Its Application*; Clarendon Press: Oxford, UK, 1982; pp. 84–129.
13. Mori, A.; Nishi, M.; Hokamoto, K. Underwater shock wave weldability window for Sn-Cu plates. *J. Mater. Process. Tech.* **2019**, *267*, 262–266.
14. Parchuri, P.; Kotegawa, S.; Yamamoto, H.; Ito, K.; Mori, A.; Hokamoto, K. Benefits of intermediate-layer formation at the interface of Nb/Cu and Ta/Cu explosive clads. *Mater. Des.* **2019**, *166*, 107610. [CrossRef]
15. Saravanan, S.; Raghukandan, K.; Hokamoto, K. Improved microstructure and mechanical properties of dissimilar explosive cladding by means of interlayer technique. *Arch. Civ. Mech. Eng.* **2016**, *16*, 563–568. [CrossRef]

16. Bataev, I.A.; Hokamoto, K.; Keno, H.; Bataev, A.A.; Balagansky, I.A.; Vinogradov, A.V. Metallic glass formation at the interface of explosively welded Nb and stainless steel. *Met. Mater. Int.* **2015**, *21*, 713–718. [CrossRef]

17. Nakajima, H. Fabrication, properties and application of porous metals with directional pores. *Prog. Mater. Sci.* **2007**, *52*, 1091–1173. [CrossRef]

18. Hyun, S.K.; Ikeda, T.; Nakajima, H. Fabrication of lotus-type porous iron and its mechanical properties. *Sci. Technol. Adv. Mat.* **2004**, *5*, 201–205. [CrossRef]

19. Sun, G.; Wang, Z.; Yu, H.; Gong, Z.; Li, Q. Experimental and numerical investigation into the crashworthiness of metal foam-composite hybrid structures. *Compos. Struct.* **2019**, *209*, 535–547. [CrossRef]

20. Heat exchangers. Available online: http://www.gunji.co.jp/products/heat_exchanger.html (accessed on 18 October 2019). (In Japanese).

21. Jung, A.; Pullen, A.D.; Proud, W.G. Strain-rate effects in Ni/Al composite metal foams from quasi-static to low-velocity impact behavior. *Compos. Part A* **2016**, *85*, 1–11. [CrossRef]

22. Itoh, S.; Kira, A.; Nagano, S.; Fujita, M. Optical study of underwater explosion of high explosive. *Kayaku Gakkaishi (Sci. Tech. Energetic Mater.)* **1995**, *56*, 188–194. (In Japanese)

Article

Joining Aluminium Alloy 5A06 to Stainless Steel 321 by Vaporizing Foil Actuators Welding with an Interlayer

Shan Su [1], Shujun Chen [1,*], Yu Mao [2], Jun Xiao [1], Anupam Vivek [2] and Glenn Daehn [2]

[1] College of Mechanical Engineering and Applied Electronics Technology, Beijing University of Technology, 100 Ping Le Yuan, Beijing 100124, China; sushans@foxmail.com (S.S.); jun.xiao@bjut.edu.cn (J.X.)

[2] Department of Materials Science and Engineering, The Ohio State University, 2041, College Road, Columbus, OH 43221, USA; mao.154@osu.edu (Y.M.); vivek.4@osu.edu (A.V.); daehn.1@osu.edu (G.D.)

* Correspondence: sjchen@bjut.edu.cn; Tel.: +86-010-67391620

Received: 27 November 2018; Accepted: 29 December 2018; Published: 5 January 2019

Abstract: Direct aluminium–stainless steel joints are difficult to create by the vaporized foil actuator welding (VFAW) method because brittle intermetallic compounds (IMCs) tend to form along the interface. The use of an interlayer as a transition layer between the two materials with vast difference in hardness and ductility was proposed as a solution to reduce the formation of the IMCs. In this work, VFAW was used to successfully weld sheet aluminium alloy 5A06 to stainless steel 321 with a 3003 aluminium alloy interlayer. Input energy levels of 6 kJ, 8 kJ, 10 kJ, and 12 kJ were used and as a trend, higher energy inputs resulted in higher impact velocities, larger weld area, and better mechanical properties. In lap-shear and peel testing, all samples failed at the interface of the interlayer and target. At 10 kJ energy input, flyer velocities up to 935 m/s, lap-shear peak load of 44 kN, and peel load of 2.15 kN were achieved. Microstructure characterization and element distribution were performed, and the results show a wavy pattern created between the flyer and interlayer which have similar properties, and the interface between the interlayer and target was dominated by element diffusion and IMCs identified mainly as FeAl$_3$ and FeAl. The results demonstrate VFAW is a suitable joining method for dissimilar metals such as aluminium alloy and stainless steel, which has a broad and significant application prospect in aerospace and chemical industry.

Keywords: dissimilar materials; interlayer; vaporizing foil actuators welding

---

1. Introduction

Aluminium–stainless steel sheet joints are of great interest for purposes of lightweighting and corrosion protection in industry applications [1]. Aluminium can be joined with stainless steel via adhesive bonding or mechanical fastening [2]. However, in order to weld aluminium to stainless steel, special techniques are required. Dissimilar metals fusion welding processes have many difficulties as a result of metallurgical incompatibility [3], the formation of brittle phases, the segregation of high- and low-melting phases due to chemical mismatch, and possibly large residual stresses from the physical mismatch. Fusion-based welding techniques such as resistance spot welding [4] (RSW) and arc and laser welding-brazing [5–8] are most common in aluminium–stainless steel welding. The joints welded through the fusion-based welding process produce brittle intermetallic compound (IMC), which easily leads to fracture due to different physical and chemical properties between dissimilar metals. Solid state welding is a suitable welding method for joining dissimilar metals and can overcome some of the disadvantages associated with the fusion-based welding process [9]. Several solid state welding methods, including explosion welding [10–15] (EXW), ultrasonic welding [16], magnetic pulse welding [17,18] (MPW), and friction stir welding [1,19,20] (FSW), have been used to weld the dissimilar materials such as aluminium alloy and stainless steel. EXW and MPW are variants of impact or collision welding, which has been established as a fast and reliable technology [15]. MPW is applied for the weld length of centimeters, while EXW is more suitable for meters. Corigliano et al. [10,11] investigated the static and fatigue bending tests of explosive welded joints of ASTM A516 low carbon steel and A5086 aluminium alloy with pure aluminium as an interlayer. Han et al. [21] use AA1050 plate as interlayer in EXW of AA5083 aluminium alloy plate and SS41 steel plate. The thin interlayer enhanced the bond strength and suppressed the formation of the brittle interfacial zone and the thickness of the generated interfacial zone increased as the thickness of the interlayer increased from 0.5 mm to 2 mm. Manikandan et al. [22] employed different thickness of interlayer to analyze the energetic conditions of the titanium/304 stainless steel joints welded by explosive welding process. The results show that a thin interlayer leads to successful welding as the use of interlayer splits the kinetic energy deposition between the two interfaces and thereby reduces the possibility of melting or formation of IMCs.

Vaporizing foil actuator welding (VFAW) is a novel solid state welding technology making use of the impulse created by vaporization of the metallic foil or wire by the passage of a high current [23]. The high-amplitude impulse accelerates the flyer workpiece toward the target workpiece. The high velocity oblique impact spits out the surface contaminants in the form of a jet and the high pressure metal-to-metal contact leads to formation of metallic bonds [24]. The technology proves to be robust to join several dissimilar metals such as Al–Mg, Al–Cu, Al–Fe, Al–Ti, etc. [25–28]. However, further development and understanding of the process is needed in order to accomplish industrial adoption. There are also certain pairs which cannot be joined directly with this process. In such cases, an interlayer can prove to be useful.

In this study, 3003 aluminium alloy was chosen as the interlayer in the 5A06 aluminium alloy and 321 stainless steel VFAW welding process. This research further studied the effects of the interlayer on the microstructure and mechanical properties of the welds. Welded joints were evaluated based on mechanical strength and microstructure. The characteristics of mechanical interlocking and metallurgical bonding areas were studied by advanced microscopy.

2. Experimental Procedure

2.1. Materials and Methods

This work follows the VFAW method described by Vivek et al. [23]. The flyer sheet was placed directly against an aluminium foil which was insulated by a 0.12 mm thick polyimide (trade name Kapton) tape and the ends of steel terminals were connected to a capacitor bank. The characteristics of the capacitor bank are shown in Table 1.

Table 1. Capacitor bank characteristics.

Capacitance	Inductance	Resistance	Maximum Charging Energy	Short Circuit Current Rise Time
426 μF	100 nH	10 mΩ	16 kJ at 8.66 kV	12 μs

As the capacitor bank was discharged, the foil vaporized in 10~15 μs under the current of 100~300 kA. The flyer sheet accelerated to a high velocity (300~1200 m/s) by the forces of the vaporized foil. 0.076 mm thick aluminium foils were used for energy input lower than 10 kJ, and 0.125 mm thick foils were used for 12 kJ energy input. The active area of the foils was 12.7 mm wide and 50.8 mm long. A schematic of the apparatus as well as the actual implementation, the sketch of the aluminium foil, and the schematic of VFAW process are shown in Figure 1. Annealed aluminium alloy type 5A06 (1.8 × 70 × 140, in mm) and 3003 (1.02 × 70 × 70, in mm) sheets were chosen as the flyer and interlayer material. 321 stainless steel sheets were cut to 4 mm × 70 mm × 140 mm. The standoff distances between the flyer and interlayer and the interlayer and target were 3 mm and 1.5 mm, respectively, in all experiments. The impacting surfaces of all materials were cleaned prior to welding with acetone after being ground by emery paper.

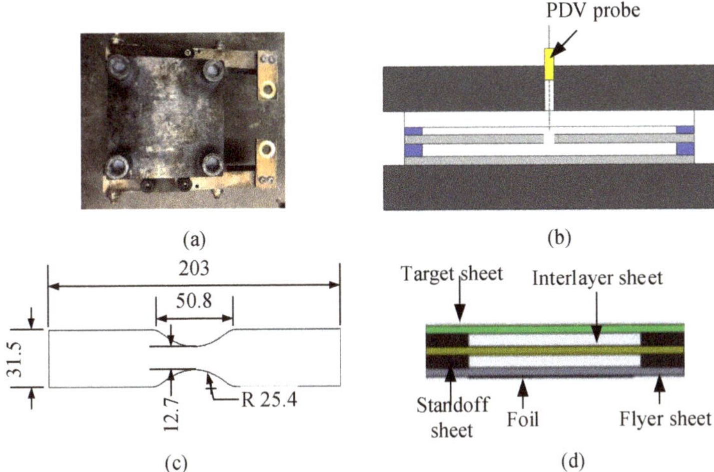

Figure 1. (**a**) actual implementation of the vaporizing foil actuator welding (VFAW) apparatus; (**b**) impact velocity test apparatus; (**c**) sketch of the aluminium foil; (**d**) schematic of the VFAW process.

2.2. Velocity Measurement

Input energies of 6 kJ, 8 kJ, 10 kJ, and 12 kJ were used in this study. Current and voltage of the foil vaporized in VFAW process were measured using a 50 kA:1 V Rogowski coil and 1000:1 high voltage probe, respectively. The velocity of flyer and interlayer impact to the target was recorded using photonic Doppler velocimetry (PDV) [29]. A hole was drilled in the center of the backing steel fixture and the interlayer using transparent acrylic as the replacement of the target sheet to allow the laser focusing probe to look at the interlayer directly. The velocity at any distance within this range can then be estimated by integration of the resulting velocity–time curve.

2.3. Strength Testing

Lap-shear and peel tests were used to analyze the joint strength. A schematic of the strength testing is shown in Figure 2. Three samples from each input energy were tested. For peel testing, the flyer sheets were bent to 90° with respect to the interlayer and target, and the target sheets were fixed to a steel die by a bolt. Testing was carried out using an MTS810 mechanical testing frame at a constant extension rate of 0.1 mm/s.

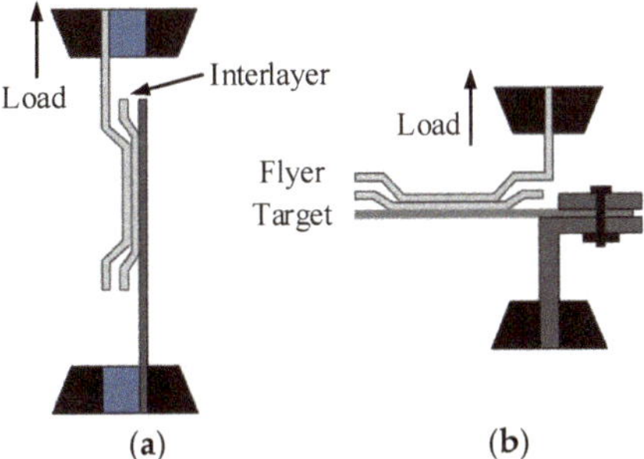

Figure 2. Mechanical testing schematic. (**a**) Peel test; (**b**) Lap-shear test.

Interfacial morphologies and element content and distribution in the interface of interlayer and target were observed by scanning electron microscopy (SEM) using a Zeiss Ultra55 equipped with a silicon drift detector for energy-dispersive X-ray spectroscopy (EDS). Optical microscopy was used to analyze the interfacial morphologies of the flyer and interlayer.

3. Results and Discussion

3.1. Velocity, Current, and Voltage Traces

The temporal evolutions of current, voltage, and velocity of the flyer sheet with 8 kJ energy input in VFAW process are shown in Figure 3.

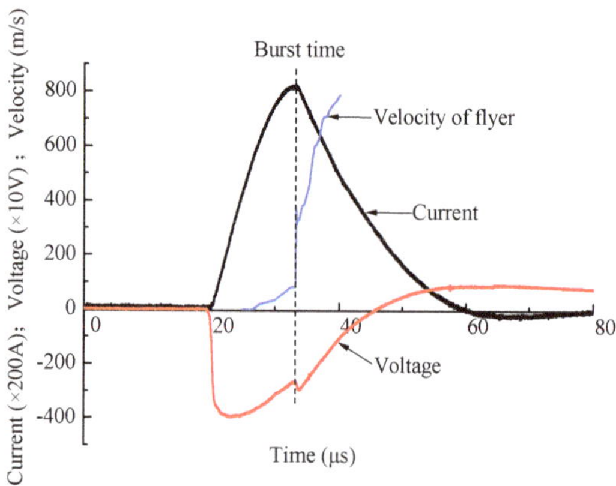

Figure 3. Voltage, current, and velocity traces with 8 kJ energy input.

The flyer sheet has a low acceleration (from 25 to 33 µs) under the electromagnetic interaction force generated by the current carrying foil [30]. The voltage has a sharp spike rise and a sudden decrease of current, while the flyer sheet accelerated to 393 m/s instant at about 33.5 µs. This moment also called burst time in the VFAW process. This voltage spike rise is a typical phenomenon in foil vaporized process due to the resistance increase when the foil is vaporized and converted into high pressure plasma. The flyer and interlayer sheet impacted at the target sheet at a velocity of 788 m/s. The velocity trace of flyer sheet with other three energies input are shown in Figure 4. The impact velocities were 635 m/s, 788 m/, 935 m/s, and 1025 m/s for input energies of 6 kJ, 8 kJ, 10 kJ, and 12 kJ, respectively.

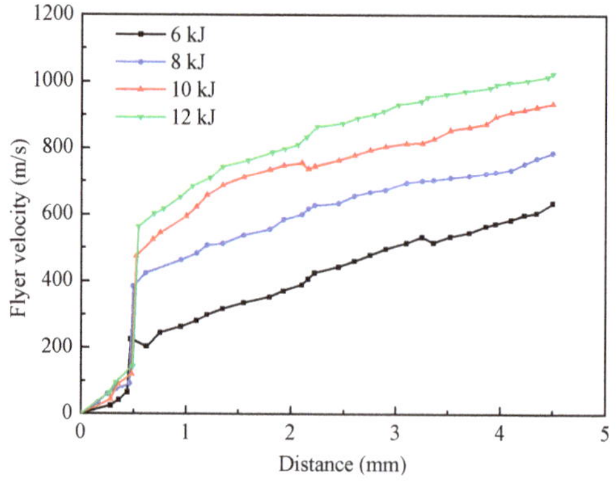

Figure 4. Flyer sheet velocity traces at 6 kJ, 8 kJ, 10 kJ, and 12 kJ input energies.

3.2. Mechanical Testing

Mechanical testing results from lap-shear and peel tests are summarized in Table 2. The peak load of lap-shear and peel tests increased with energy input increased from 6 kJ to 10 kJ. The average lap-shear loads were 29.9 kN, and 35 kN, 44 kN, and 42.6 kN for the 6 kJ, 8 kJ, 10 kJ, and 12 kJ input energies, respectively. The average peak peel loads were 1.62 kN, 1.65 kN, 2.15 kN, and 2.03 kN, respectively.

Table 2. Lap-shear and peel test results.

Energy Input	Lap-Shear Average	Test 1	Test 2	Test 3	Peel Average	Test 1	Test 2	Test 3
(kJ)	(kN)	(kN)	(kN)	(kN)	(kN)	(kN)	(kN)	(kN)
6	29.9	28.8	31.9	28.9	1.62	1.59	1.61	1.66
8	35	34.2	36.2	34.7	1.65	1.69	1.65	1.62
10	44	43.5	45.4	43.1	2.15	2.12	2.18	2.16
12	42.6	42.9	42.2	42.6	2.03	2.01	2.04	2.03

The load–displacement curves of the lap-shear samples are shown in Figure 5. The lap-shear test fractured surface are shown in Figure 6. The failure zones of all specimens for lap-shear and peel test are located at the interface between the interlayer and target plate. As shown in the cross-sectional morphology of the specimen for tensile test, the good weld area between the flyer plate and interlayer is smaller than that between the interlayer and target plate. The diameter of the failure zone increases from 16 mm to 28.8 mm with the input energy from 6 kJ to 10 kJ, thereby increasing the peak tensile force sustained by the joint. The results show that the welded area of the flyer and interlayer was smaller than that of interlayer and target, and the bonding strength of interlayer and target near the center is smaller than that of outside. The samples welded at higher energies had larger welding interface of interlayer and flyer, and these correlated with higher bonding strength. However, the increase of the input energy slightly reduced the strength of the joining between the interlayer and target with the failure load decreasing from 44 kN to 42.6 kN.

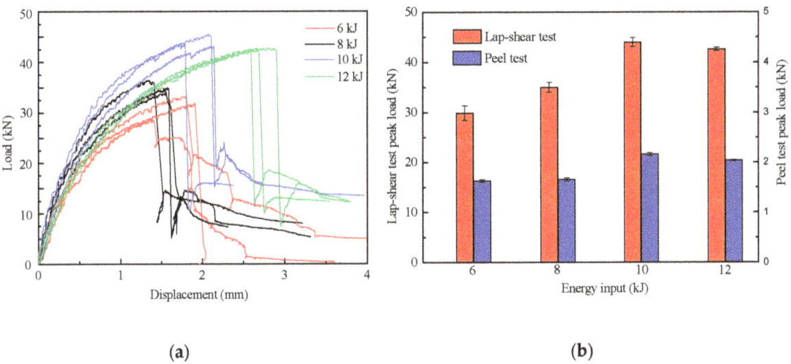

(**a**)   (**b**)

Figure 5. Lap-shear and peel test results at 6 kJ, 8 kJ, 10 kJ, and 12 kJ energies input. (**a**) Load–displacement curves of lap-shear test; (**b**) Lap-shear and peel average load.

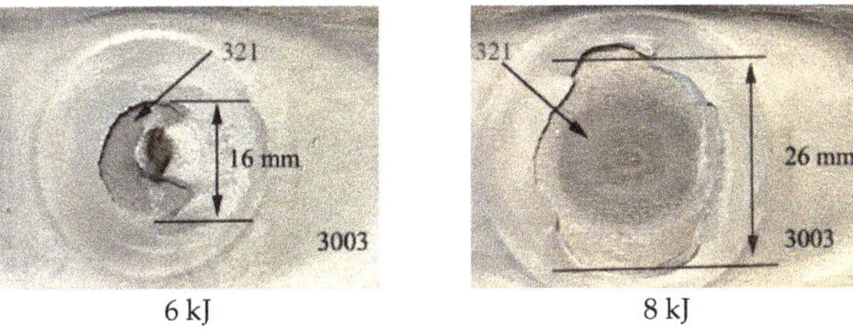

Figure 6. *Cont.*

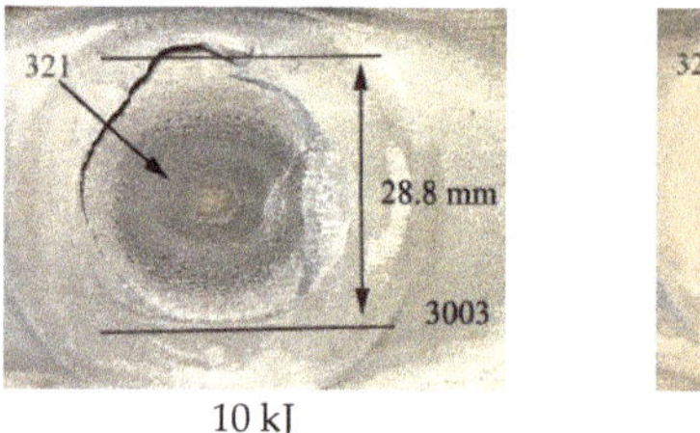
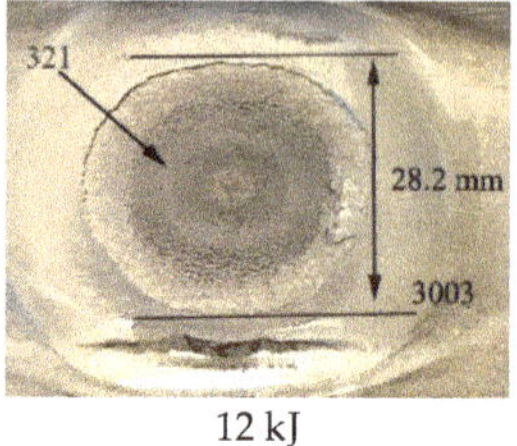

Figure 6. Lap-shear tested samples failed area at 6 kJ, 8 kJ, 10 kJ, and 12 kJ input energies.

3.3. Interfacial Morphologies

The interface between 3003 and 5A06 is dominated by a wave-like bond, whereas AA3003 and SS321 are bonded with IMC possibly caused by element diffusion and mixing. The wavy-bonding regions with the input energy of 6, 8, 10, and 12 kJ are shown in the red frame of Figure 7. A symmetrically distributed wavy-bonding region is formed at the Al–Al interface area at approximately 5 mm from the central area. The metals of the interlayer and the flyer plate are embedded with each other, the crest height gradually increases from the center to the sides, and the distance between the wavy bonding region and the center decreases with increasing energy. When the input energy increases, the tendency of the crest tilting towards the center increases.

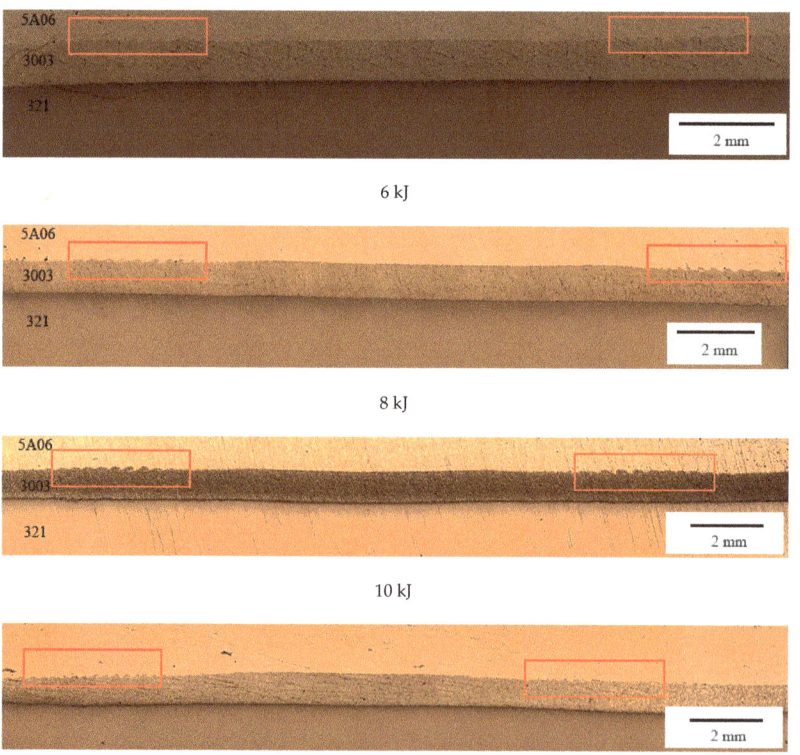

Figure 7. Interface morphologies of flyer and interlayer at 6 kJ, 8 kJ, 10 kJ, and 12 kJ input energies.

The distance between the crest height, λ, and the distance, L, between two waves are given in Figure 8. The crest height and inter-wave interval increase with the input energy increase. The crest heights are 96, 116, 120, and 124 µm, whereas the inter-wave intervals are 332, 358, 471, and 493 µm.

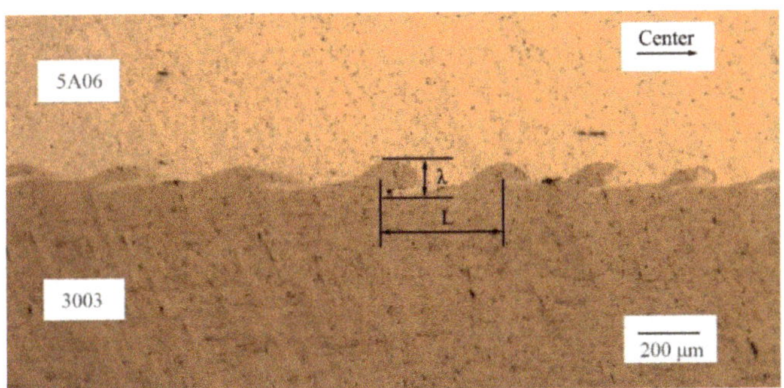

Figure 8. Interface morphologies of wave bonding area with 6 kJ energy input.

Figure 9 shows the distribution pattern at the interface between aluminium alloy 3003 and stainless steel 321 when the input energy is 8 kJ. The graph shows that the bonding region at the interface between aluminium alloy 3003 and stainless steel 321 consists of the mechanical-clinch area, IMC bonding area, and an unwelded area. The unwelded area is located at the center of the joint, the mechanical-clinch zone by element diffusion is outside the IMC bonding zone, and the IMC bonding zone is flat near the aluminium alloy and irregularly undulated near the stainless steel.

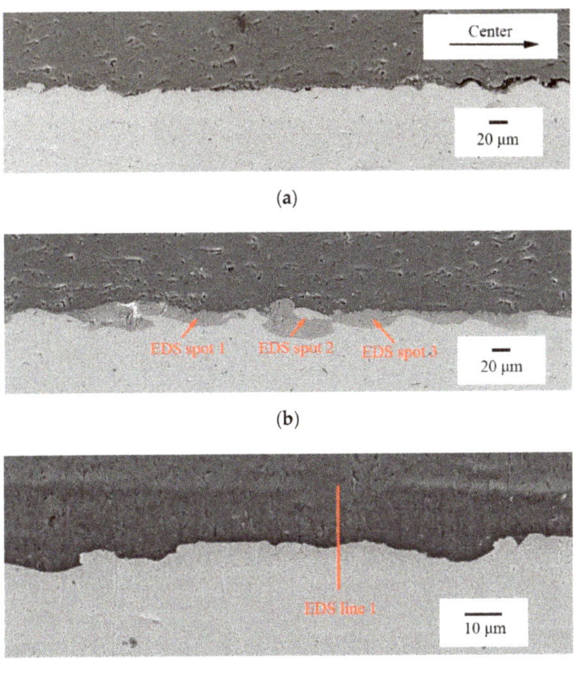

Figure 9. Interface morphologies of 3003 aluminium and 321 stainless steel at 8 kJ energy input. (**a**) unwelded area. (**b**) IMC bonding area. (**c**) Mechanical-clinch area.

The morphologies of the IMC bonding area at energies input 6, 10, and 12 kJ are shown in Figure 10. The length and thickness of the IMC bonding zone at the interface increases and decreases with the input energy, respectively. When the input energy is 6 kJ, the largest thickness of IMC at 50 μm is achieved, whereas its length is approximately 300 μm. When the input energy increases to 10 kJ, the thickness of the interface is reduced to 30 μm, and the IMC is discontinuously distributed with approximately 2 mm in length. The thickness of IMC reduced to 26 μm with energy input increased to 12 kJ.

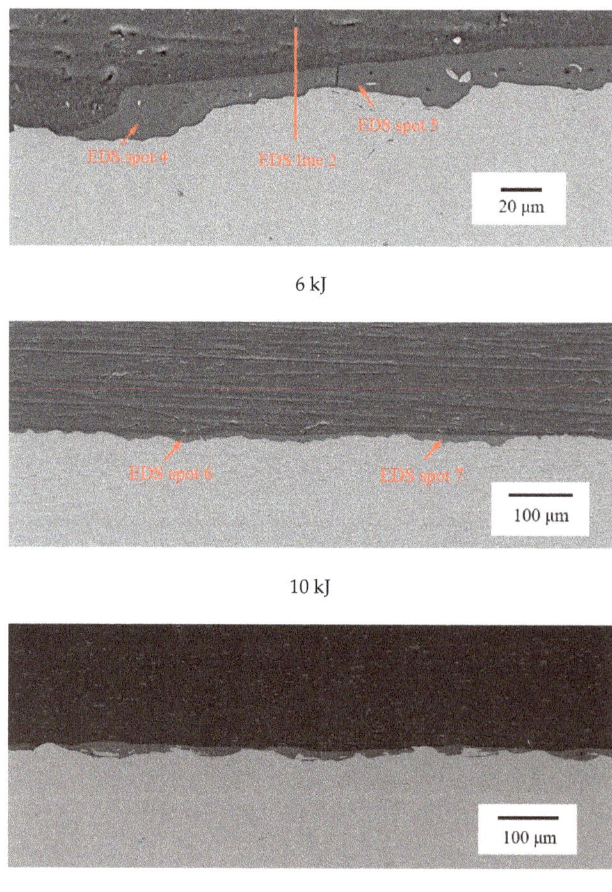

Figure 10. Interface morphologies with 6 kJ, 10 kJ, and 12 kJ input energies.

The interface morphologies of the center area with 12 kJ energy input is shown in Figure 11. At high impact velocities and low impact angles, overheated regions were formed at the center of the interface between the interlayer and target at the input energy level of 12 kJ causing damage on the interlayer, and therefore the bonding strength was impaired, with the tensile load decreased to 42.6 kN.

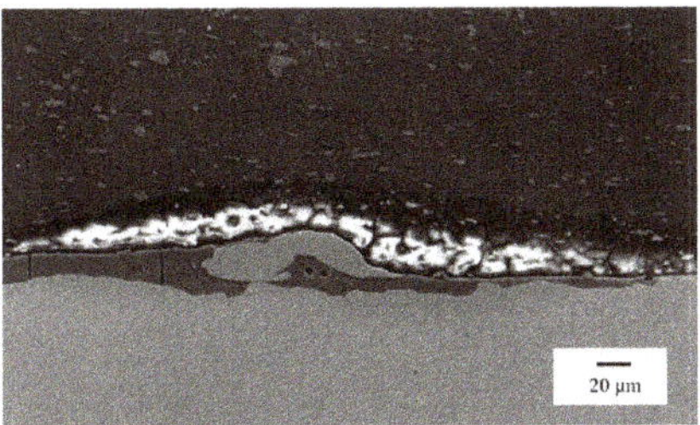

Figure 11. Interface morphologies of center area with 12 kJ energy input.

As shown from the failure areas of the specimens after the lap-shear and peel tests, the weld area at the interface between aluminium alloy 3003 and stainless steel 321 is significantly larger than that between aluminium alloy 3003 and aluminium alloy 5A06. The failure zone of each joint is located at the central area of the interface between aluminium alloy 3003 and stainless steel 321. In the VFAW process, which is driven by the energy of aluminium foil vaporization, the flyer plate initially collides with the interlayer. The flyer plate and interlayer then impact the target plate together, thus producing a high collision angle between the flyer plate and external side of the interlayer. This collision prevents the formation of a reliable weld. The collision angle between the interlayer and central area of the target plate increases externally, thus rendering the bonding strength of the region smaller than that of the external side.

The element line scanning results at the interface between aluminium alloy 3003 and stainless steel 321 are given in Figure 12. The welded areas at the interface between aluminium alloy 3003 and stainless steel 321 and at the central area close to the joint are mainly characterized by IMC. Within a certain distance from the center of the joint, the collision angle between the interlayer and target plate increases. At the interface, a 3 μm thick bonding region is formed. The intensity is significantly larger than that in the central area. The EDS scan results of Points 1–7 are shown in Table 3. The IMCs along the interface between aluminium alloy 3003 and stainless steel 321 are FeAl and FeAl$_3$.

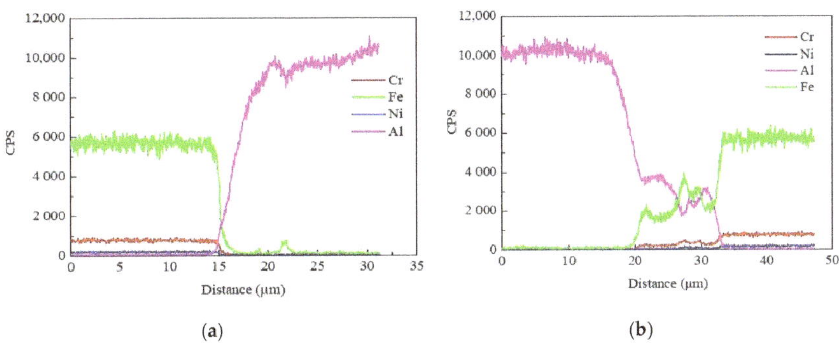

(a)   (b)

Figure 12. Element distributions of the 3003–321 interface at different areas. (**a**) EDS line scan of line 1; (**b**) EDS line scan of line 2.

Table 3. Point EDS compositional result at the interface of the interlayer and target sheet (at. %).

	EDS Spot 1	EDS Spot 2	EDS Spot 3	EDS Spot 4	EDS Spot 5	EDS Spot 6	EDS Spot7
Al	42.3	0.77	42.49	62.99	67.16	66.79	67.17
Fe	42.56	69.79	42.47	21.61	22.84	22.54	22.25

4. Conclusions

5A06 aluminium alloy and 321 stainless steel were welded by VFAW using an AA3003 interlayer and joints with good structure and properties were obtained. With input energies of 6 kJ, 8 kJ, and 10 kJ, the shear and tensile strengths are increased due to increased welded area. The joints with the best mechanical properties were obtained at the input energy of 10 kJ, which had a failure load of 44 kN in lap shear mode and 2.15 kN in peel mode of loading. Microstructure characterization revealed a wavy pattern between the flyer and interlayer which have similar properties. The interface between the interlayer and target showed the presence of IMCs, identified as Al_3Fe and $FeAl$, which were continuously or intermittently distributed along the interface depending on the input energy. The results show that the use of an interlayer is a viable solution for impact welding of materials which are otherwise difficult to weld.

Author Contributions: Data curation, S.S.; Funding acquisition, S.C. and G.D.; Investigation, S.S. and A.V.; Methodology, A.V.; Project administration, S.C.; Software, J.X.; Supervision, G.D.; Writing—original draft, S.S.; Writing—review & editing, Y.M.

Funding: This research was funded by National Nature Science Foundation of China (Grant Number 51575012). The OSU work was funded by the Department of Energy, Office of Energy Efficiency and Renewable Energy (EERE), Vehicle Technology Office under award number DE-0007813. The author gratefully acknowledges financial support from China Scholarship Council.

Conflicts of Interest: The authors declare no conflict of interest.

Disclaimer: Neither the U. S. Government, nor any of its employees, makes any warranty, express or implied, or assumes any legal liability or responsibility for the accuracy or completeness of any information, product, or process disclosed, or represents that its manufacture or use would not infringe privately owned rights. Any reference to a commercial product, process, or service does not constitute an endorsement or favoring by the U. S. Government. The views and opinions of the authors stated do not necessarily reflect those of the U. S. Government.

References

1. Sahin, M. Joining of stainless-steel and aluminium materials by friction welding. *Int. J. Adv. Manuf. Technol.* **2009**, *41*, 487–497. [CrossRef]
2. Sun, Z.; Karppi, R. The application of electron beam welding for the joining of dissimilar metals: An overview. *J. Mater. Process. Technol.* **1996**, *59*, 257–267. [CrossRef]
3. Jafarian, M.; Khodabandeh, A.; Manafi, S. Evaluation of diffusion welding of 6061 aluminum and AZ31 magnesium alloys without using an interlayer. *Mater. Des.* **2015**, *65*, 160–164. [CrossRef]
4. Zhang, W.; Sun, D.; Han, L.; Liu, D. Interfacial microstructure and mechanical property of resistance spot welded joint of high strength steel and aluminium alloy with 4047 AlSi12 interlayer. *Mater. Des.* **2014**, *57*, 186–194. [CrossRef]
5. Song, J.L.; Lin, S.B.; Yang, C.L.; Fan, C.L. Effects of Si additions on intermetallic compound layer of aluminum–steel tig welding–brazing joint. *J. Alloys Compd.* **2009**, *488*, 217–222. [CrossRef]
6. Wang, H.Y.; Zhang, Z.D.; Liu, L.M. The effect of galvanized iron interlayer on the intermetallics in the Laser weld bonding of Mg to Al fusion zone. *J. Mater. Eng. Perform.* **2013**, *22*, 351–357. [CrossRef]
7. Wang, H.; Liu, L.; Liu, F. The characterization investigation of laser-arc-adhesive hybrid welding of Mg to Al joint using Ni interlayer. *Mater. Des.* **2013**, *50*, 463–466. [CrossRef]
8. Balakrishnan, M.; Balasubramanian, V.; Madhusudhan Reddy, G. Effect of hardfaced interlayer thickness on ballistic performance of armour steel welds. *Mater. Des.* **2013**, *44*, 59–68. [CrossRef]
9. Date, H.; Kobayakawa, S.; Naka, M. Microstructure and bonding strength of impact-welded aluminium–stainless steel joints. *J. Mater. Process. Technol.* **1999**, *85*, 166–170. [CrossRef]
10. Corigliano, P.; Crupi, V.; Guglielmino, E.; Mariano Sili, A. Full-field analysis of Al/Fe explosive welded joints for shipbuilding applications. *Mar. Struct.* **2018**, *57*, 207–218. [CrossRef]
11. Corigliano, P.; Crupi, V.; Guglielmino, E. Non linear finite element simulation of explosive welded joints of dissimilar metals for shipbuilding applications. *Ocean Eng.* **2018**, *160*, 346–353. [CrossRef]
12. Topolski, K.; Szulc, Z.; Garbacz, H. Microstructure and properties of the Ti6Al4V/Inconel 625 bimetal obtained by explosive joining. *J. Mater. Eng. Perform.* **2016**, *25*, 3231–3237. [CrossRef]

13. Findik, F. Recent developments in explosive welding. *Mater. Des.* **2011**, *32*, 1081–1093. [CrossRef]

14. Xie, M.; Shang, X.; Zhang, L.; Bai, Q.; Xu, T. Interface characteristic of explosive-welded and hot-rolled Ta1/X65 bimetallic plate. *Metals* **2018**, *8*, 159. [CrossRef]

15. Kaya, Y. Microstructural, mechanical and corrosion investigations of ship steel-aluminum bimetal composites produced by explosive welding. *Metals* **2018**, *8*, 544. [CrossRef]

16. Balasundaram, R.; Patel, V.K.; Bhole, S.D.; Chen, D.L. Effect of zinc interlayer on ultrasonic spot welded aluminum-to-copper joints. *Mater. Sci. Eng. A* **2014**, *607*, 277–286. [CrossRef]

17. Berlin, A.; Nguyen, T.C.; Worswick, M.J.; Zhou, Y. Metallurgical analysis of magnetic pulse welds of AZ31 magnesium alloy. *Sci. Technol. Weld. Join.* **2013**, *16*, 728–734. [CrossRef]

18. Zhang, Y.; Babu, S.S.; Prothe, C.; Blakely, M.; Kwasegroch, J.; LaHa, M.; Daehn, G.S. Application of high velocity impact welding at varied different length scales. *J. Mater. Process. Technol.* **2011**, *211*, 944–952. [CrossRef]

19. Yılmaz, M.; Çöl, M.; Acet, M. Interface properties of aluminum/steel friction-welded components. *Mater. Charact.* **2002**, *49*, 421–429. [CrossRef]

20. Madhusudhan Reddy, G.; Venkata Ramana, P. Role of nickel as an interlayer in dissimilar metal friction welding of maraging steel to low alloy steel. *J. Mater. Process. Technol.* **2012**, *212*, 66–77. [CrossRef]

21. Han, J.H.; Ahn, J.P.; Shin, M.C. Effect of interlayer thickness on shear deformation behavior of AA5083 aluminum alloy/SS41 steel plates manufactured by explosive welding. *J. Mater. Sci.* **2003**, *38*, 13–18. [CrossRef]

22. Manikandan, P.; Hokamoto, K.; Fujita, M.; Raghukandan, K.; Tomoshige, R. Control of energetic conditions by employing interlayer of different thickness for explosive welding of titanium/304 stainless steel. *J. Mater. Process. Technol.* **2008**, *195*, 232–240. [CrossRef]

23. Vivek, A.; Hansen, S.R.; Liu, B.C.; Daehn, G. Vaporizing foil actuator: A tool for collision welding. *J. Mater. Process. Technol.* **2013**, *213*, 2304–2311. [CrossRef]

24. Vivek, A.; Hansen, S.; Benzing, J.; He, M.; Daehn, G. Impact welding of aluminum to copper and stainless steel by vaporizing foil actuator: Effect of heat treatment cycles on mechanical properties and microstructure. *Metall. Mater. Trans. A* **2015**, *46*, 4548–4558. [CrossRef]

25. Liu, B.; Vivek, A.; Presley, M.; Daehn, G. Dissimilar impact welding of 6111-T4, 5052-H32 aluminum alloys to 22MnB5, Dp980 steels and the structure–property relationship of a strongly bonded interface. *Metall. Mater. Trans. A* **2018**, *49*, 899–907. [CrossRef]

26. Liu, B.; Vivek, A.; Daehn, G.S. Joining sheet aluminum AA6061-T4 to cast magnesium AM60B by vaporizing foil actuator welding: Input energy, interface, and strength. *J. Manuf. Process.* **2017**, *30*, 75–82. [CrossRef]

27. Nassiri, A.; Zhang, S.; Lee, T.; Abke, T.; Vivek, A.; Kinsey, B.; Daehn, G. Numerical investigation of Cp-Ti & Cu110 impact welding using smoothed particle hydrodynamics and arbitrary lagrangian-eulerian methods. *J. Manuf. Process.* **2017**, *28*, 558–564.

28. Chen, S.; Huo, X.; Guo, C.; Wei, X.; Huang, J.; Yang, J.; Lin, S. Interfacial characteristics of Ti/Al joint by vaporizing foil actuator welding. *J. Mater. Process. Technol.* **2019**, *263*, 73–81. [CrossRef]

29. Johnson, J.R.; Taber, G.; Vivek, A.; Zhang, Y.; Golowin, S.; Banik, K.; Fenton, G.K.; Daehn, G. Coupling experiment and simulation in electromagnetic forming using photon doppler velocimetry. *Metal Form.* **2009**, *80*, 359–365.

30. Vivek, A.; Hansen, S.R.; Daehn, G. High strain rate metalworking with vaporizing foil actuator: Control of flyer velocity by varying input energy and foil thickness. *Rev. Sci. Instrum.* **2014**, *85*, 75101. [CrossRef] [PubMed]

Article

Characteristics of Flyer Velocity in Laser Impact Welding

Huimin Wang [1] and Yuliang Wang [2,3,*]

[1] National Center for Materials Service Safety, University of Science and Technology Beijing, Beijing 100083, China; wanghuimin@ustb.edu.cn
[2] School of Mechanical Engineering and Automation, Beihang University, Beijing 100191, China
[3] Beijing Advanced Innovation Center for Biomedical Engineering, Beihang University, Beijing 100191, China
* Correspondence: wangyuliang@buaa.edu.cn; Tel.: +86-18612525756

Received: 11 January 2019; Accepted: 26 February 2019; Published: 1 March 2019

Abstract: The flyer velocity is one of the critical parameters for welding to occur in laser impact welding (LIW) and plays a significant role on the welding mechanism study of LIW. It determines the collision pressure between the flyer and the target, and the standoff working distance. In this study, the flyer velocity was measured with Photon Doppler Velocimetry under various experimental conditions. The laser energy efficiency was compared with measured flyer velocity for various laser energy and flyer thickness. In order to reveal the standoff working window, the peak flyer velocity and flyer velocity characteristic before and after the peak velocity and the flyer velocity was measured over long distance. In addition, the rebound behavior of the flyer was captured to confirm the non-metallurgical bonding in the center of the weld nugget in LIW. Furthermore, the flyer size and confinement layer effect on the flyer velocity were investigated.

Keywords: flyer velocity; energy efficiency; peak velocity; flyer rebound; flyer size; confinement layer

---

1. Introduction

Laser impact welding (LIW) is a relatively novel welding technique. It is one of the solutions for dissimilar materials welding due to its solid state welding nature. Different from explosive welding and magnetic pulse welding, LIW focuses on solving problems in areas like electronics and medical devices. The first U.S. patent of LIW was filed in 2009 by Daehn and Lippold from The Ohio State University [1]. Wang et al. proposed the possible industrial application setup in 2015 for the first time [2]. In recent years, LIW has been studied in joining materials, like aluminum [3–5], titanium [3], steel [6], and copper [4,7]. The feasibility and ability of LIW in welding dissimilar materials has been demonstrated.

In the LIW process, a laser goes through the transparent confinement layer and ablates the ablative layer into high temperature and high pressure plasma. The expanding plasma pushes the flyer to move at a high velocity until it collides on the target, where the metallurgical bonding occurs at the collision interface with a proper impact angle. The flyer velocity is one of the critical parameters in LIW [8]. The Hugoniot theory states that the impact velocity (final flyer velocity) determines the impact pressure between the flyer and the target [9–12]. For LIW, metallurgical bonding occurs when the nascent surfaces are brought within atomic distance. Thus, the collision pressure is required to be sufficient to remove the surface contaminants, oxides, asperities. At the same time, the nascent surface should be pushed within atomic distance by the impact pressure. However, too much pressure will result in melting at the interface [13] and spallation on materials [14]. There are difficulties in measuring the pressure from the experiments directly for impact welding due to the fact that the measurement is destructive. Polyvinylidene Flouride (PVDF) gage was applied in the pressure measurement in impact welding, which is a film type polymer sensor. The measurement is sensitive to the size and location of the gage [15]. The flyer velocity is usually measured directly [16,17], which is a non-destructive method.

In LIW, the current general welding configuration for LIW is spot welding [1]. The central part of the weld nugget does not get bonded between the flyer and the target. The bonding occurs along the edges of the weld nugget [7].The flyer velocity/movement for LIW should be studied to solve this common issue in LIW. Besides that, the experimental setup can be optimized for better laser energy efficiency, which is calculated with the flyer velocity, as shown in Equation (1) (t is the thickness of the flyer, D is the diameter of the laser spot size, v is the flyer velocity, E is the laser energy, ρ is the density of the flyer).

$$\eta = \frac{0.5(\pi(D/2)^2 t\rho)v^2}{E} \tag{1}$$

In the study of the laser-driven flyer, the flyer velocity was investigated by Brown et al. [10]. They studied the variation of flyer velocity with a different confinement layer, ablative layer and adhesive. Their study was within a time of 200 ns. In the study of Zhao et al. [18], the effect of laser energy on the flyer velocity was studied, which was also investigated by Shaw-Stewart et al. [19]. The flyer velocity as a function of time was displayed in the study by Paisley et al. [11]. The measured flyer velocity was lower than 500 m/s and the energy efficiency was not calculated for laser energy up to 52 J. In the study of LIW, Wang et al. showed the flyer velocity as a function of time with various laser energy and flyer thickness [2,3]. The laser energy efficiency has not been studied with various laser energy and flyer thickness.

This study investigated the characteristics of the flyer velocity under various conditions with Photon Doppler Velocimetry (PDV), which is much more practical and easier for surface velocity measurement than any other competing techniques [20]. The laser energy efficiency was investigated for various laser energy and flyer thickness by measuring the flyer velocity under those conditions. The study of flyer velocity over long travel distance would propose the proper standoff distance and reveal the plasma working distance. The investigation of flyer rebound behavior should reveal the flyer behavior before, at and after collision with the target. Other than that, the parameters effect on the flyer velocity would provide instructions for the optimization of the current experimental setup.

2. Experimental Setup

The schematic experimental setup is shown in Figure 1. The laser beam passes through the confinement layer and connection layer (double-sticky tape in this study), and vaporizes the ablative layer into hot plasma. The expanding of the plasma separates the confinement layer and flyer. Due to the conservation of momentum, the flyer moves much faster than the more massive confinement layer in the laser incident direction, which is also the flyer thickness direction. PDV measures the flyer velocity during its flying process. The polycarbonate (1 mm thick) between the PDV probe and flyer prevents the direct contact between the flyer and the PDV probe to protect the probe from damage. The pre-set spacer provides the acceleration distance (travel distance) for the flyer velocity to get to a certain value. The distance between the flyer and the polycarbonate (1 mm thick) is called standoff distance/preset distance. The ablative layer was painted on the flyer surface.

In this study, the laser system was a commercially available Nd:YAG laser (Continuum, San Jose, FL, USA), which was a Continuum PowerliteTM Precision II Scientific System with a maximum laser energy of 3.1 J and a pulse width of 8 ns. More information about the laser system is provided in [2]. The flyer material was Al1100. Other experimental parameters are listed in Table 1. In this study, the confinement layers included 0.5 mm polycarbonate, 1 mm polycarbonate, 3 mm glass and 5 mm glass. The ablative layer was a commercial black spray paint (RUST-OLEUM^{TM}, Enamel). 1 mm thick polycarbonate was used as the target. Laser beam energies of 3.1 J, 1.9 J and 0.8 J were used. The diameter of the laser beam spot was adjusted by changing the distance between the experimental setup and focus lens, as shown in Figure 2. The power and energy density are shown in Table 2 for laser energies of 3.1 J, 1.9 J and 0.8 J, respectively. Parameters used in this study were defined as follows: E: Laser energy; D: Diameter of laser spot; t: Flyer thickness; v: Flyer velocity; L: Travel distance.

Table 1. Experimental parameters.

Confinement Layer	0.5 mm Polycarbonate; 1 mm Polycarbonate; 3 mm Glass; 5 mm Glass
Ablative Layer	Commercial black spray paint (RUST-OLEUM^{TM}, Enamel)
Laser Beam Energy (J)	3.1, 1.9 and 0.8
Flyer Material	Al1100

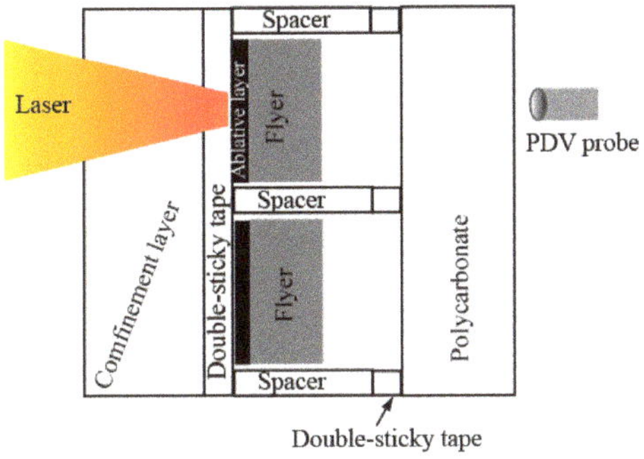

Figure 1. Schematic experimental setup of the flyer velocity measurement for laser impact welding (LIW).

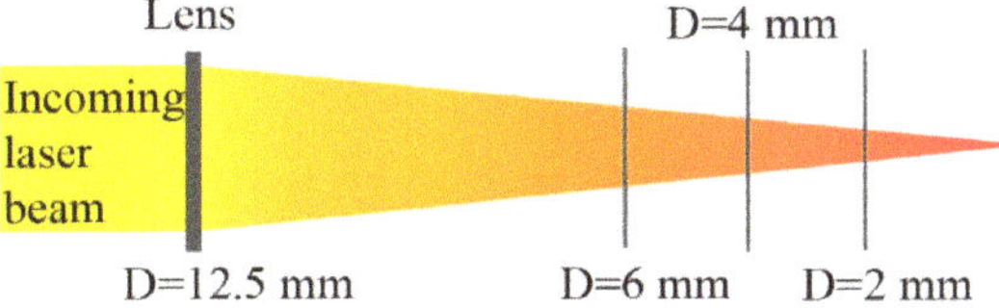

Figure 2. Laser spot size variation by changing the distance between the focus lens and the sample.

Table 2. Laser beam power density and energy density at various laser spot sizes.

Spot Size / Large Energy	Energy Density ($\times 10^5$ J/m^2)			Power Density ($\times 10^{13}$ W/m^2)		
	3.1	1.9	0.8	3.1	1.9	0.8
2 mm	9.87	6.05	2.55	12.34	7.56	3.19
4 mm	2.47	1.51	0.64	3.09	1.89	0.80
6 mm	1.10	0.67	0.28	1.38	0.84	0.35

3. Photon Doppler Velocimetry

Photon Doppler Velocimetry (PDV) is an advanced technique, which is used to measure the flyer velocity [20]. In this study, the PDV system used an erbium fiber laser with a wavelength of 1550 nm. It was designed such that the original laser beam was divided into two equal beams. One beam was directly sent back to the detector while the other beam was delivered to point to the surface of a moving target. This beam was reflected by the moving target surface and was also sent to the detector. By mixing the original and reflected beams, a beat frequency was generated by the detector. The signal with beat frequency was displayed with an oscilloscope (digitizer). The velocity of the moving target v was equal to the beat frequency f_b times a half of the laser wavelength λ, as shown in

Equation (2). Detailed information about the PDV system used in this study was discussed in [21]. In this study, during the experiment, the PDV system was triggered by the Q-switch of the laser system to start recording.

$$v = \frac{\lambda f_b}{2} \tag{2}$$

4. Results and Discussion

4.1. Laser Energy Efficiency at Various Laser Energy and Flyer Thickness

The flyer velocity for various laser energy and flyer thickness was measured. Some of the results have been reported in [2,3]. In LIW, the flyer velocity reached above 1000 m/s within 0.2 μs (E = 3.1 J, t = 25 μm, D = 2 mm). The laser energy efficiency is one of the main factors for the industrial application of LIW. In this study, the laser energy efficiency was calculated for various laser energies (3.1 J, 1.9 J and 0.8 J) and flyer thicknesses (25 μm, 75 μm, 100 μm, 125 μm, and 150 μm) at a flyer travel distance of 100 μm and 2 mm laser spot size. Table 3 lists the flyer velocities used for the laser energy efficiency calculation at the flyer travel distance of 100 μm and 2 mm laser spot size. The flyer velocity increased with laser energy and decreased with flyer thickness. The laser energy efficiency was calculated with Equation (1), as shown in Figure 3. The laser energy efficiency decreased with laser energy for all thickness flyers. This indicated that there was more laser energy wasted by increasing the laser energy when the laser energy density was above 2.55×10^5 J/m^2. The previous study also revealed that the flyer velocity increased dramatically when the laser energy density was below 2.55×10^5 J/m^2. Above 2.55×10^5 J/m^2, the flyer velocity formed a plateau [3]. Therefore, the absorption limit of the current ablative layer should be close to 2.55×10^5 J/m^2. This resulted in the laser energy efficiency decreasing by increasing the laser energy. Laser energy efficiency was not affected by flyer thickness.

Table 3. Flyer velocities (m/s) with various laser energy and flyer thickness (D = 2 mm, L = 100 μm).

Flyer Velocity (m/s)	25 μm	75 μm	100 μm	125 μm	150 μm
3.1 J	1036	613	416	397	404
1.9 J	895	596	414	365	383
0.8 J	837	548	368	351	381

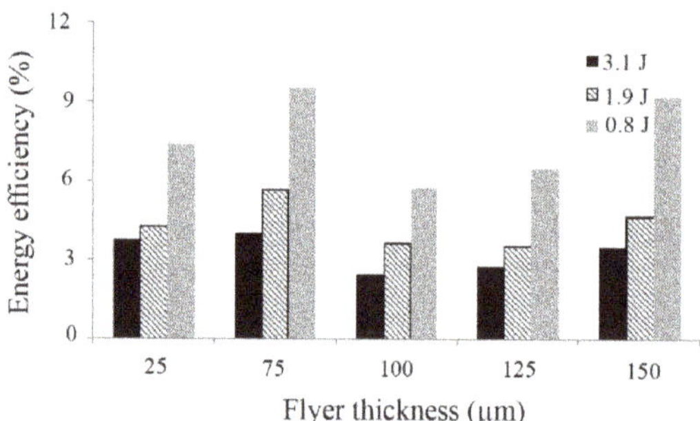

Figure 3. Energy efficiency with various laser energy and flyer thickness (D = 2 mm, L = 100 μm).

4.2. Flyer Velocity Over a Long Travel Distance

The preset standoff distance provided the acceleration distance for the flyers. This study would reveal the proper standoff distance for LIW. In this experiment, the preset standoff distance was 32.5 mm. The experiment was carried out with 114 μm thick flyer at a laser energy of 3.1 J and laser spot size of 5.5 mm. Figure 4 shows the flyer velocity and travel distance as a function of time for the first 12 μs travel time. The flyer velocity had a dramatic increase at the beginning of the flying time and then gradually increased to the peak velocity at a travel distance of 4 mm. Therefore, a standoff distance up to 4 mm should work under these experimental conditions. After the peak velocity, due to aerodynamic resistance, the flyer velocity started to reduce. Before the flyer reached the peak velocity, the pressure from plasma was higher than aerodynamic resistance. Based on the theory of aerodynamics, the aerodynamic resistance increases with the objective speed [22]. During the acceleration process, the aerodynamic resistance kept increasing while the pressure from the plasma kept decreasing due to its expansion. When the pressure from the plasma was lower than aerodynamic resistance, the flyer velocity started to decrease. Thus, the aerodynamic resistance started to decrease. From the beginning to a travel distance of 257 µm, over 80% of the peak velocity was reached. Therefore, the standoff working window was wide. During long distance flying (over 4 mm), the shape of the flyer gradually changed from a disc to an irregular shape due to its interaction with air. With the peak velocity, the energy efficiency was 27%.

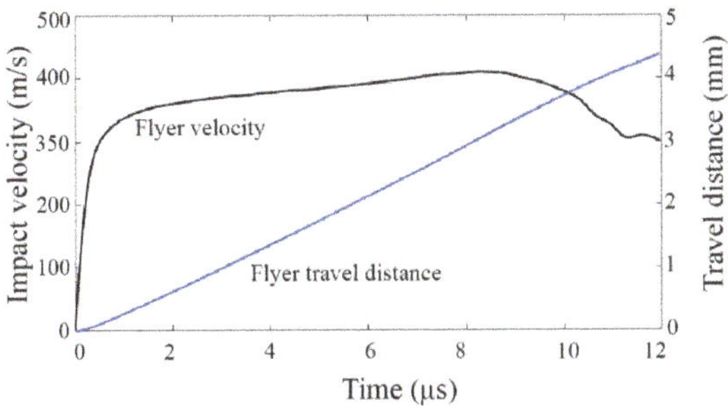

Figure 4. Flyer velocity and travel distance as a function of time (E = 3.1 J, D = 5.5 mm, t = 114 µm).

4.3. Flyer Rebound Behavior

In LIW, the current weld configuration is spot weld. The size of the nugget is the laser spot size. The cross section of the nugget shows that the metallurgical bonding occurs along the edge of the nugget. There is no metallurgical bonding in the center of the nugget. At the non-welded region within the nugget, the flyer and the target is separated from each other. On the separated flyer and target, both metallurgical bonding trace (Gaussian laser) [7] and non-metallurgical bonding (top-hat laser) [5] were observed. The metallurgical bonding trace on the separated flyer and target indicated that the metallurgical bonding was teared after its formation. The non-metallurgical bonding was due to the fact that the welding parameters were not suitable for metallurgical bonding to occur. Wang et al. studied the flyer flying behavior through a high speed camera [2]. The high speed camera video showed that the center part of the flyer travelled at the highest speed. Therefore, the center part of the flyer contacted the target first. Both the high speed camera result and metallurgical bonding confirmed the contact between the central part of the flyer and target. The rebound of the flyer must have occurred after contact, which resulted in the separation of the flyer and target. This study was carried out in order to reveal the flyer behavior before, at and after collision with the target. In this experiment, the flyer thickness was 250 µm with a laser energy of 3.1 J and laser spot size of 2 mm.

Figure 5a is the output of Photon Doppler Velocimetry, from which Figure 5b was derived through Fourier-transformation. Figure 5b shows a sudden change in the beat frequency around 1.3 s. This change should be caused by the sudden movement of the flyer, which directly affected the beat frequency. At this point, the flyer travel distance reached the standoff distance, which was 350 µm. Therefore, the continuum movement of the flyer should be after collision with the target and in the opposite direction. For the following movement, the velocity started to decrease due to the drag force from the movement of other part of the flyer in the opposite direction until it stopped. By calculation, Figure 5c,d showed the flyer velocity and travel distance before and after collision with the target. The rebound velocity reached 80% of the flyer impact velocity. The rebound travel distance was about

150 µm, which was the separation distance between the flyer and the target. This observation is valuable for future accurate measurement of flyer rebound behavior and its elimination. The rebound behavior was due to the shock wave generated within the flyer and target after their collision. Actions should be taken to absorb or reduce the reflected shock wave from the collision interface. Increasing the metallurgical bonding strength between the flyer and target may be a good way to prevent the tearing. In explosive welding, the rebound phenomenon has not been reported widely while good metallurgical bonding has been reported along the collision interface [13,23,24]. There are also shock waves on the collision interface in explosive welding. The metallurgical bonding was not teared by the reflected shock waves, which indicated that the welding strength was high enough to prevent the tearing.

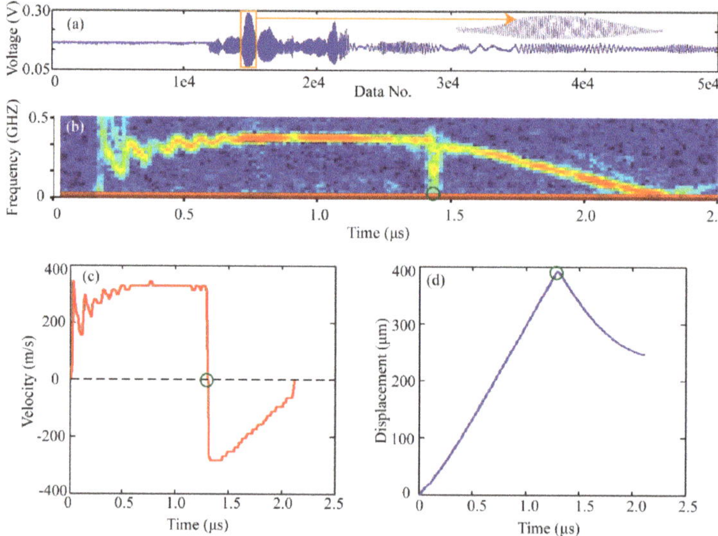

Figure 5. (**a**) Data from the detector of Photon Doppler Velocimetry, (**b**) beat frequency of the flyer, (**c**) flyer velocity, (**d**) flyer travel distance, (E = 3.1 J, D = 2 mm, t = 250 µm).

4.4. Parameter Effect on the Flyer Velocity

4.4.1. Flyer Size Effect

The flyer thickness effect on the flyer velocity has been studied by Wang et al. [3]. Their results showed that the flyer thickness had a significant effect on the flyer velocity. It decreased with flyer thickness increase. The width and length effect on the flyer velocity was investigated in this study. If the flyer size was larger than the laser spot size, when the laser incident part of the flyer started to move, the connection between the rest part of the flyer and the confinement layer applied a drag force to the flying part. This drag force was resistance to the flyer movement. In another way, the hot plasma should also be confined by the connection between the rest part of the flyer and the confinement layer. If the flyer had the same size with the laser spot, at the time the flyer started to move, the plasma escaped from the circumferential direction immediately. Thus, the pressure from the plasma should decrease. This study was conducted to study the flyer size effect on the flyer velocity due to the dual effect of the flyer size: Drag effect and confinement effect.

The experiment was conducted under the following conditions: 3.1 J laser energy, 4 mm laser spot size, 75 µm thick flyer, 5 mm thick glass (confinement layer). The dimension of the flyers were 12 mm × 12 mm, 7 mm × 7 mm and 4 mm × 4 mm. Due to fact that the laser spot size was 4 mm, for 4 mm × 4 mm flyer, there was no drag effect and confinement effect. After laser shot, the whole piece of flyer moved at the same time instantaneously. For flyers with a dimension of

12 mm $\times$ 12 mm and 7 mm $\times$ 7 mm, the dual effect from the flyer size started to play their roles. Figure 6 shows the measured flyer velocity for flyers. Experiments were repeated four times for each experimental condition. The results showed good repeatability in terms of acceleration and final velocity. The average final velocity and travel distance were displayed in Figure 7. The average velocities were 686 $\pm$ 35 m/s, 664 $\pm$ 16 m/s and 666 $\pm$ 20 m/s for 12 mm $\times$ 12 mm, 7 mm $\times$ 7 mm and 4 mm $\times$ 4 mm, respectively. The travel distances were 472 $\pm$ 32 μm, 458 $\pm$ 13 μm and 462 $\pm$ 14 μm for 12 mm $\times$ 12 mm, 7 mm $\times$ 7 mm and 4 mm $\times$ 4 mm, respectively. Therefore, there was no dramatic change of the flyer velocity when the flyer dimension increased from 4 mm $\times$ 4 mm to 12 mm $\times$ 12 mm. With the flyer size increase, both drag effect and confinement effect would increase. The 3% increase of the flyer velocity from 4 mm $\times$ 4 mm to 12 mm $\times$ 12 mm may indicated that the confinement effect started to dominate. Before the domination, there must be a flyer size with which the dual effect cancelled each other. However, if the flyer size was infinitely large, the movement of the central part of the flyer would result in plastic deformation (instead of peeling the rest of the flyer from the confinement layer) on the flyer in the region next to the moving part. This deformation would reduce the flyer velocity by consuming some flyer dynamic energy.

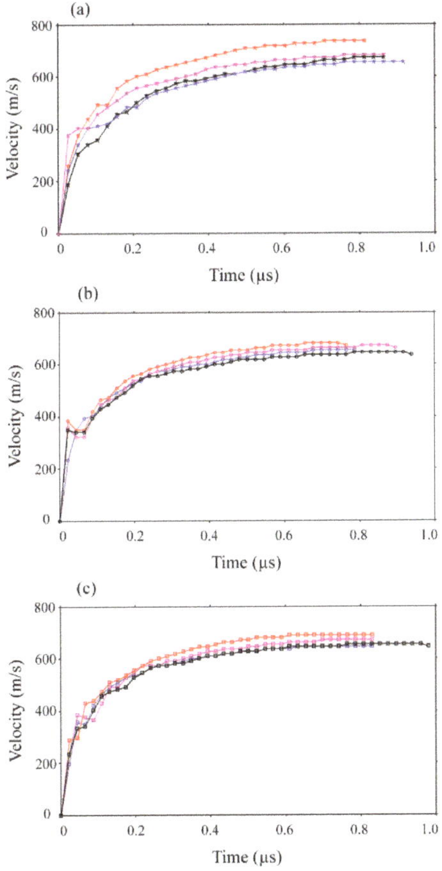

Figure 6. Flyer Velocity for flyers with dimensions of (**a**) 12 mm $\times$ 12 mm, (**b**) 7 mm $\times$ 7 mm, (**c**) 4 mm $\times$ 4 mm (E = 3.1 J, D = 4 mm, t = 75 μm, L = 525 μm).

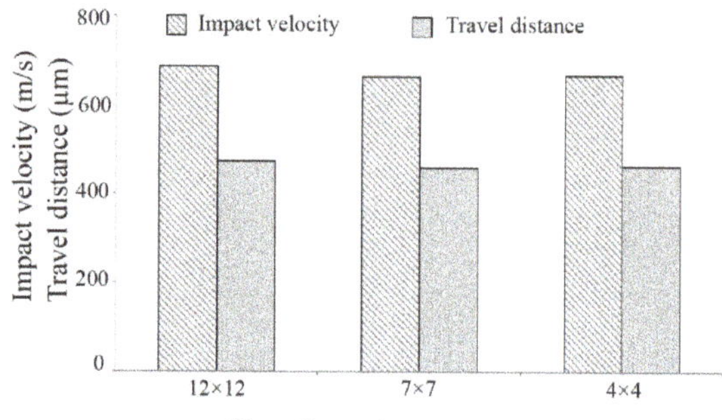

Figure 7. Average flyer velocity and travel distance for flyers with dimensions of 12 mm × 12 mm, 7 mm × 7 mm and 4 mm × 4 mm (E = 3.1 J, D = 4 mm, t = 75 µm, L = 525 µm).

4.4.2. Confinement Layer Effect

Confinement layer effect on laser energy efficiency was studied by Wang et al. [2] in means of measuring the dimple height, which was the result of flyer deformation after laser shot. They studied 1mm thick glass, 0.5 mm thick polycarbonate and 0.07 mm thick tape. Their results showed that glass was the one that provided the highest energy efficiency. In this study, the confinement layer effect on flyer velocity was investigated. The studied confinement layers in this work included 0.5 mm thick polycarbonate (0.5mmP), 1 mm thick polycarbonate (1mmP), 1 mm thick glass (1mmG) and 5 mm thick glass (5mmG). The experiments were carried out with a 25 µm thick flyer at a laser energy of 3.1 J and laser spot size of 4 mm with a preset standoff distance of 325 µm. The flyer velocity as a function of time was shown in Figure 8a. The average flyer velocity was shown in Figure 8b. The flyer travel distance as a function of time was shown in Figure 8c. The flyer velocity increased with time until the preset standoff distance was reached. One of the two experiments (under the same experimental conditions) with polycarbonate as the confinement layer travelled a much longer time/distance than the other experiment. The average flyer velocities were 727 m/s, 710 m/s, 759 m/s and 759 m/s for confinement layers of 0.5mmP, 1mmP, 1mmG and 5mmG, respectively. The average travel distances were 441 µm, 415 µm, 362 µm and 354 µm for confinement layers of 0.5mmP, 1mmP, 1mmG and 5mmG, respectively. Therefore, the confinement layers had little effect on flyer velocity and a significant effect on travel distance.

As shown in Figure 8a, the acceleration process was very short. Within 0.2 µs, over 80% of the final velocity had been reached. Within 0.2 µs, the flyer travelled around 100 µm. Therefore, even though flyers with a confinement layer of 0.5mmP_62 and 1mmP_59 travelled much longer than 0.5mmP_61 and 1mmP_58, the final flyer velocities were not affected significantly. As shown in Figure 8c, the travel distance with polycarbonate had a large variation between experiments while travel distance with glass had much better repeatability. Therefore, stiffness of confinement layer played a significant role in controlling the repeatability of experiments. As Figure 1 shows, the confinement layer was shared by flyers. For a low stiffness material, laser shot may cause its deformation. This deformation would affect the preset standoff distance for the nearby flyers. Therefore, if the travel distance had a strict requirement, the following procedures should be followed. First of all, the confinement layer would be replaced by a more rigid plate, like glass. Super glue would be used to connect the standoff and target instead of sticky tape to prevent the nearby polycarbonate deformation effect on the preset standoff distance.

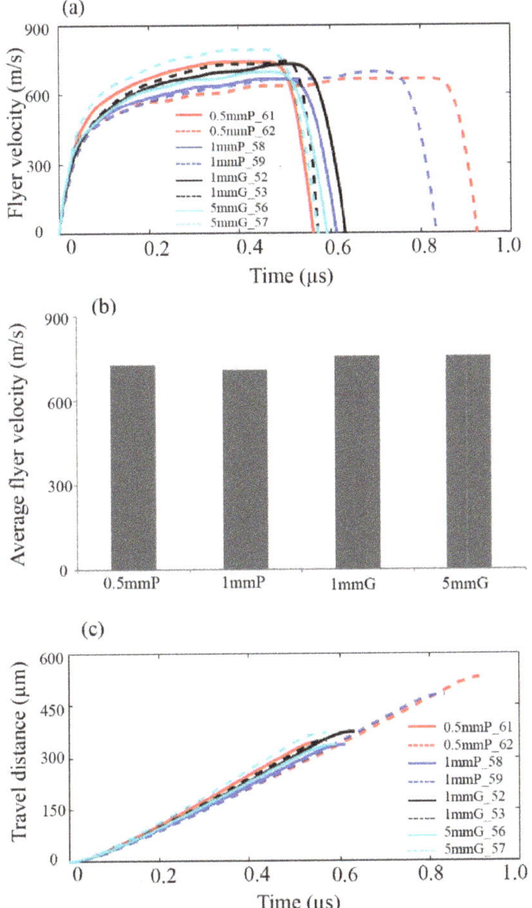

Figure 8. (**a**) Flyer velocity as a function of time, (**b**) average flyer velocity with various confinement layers, (**c**) flyer travel distance as a function of time (E = 3.1 J, D = 4 mm, t = 25 μm, 0.5mmP_61: 0.5 mm thick polycarbonate as the confinement layer for experiment number 61, 1mmG_52: 1 mm thick glass as the confinement layer for experiment number 52).

5. Conclusions

In this study, the flyer velocity was measured with Photon Doppler Velocimetry under various experimental conditions. The flyer characteristics were revealed through the measurement, which was analyzed and summarized below.

- The energy efficiency was calculated with the measured flyer velocity. Above a laser energy density of 2.55 × 10^5 J/m^2, the energy efficiency decreased with laser energy increase. Flyer thickness had no apparent effect on laser energy efficiency.
- In LIW, the flyer velocity had a dramatic increase at the beginning of the flying, then gradually increased to the peak velocity with a travel distance of 4 mm. From the beginning to a travel distance of 257 μm, over 80% of the peak velocity was reached. Therefore, the standoff working window is wide for LIW.

- The flyer behavior of before, at and after collision with the target was captured with PDV measurement. The results confirmed the rebound behavior of the flyer, which resulted in the tearing of the metallurgical bonding in the center of the weld nugget in LIW.
- When the flyer size was within three times larger than the laser spot size, flyer width and length had little effect on the flyer velocity due to the dual effect of the drag force and confinement. The confinement layer had little effect on the final flyer velocity. With rigid confinement layer, the experiment has better repeatability on the flyer travel distance.

Author Contributions: Conceptualization, H.W. and Y.W.; methodology, H.W.; formal analysis, H.W. and Y.W.; investigation, H.W.; writing—original draft preparation, H.W.; writing—review and editing, Y.W.

Funding: This work was sponsored by the Fundamental Research Funds for the Central Universities (Grant No. 06500107).

Acknowledgments: Thanks to Professor Glenn Daehn for providing laboratory access for the experimental work at The Ohio State University.

Conflicts of Interest: The authors declare no conflict of interest.

References

1. Daehn, G.S.; Lippold, J.C. Low Temperature Spot Impact Welding Driven without Contact. U.S. Patent WO/2009/111774 A2, 9 November 2009.
2. Wang, H.; Taber, G.; Liu, D.; Hansen, S.; Chowdhury, E.; Terry, S.; Lippold, J.C.; Daehn, G.S. Laser impact welding: Design of apparatus and parametric optimization. *J. Manuf. Process.* **2015**, *19*, 118–124. [CrossRef]
3. Wang, H.; Vivek, A.; Wang, Y.; Taber, G.; Daehn, G.S. Laser impact welding application in joining aluminum to titanium. *J. Laser Appl.* **2016**, *28*, 032002. [CrossRef]
4. Wang, X.; Gu, C.X.; Zheng, Y.Y.; Shen, Z.B.; Liu, H.X. Laser shock welding of aluminum/aluminum and aluminum/copper plates. *Mater. Des.* **2014**, *56*, 26–30. [CrossRef]
5. Wang, H.; Wang, Y. Laser-driven flyer application in thin film dissimilar materials welding and spalling. *Opt. Lasers Eng.* **2017**, *97*, 1–8. [CrossRef]
6. Zhang, Y.; Babu, S.S.; Prothe, C.; Blakely, M.; Kwasegroch, J.; Laha, M.; Daehn, G.S. Application of high velocity impact welding at varied different length scales. *J. Mater. Process. Technol.* **2011**, *211*, 944–952. [CrossRef]
7. Liu, H.; Gao, S.; Yan, Z.; Li, L.; Li, C.; Sun, X.; Sha, C. Investigation on a novel laser impact spot welding. *Metals* **2016**, *6*, 179. [CrossRef]
8. Wang, H. *Laser Impact Welding and High Strain Rate Embossing*; The Ohio State University: Columbus, OH, USA, 2013.
9. Fujiwara, H.; Brown, K.E.; Dlott, D.D. A thin-film Hugoniot measurement using a laser-driven flyer plate. *AIP Conf. Proc.* **2012**, *1426*, 382–385.
10. Brown, K.E.; Shaw, W.L.; Zheng, X.; Dlott, D.D. Simplified laser-driven flyer plates for shock compression science. *Rev. Sci. Instrum.* **2012**, *83*, 103901. [CrossRef] [PubMed]
11. Paisley, D.L.; Luo, S.N.; Greenfield, S.R.; Koskelo, A.C. Laser-launched flyer plate and confined laser ablation for shock wave loading: Validation and applications. *Rev. Sci. Instrum.* **2008**, *79*, 023902. [CrossRef] [PubMed]
12. Marsh, S.P. *LASL Shock Hugoniot Data*; University of California Press: Berkeley, CA, USA, 1980.
13. Mousavi, S.; Al-Hassani, S.T.S.; Atkins, A.G. Bond strength of explosively welded specimens. *Mater. Des.* **2008**, *29*, 1334–1352. [CrossRef]
14. Jarmakani, H.; Maddox, B.; Wei, C.T.; Kalantar, D.; Meyers, M.A. Laser shock-induced spalling and fragmentation in vanadium. *Acta Mater.* **2010**, *58*, 4604–4628. [CrossRef]
15. Dung, C.V.; Sasaki, E. Numerical simulation of output response of PVDF sensor attached on a cantilever beam subjected to impact loading. *Sensors* **2016**, *16*, 601. [CrossRef] [PubMed]
16. Stahl, D.B.; Gehr, R.J.; Harper, R.W.; Rupp, T.D.; Sheffield, S.A.; Robbins, D.L. Flyer velocity characteristics of the laser-driven MiniFlyer system. *AIP Conf. Proc.* **2000**, *505*, 1087–1090.
17. Watson, S.; Field, J.E. Measurement of the ablated thickness of films in the launch of laser-driven flyer plates. *J. Phys. D Appl. Phys.* **2000**, *33*, 170–174. [CrossRef]

18. Zhao, X.H.; Zhao, X.; Shan, G.C.; Gao, Y. Fiber-coupled laser-driven flyer plates system. *Rev. Sci. Instrum.* **2011**, *82*, 043904. [CrossRef] [PubMed]

19. Shaw-Stewart, J.; Lippert, T.; Nagel, M.; Nuesch, F.; Wokaun, A. A simple model for flyer velocity from laser-induced forward transfer with a dynamic release layer. *Appl. Surf. Sci.* **2012**, *258*, 9309–9313. [CrossRef]

20. Strand, O.T.; Goosman, D.R.; Martinez, C.; Whitworth, T.L.; Kuhlow, W.W. Compact system for high-speed velocimetry using heterodyne techniques. *Rev. Sci. Instrum.* **2006**, *77*, 083108. [CrossRef]

21. Johnson, J.R.; Taber, G.; Vivek, A.; Zhang, Y.; Golowin, S.; Banik, K.; Fenton, G.K.; Daehn, G.S. Coupling experiment and simulation in electromagnetic forming using Photon Doppler Velocimetry. *Steel Res. Int.* **2009**, *80*, 359–365.

22. Carafoli, E.; Street, R.E. High-Speed Aerodynamics. *Phys. Today* **1958**, *11*, 48–50. [CrossRef]

23. Akbari-Mousavi, S.A.A.; Barrett, L.M.; Al-Hassani, S.T.S. Explosive welding of metal plates. *J. Mater. Process. Technol.* **2008**, *202*, 224–239. [CrossRef]

24. Chizari, M.; Al-Hassani, S.T.S.; Barrett, L.M. Effect of flyer shape on the bonding criteria in impact welding of plates. *J. Mater. Process. Technol.* **2009**, *209*, 445–454. [CrossRef]

Article

Influence of Surface Preparation on the Interface of Al-Cu Joints Produced by Magnetic Pulse Welding

Omid Emadinia [1,2,*], Alexandra Martins Ramalho [1], Inês Vieira de Oliveira [1], Geoffrey A. Taber [3] and Ana Reis [1,2]

1   Faculty of Engineering, University of Porto (FEUP), 4200-465 Porto, Portugal;
    alexandramartinsramalho@gmail.com (A.M.R.); prodem0606954@fe.up.pt (I.V.d.O.); areis@inegi.up.pt (A.R.)
2   Institute for Mechanical Engineering and Industrial Management (LAETA/INEGI), 4200-465 Porto, Portugal
3   Department of Materials Science and Engineering, The Ohio State University, Columbos, OH 43221, USA;
    taber.17@osu.edu
*   Correspondence: omid@fe.up.pt or omid.emadinia@fe.up.pt; Tel.:+351-22-041-31-11

Received: 29 June 2020; Accepted: 21 July 2020; Published: 24 July 2020

Abstract: Magnetic pulse welding can be considered as an advanced joining technique because it does not require any shielding atmosphere and input heat similar to conventional welding techniques. However, it requires comprehensive evaluations for bonding dissimilar materials. In addition to processing parameters, the surface preparation of the components, such as target material, needs to be evaluated. Different surface conditions were tested (machined, sand-blasted, polished, lubricated, chemically attacked, and threaded) using a fixed gap and standoff distance for welding. Microstructural observations and tensile testing revealed that the weld quality is dependent on surface preparation. The formation of waviness microstructure and intermetallic compounds were verified at the interface of some joints. However, these conditions did not guarantee the strength.

Keywords: magnetic pulse welding; dissimilar metals; surface preparation; interface; microstructure

---

1. Introduction

The use of multi-material structures for complex lightweight applications is noticeable; however, this requires the implementation of joining technologies for dissimilar materials. These joints are produced by different techniques such as mechanical fastening, fusion joining, and adhesive bonding, etc., each technique can provide different levels of strength. The joint strength produced by a fusion welding process can surpass those achieved by the application of adhesives or fasteners; however, a dissimilar joint interface might be very susceptible to an early failure caused by the formation and propagation of cracks. The welding of dissimilar materials, e.g., Aluminium (Al) to copper (Cu) is a challenge because these materials have different melting points, thermal conductivities, volumetric specific heats, and thermal expansion coefficients [1,2]. Cracking can occur at the welded interface during solidification, and can also be affected by stresses induced by the difference of coefficients of thermal expansion of the base materials, or due to the presence of brittle phases such as intermetallic phases, etc. Considering the increase of performing dissimilar joints for sufficient ductile materials, the application of joining by plastic deformation is needed, and this manufacturing approach does not require an external heat supply as, instead, pressure is applied on the workpiece [3]. Plastic deformation joining can be categorized as metallurgical and mechanical processes, the former type involves a temperature increase caused by a severe plastic deformation effect, e.g., during the friction welding process, the mechanical type can be achieved without a thermal effect such as riveting. However, there are other techniques such as explosive welding and magnetic pulse welding (MPW), and these processes also involve a high velocity plastic deformation and collision effects [4]. The pressure is provided by an explosive material for the former technique. Regarding MPW, a repulsive force is generated when two opposite magnetic fields meet each other, this supplies the pressure required for joining [3,5], i.e., the generated pressure accelerates the outer workpiece (known as flyer) towards the inter one (called target), this causes a collision which is associated with plastic deformation. The MPW is considered as a solid-state joining process. However, some studies reported the formation of a liquid phase [6] which is influenced by the processing design, defects such as voids, cracks, and melted zones may occur if the parameters exceed the processing window [7].

In the MPW process, two opposite currents, one passes via a coil which induces an eddy current in the flyer, this generates a magnetic field between the coil and flyer, this interaction causes Lorentz Forces by which a repelling pressure expels the flyer to the target. This pressure should provide enough energy for the occurrence of plastic deformation at the flyer material and also enough kinetics for bonding. This process is fulfilled within some seconds [8]. What regards the processing conditions, the velocity, and the angle of impact are vital parameters, the former is a characteristic of the equipment (charged electrical energy, frequency of discharge current, and inductance of the coil) and of the materials' properties such as strength and conductivity. Regarding the impact angle, this is affected by the geometries of the materials being welded. Moreover, material type, dimensions of the materials such as flyer thickness, the gap between the materials (standoff gap), electric conductivity, strength, elongation, and surface conditions are also effective parameters for MPW [5,9].

Impact welding processes such as MPW are associated with a jet phenomenon that consists of thin layers of the oxide, of the flyer, and of the target metals and also contaminants from the colliding surfaces. This jet is controlled by the collision angle and must fly away otherwise it is trapped at the collision interfaces. Thus, a clean surface is provided, and virgin metals are contacted under extreme pressure resulting in the formation of atomic bonds of flyer and target [4].

What regards the morphology of the joint interface produced by MPW, a wavy microstructure is expected. However, this morphology depends on the collision energy, impact angle, and the geometry of the joint [10]. MPW also exhibits a negligible heat affected zone at the interface [7,11]. Moreover, micro or sub micro scale local melting and solidification can also be considered as the bonding mechanism for MPW, however, this depends on the level of the input energy [11]. In addition to the processing parameters, geometrical designs, and materials characteristics, some researches were conducted to evaluate the influence of the surface preparation on the bonding quality for the MPW process [12–15] as summarized in Table 1. However, there are two studies that clearly presented the influence of pre-treatments on the joint quality [12,14], and the other two works mentioned that the processing conditions such as impact angle strongly determine the weld quality [13,15].

Table 1. A summary of related studies on the influence of surface preparation for MPW process.

Materials	Preparation Conditions	Results	Ref.
AA6063-O Al (flyer) to C110 Cu (target) tubes	A: tangential scratches over Cu length made by lathe B: axial scratches along the Cu length by 200-grit C: A + silicon-based high-viscosity lubricant oil	A: was in favour C: failed	[12]
EN AW-1050 Al (flyer) to S235 JR steel (target) sheets	Belt grinding Laser ablation	Not clearly explained	[13]
Al5182 (flyer) to HC340LA steel	Grinding steel parallel to welding (PW) Grinding steel vertical to welding (VW)	VW caused a wavy-shape interface and better mechanical properties PW caused straight interface with less elemental diffusion	[14]
EN AW-6016-T6 Al (flyer) to DC04 steel (target) sheets	Untreated surface Polished surface Laser ablation	Surface pre-treatment was not essential Processing window was the most important factor	[15]

Thus, the main objective of this study includes further evaluations on the influence of surface roughness on the welding quality such as microstructure (aiming at evaluating the formation of intermetallic phases and waviness structure) and tensile strength. Two metallic systems, namely Al and Cu, were selected, this bimetallic system provides a lighter assembly than the Cu/Cu. Since the cross-section area of a conductor supports the current passage, the use of a lighter material can lead to enlarge the surface area for a higher current. However, the formation of intermetallic compounds strongly influences the electrical and mechanical properties of the Al/Cu joint and also resulting in a lack of long-term electrical stability [16]. Therefore, further investigations on advanced techniques such as MPW are still needed in order to evaluate the influence of surface preparation on the weld.

2. Materials and Methods

The base materials for this study included AA 6063-T5 commercial Al alloy (98.9% Al, 0.7 Mg and 0.4 Si (in wt.%) declared by AZO MATERIALS) and copper R300 (99.99% Cu, 0.01% O_2 (in at.%) as flyer and target, respectively. The physical and mechanical properties of the aluminium alloy and of the copper are presented in Table 2. The Al was a tube of 100 mm length, 1 mm wall thickness, and an outer diameter of 20 mm. The copper was a rod with a diameter adapted from a related study [17] based on the gap distance that produced the best results (ϕ = 16 mm). Several copper rods were surface treated to obtain different surface roughness, these treatments included machining (appeared as M in this study), sandblasting (G), polishing (P), chemical attacking (Q), lubricating (O), and threading (R). The application of a greasy surface aimed at studying the MPW process when it is applied for a very contaminated welding surface, this can clarify if surface cleaning before welding is relevant to a solid joint and if the occurrence of jet phenomenon can effectively remove the contamination layer and creates a weld.

Table 2. Physical and mechanical properties of the Al flyer and of the Cu target used in this study.

Material	Density (kg/m³)	Youg's Modulus (GPa)	Yield Strength (MPa)	Shear Modulus (GPa)	Tensile Strength (MPa)	Hardness (HV)	Fracture Toughness (MPa m)	Melting Point (°C)	Electrical Resistivity (μΩcm)
Cu-R300	8940–8950	127	250	45–50	290–360	90–110	43.2–57.6	1083	1.70–1.74
AA 6063-T5	2660–2710	67.2–70.7	113–125	25.3–26.6	158–175	61.8–68.3	30–36	615–655	3.08–3.21

What regards the surface properties of the base materials and the nomenclature of the Al/Cu couples used in this study, the Al flyer was an extruded tube expecting a surface roughness of almost 4 μm, in average. Regarding the target, copper rods were machined by a lathe and coupled with Al tubes named Al/Cu-M; the machined rods were sandblasted after and the coupled named Al/Cu-G, the goal was to replace the machined surface by a new topography for study; other machined Cu rods were polished by emery papers up to 1200-grit sandpaper to remove the lathe traces, these rods were coupled and called Al/Cu-P. Some machined surfaces were chemically attacked in a solution of 50 mL HNO_3, 10 mL H_2N_2, and 50 mL H_2O, for 90 s, these attached rods were coupled with Al tube and called Al/Cu-Q; the greasy samples were prepared by using a high-viscosity lubricant oil on the machined Cu surfaces and called Al/Cu-O; and finally, the threaded rods were created by a 1 mm pitch on the surfaces of machined rods coupled with Al tube and named Al/Cu-R. The roughness of these copper workpieces was measured using an optical non-contact profiler. For each condition, a total of three measurements were done and the average results are presented in Table 3, Ra is the mean roughness value and Rq is the mean square roughness. Thus, the highest value of roughness belongs to the shot blasted specimens with an average roughness value of 4.8 μm, and the polished specimens comprise the least roughness, nearby 0.6 μm.

The copper and aluminium parts were subjected to an ultrasonic cleaning device with acetone for five minutes (before and after any surface treating) to properly clean the surface and assure that there was not any kind of contaminant which could influence the joining process. However, the greasy sample was excluded from this cleaning procedure.

Table 3. Roughness measurements of the copper R300 rods.

Sample Indication	M	G	P	Q
Ra (μm)	1.5	4.8	0.6	2.1
Rq (μm)	1.9	6.1	0.7	2.5

Welding experiments were carried out using a 25/25 magnetic pulse system, it consisted of a capacitor bank and a high voltage cabinet for charging the capacitors, capable of generating 25 kJ at a charging voltage of 25 kV. The bank of capacitors possessed a total capacitance of 80 μF and a total inductance of 0.1 μH, the internal resistivity was 19 mK providing a current up to 400 kA. The electromagnetic actuator was a single turn coil made of a 40CrMnNiMo 7 steel with an outer diameter of 20 mm which is suitable for welding cylindrical parts. Figure 1 shows the setup of the MP tool which includes a single turn coil (part A in Figure 1a) installed on the station (B in Figure 1a) of the MPW equipment and surrounded by a shielding frame. A schematic of the weld assembly is also illustrated in Figure 1b.

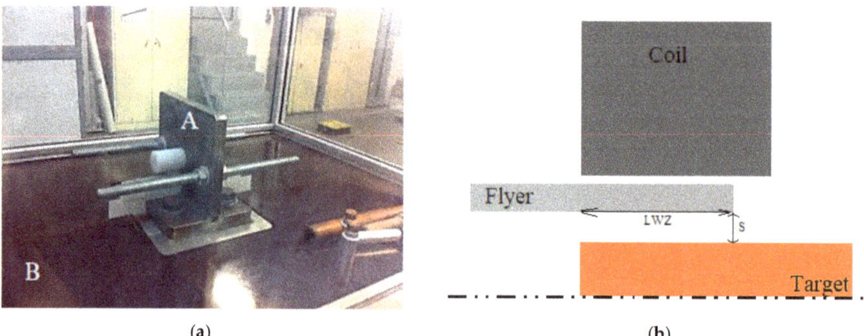

(a)   (b)

Figure 1. The setup applied for the performance of MPW in this study: (**a**) the coil assembly (A) on the station of the equipment, (**b**) the arrangement of coil-flyer-target (that were kept constant in this study).

Figure 2 illustrates the welding window from which the welding parameters were selected for joining the aluminium tube to copper rod, as presented in Table 4. A welding window can give us an indication of the welding range, this window was obtained from the results of the compression tests of similar aluminium tubes and copper rods in a previous study [18] taking into account the effect of three main parameters such as the charging voltage, the air gap width (S) and overlap distance (LWZ), this was possible to monitor these parameters during the MPW process. The experimental conditions that led to successful welds constitute the two planes shown in this window (Figure 2). However, the combination of different gaps and overlap distances will result in successful weld if the selected energy provides an appropriate impact velocity. Moreover, it should be noted that the welding window shape can change by the application of different input parameters, thus each experiment can have its own welding window depending on the values of energy, overlap distance (LWZ), and gap used for weld performance.

Regarding microstructure observations, the samples were cross sectioned (on the middle of the weld seem and in front of the slot of the coil) and polished through standard metallurgical procedures, the polished samples were not etched, and then were observed by scanning electron microscopy (SEM), a FEI QUANTA 400 FEG equipment (Hillsboro, OR, USA), using the electron backscatter imaging (EBI) mode. Moreover, the chemical composition distribution across the interfaces was analysed with an electron dispersive spectroscopy (EDS) probe, Oxford Instrument (Oxfordshire, UK), then these results were compared with the equilibrium Cu-Al binary phase diagram [2].

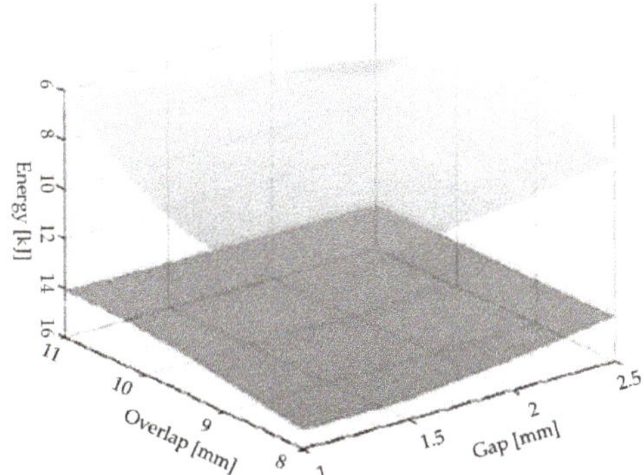

Figure 2. The processing window applied for welding in this study (adapted from [18]).

Table 4. Welding parameters applied in this study that were kept constant [17].

Energy [kJ]	Voltage [kV]	LWZ [mm]	Air Gap [mm]
10.24	16	8	1

Regarding the mechanical properties of the welded experiments, a uniaxial tensile test was conducted by a machine illustrated in Figure 3a (Instron model 4507) using a strain rate of 2 mm/min. The parameters of the tensile tests were designed in accordance with ASTM A 370 as presented in Table 5. Moreover, it was necessary to prepare new grips for the stabilization of the cylindrical samples in order to perform the tensile test as illustrated in Figure 3. Since the tensile specimens have tubular sections, snug-fitting metal plugs were inserted far enough at the end of the tubes. This assembly was designed in accordance with the plug design mentioned in ASTM A 370 and applied to allow the jaws gripping the specimens properly. This assembly guaranteed sample holding without the tube walls being crushed. Moreover, the plugs did not extend into that part of the specimen where the elongation was measured.

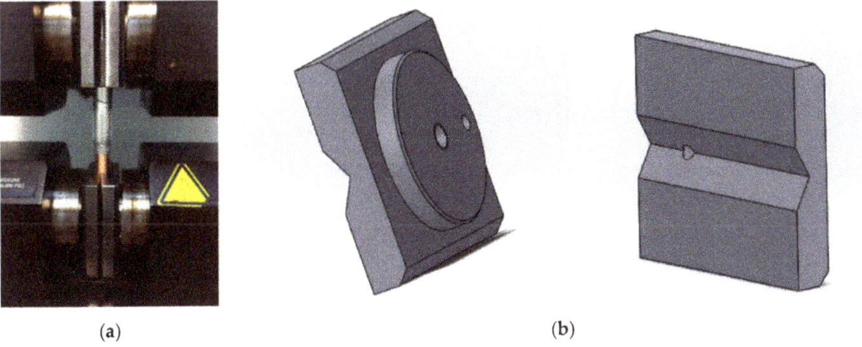

(**a**)　　　　　　　　　　　　　　　　　　　(**b**)

Figure 3. (**a**) The placement of the welded specimen between the grips of the tensile test, (**b**) the grips used for holding samples (a SOLIDWorks® design).

Table 5. Tensile tests parameters according to ASTM A 370.

Cell	Type of Grip	Test Speed
200 kN	Hydraulic	2 mm/min

3. Results and Discussions

3.1. Microstructure of Interfaces

What regards the formation of waviness structure at the joint interfaces as a general aspect of joints produced by MPW technique, the morphology of the interfaces, which are seen on the polished cross sections (Figure 4), revealed the formation of a weak waviness pattern for the joints produced by the use of machined specimens (Figure 4a), sandblasted (Figure 4b), polished (Figure 4c), and chemically attached (Figure 4d) specimens. The two other specimens, lubricated and threaded, do not show this effect. Moreover, it is seen that the waviness pattern observed in the chemically treated specimen is the strongest one. This depicts that neither a very smooth surface (produced by polishing) nor a very rough surface (produced by sandblasting) was effective, therefore the roughness should be a moderated value. Regarding the morphology of the joint interfaces obtained in the lubricated and threaded specimens (Figure 4e,f), the wavy pattern is not observed in any of these conditions. This means that the jet formation during the MPW process was influenced by the presence of oil and threads. Regarding this effect on the lubricated surface, it is consistent with another research [12]. The energy has not been sufficient to remove the contamination and to produce a wavy pattern. This shortcoming might be attributed to the welding windows applied in this study (Figure 2). Moreover, the jet has also not been robust to form any wavy pattern on the threaded surface, and this means that the jet formation was blocked by the topography of the threaded surface.

Regarding the formation of metallurgical bonds at the joint interfaces, as a guarantee for bonding, Figure 4 provides some information about the microstructure of the joint interfaces. The MPW is considered as a solid-state welding process, but the contrasts illustrated in these images confirm the presence of phases which have different atomic contrast from the Al and Cu base materials (indicated by white arrows in the images illustrated in Figure 4a–d). Moreover, a wavy pattern has formed on interfaces of these joints. Table 6 presents the average thickness values of these layers which are seen in these microstructures (Al/Cu-M, G, P, and Q joints as illustrated in Figure 4). This means that diffusion or melting has occurred on the surfaces of these flyers and targets during welding. However, the bonding time is so short in MPW that the hypothesis of diffusion is not achievable, therefore melting is confirmed. The supplied energy in conjunction with these four surfaces (M, G, P, and Q) generated jet and locally heated surfaces sufficiently to melt Al and/or Cu base materials in tiny zones by which they reacted and then solidified [6,7]. Moreover, these interfaces (Figure 4a–d) also consist of dark intermediate layers between the reacted zones which seem that the Al/Cu-G, and Al/Cu-Q joints contain the smallest layers. This intermediate layer is composed of oxide particles that mean that the jet formation has not been strong enough to reject these compounds out of the mating surface; this also indicates that the welding windows have not been adequate for the geometries used in this study. Moreover, it appears that the latter interface comprises porous reacted zones. These microscopic observations also revealed that the reaction zones formed at the copper base material appear almost continuous at the copper side and the thinnest layer has formed in the sandblasted specimen. However, some pocket-like zones have formed on the aluminium base.

Table 6. The average thickness (in μm) of intermetallic layers formed at the interface of the joints obtained in this study.

Joint	Aluminium Side	Copper Side
Al/Cu-M	0 to 12 ± 8	23 ± 22
Al/Cu-G	17 ± 15	27 ± 15
Al/Cu-P	0 to 13 ± 7	11 ± 7
Al/Cu-Q	0 to 30 ± 13	18 ± 15

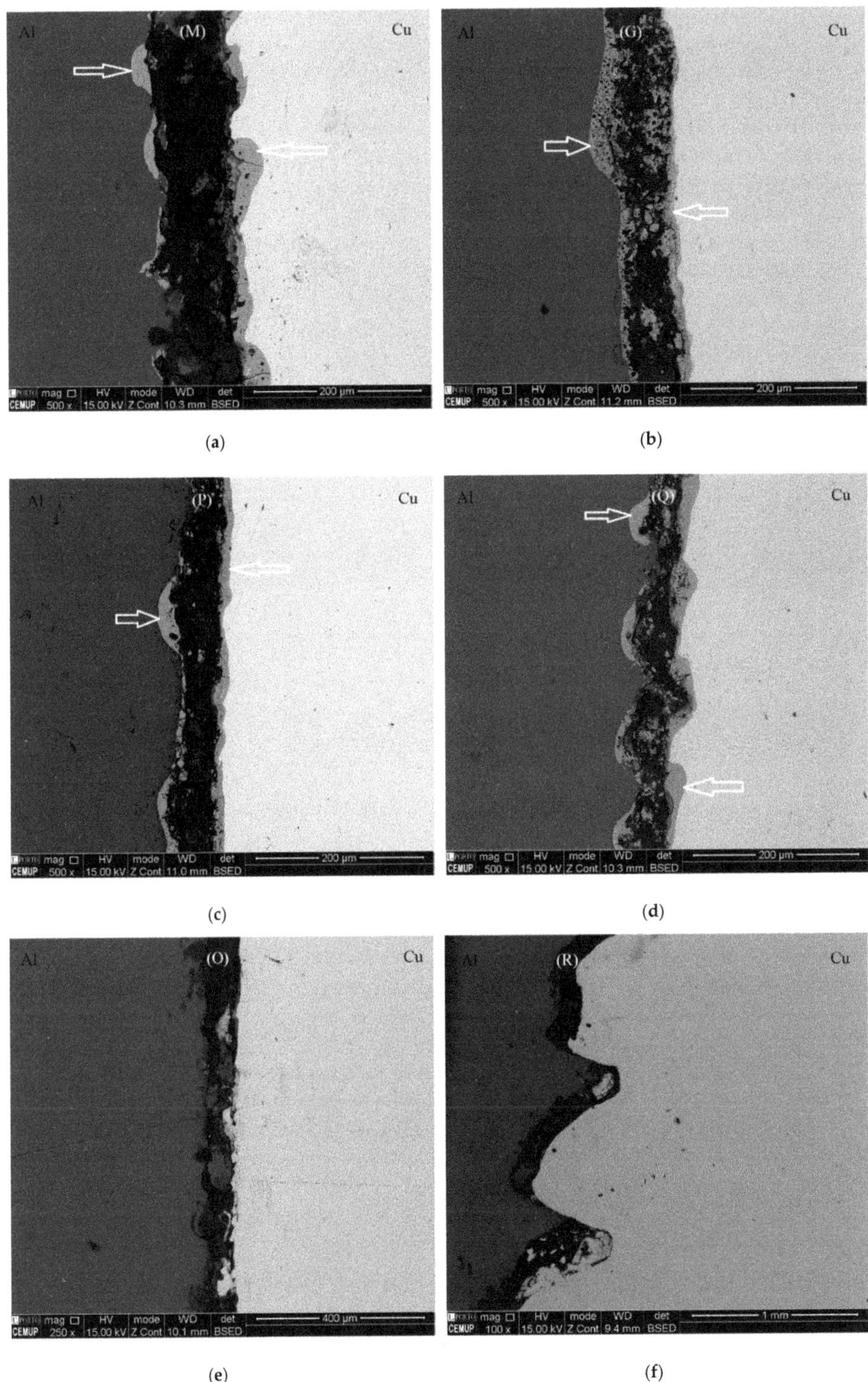

Figure 4. SEM/BSE images from the (**a**) Al/Cu-M, (**b**) Al/Cu-G, (**c**) Al/Cu-P, (**d**) Al/Cu-Q, (**e**) Al/Cu-O, and (**f**) Al/Cu-R joints produced by MPW technique.

The formation of intermetallic phases at the lubricated or threaded bases was not observed (Figure 4e,f). It seems the presence of oil and threaded topography interfered in the impact angle, which determines the impact energy/velocity, and consequently led to the dissipation of kinetic energy at the mating surfaces.

In what concerns the type of reacted zones formed at the interface of these joints, the EDS analysis was performed on the zones of interests as illustrated in Figure 5, the semi-quantitative results are presented in Table 7. Since the major elements are copper and aluminium, these results were correlated with the Al-Cu phase diagram [2] though these values were obtained in non-equilibrium conditions. Table 5 depicts that, independent on the preparation techniques, these reacted zones were generally $CuAl_2$ intermetallic, also known as θ phase. However, this evaluation has been influenced by the volume interaction, which is a characteristic phenomenon of the SEM/EDS analysis technique. This requires further assessments (as future works), such as transmission electron microscopy and electron backscatter diffraction analyses.

Table 7. The SEM/EDS results obtained from the zones of interest indicated in the images in Figure 5.

Zone	Al (at.%)	Cu (at.%)	Others (at.%)	Probable Coumpound
Z1	63	37	-	θ
Z2	72	28	-	θ
Z3	62	38	-	θ
Z4	70	30	-	θ
Z5	66	34	-	θ
Z6	65	35	-	θ
Z7	67	33	-	θ
Z8	2.8	1.0	19.3 O_2, 76.4 C, 0.5 S	-
Z9	1.5	26.3	5.6 O_2, 66.6 C	-
Z10	93		7 O_2	-
Z11	68	-	25 C, 7 O_2	-
Z12	-	88	10 C, 2 O_2	-

A related study that used similar processing windows [17], the formation of other intermetallic compounds (which contain a higher quantity of copper) were reported, therefore the interaction of copper and aluminium atoms has changed. The sample length used for this research was longer than that of the related study which likely influenced the welding window and consequently the heat and atomic interactions. Thus, the re-optimization of the processing parameters for this length of the sample is suggested for future works. Moreover, some microcracks are seen in the intermetallic compounds, perpendicular to the joint interface and not parallel with it, formed in the Al/Cu-M and Al/Cu-P systems on the copper side (yellow arrows illustrated in Figure 5a,c). The presence of residual stresses, due to the solidification shrinkage and/or hardness differences between the intermetallic and the surrounding region, can contribute to the formation of this defect.

What also regards the influence of localized melting, which is associated with the heated generated above the melting point of either of Al or Cu and following by a very quick solidification process, this can contribute to the formation of microvoids observed in the intermetallic zones (Figure 5) which is very concentrated for the joint sandblasted surface.

In addition, the microstructure of these interfaces can be influenced by the superficial energy of these mating surfaces which were prepared by different techniques.

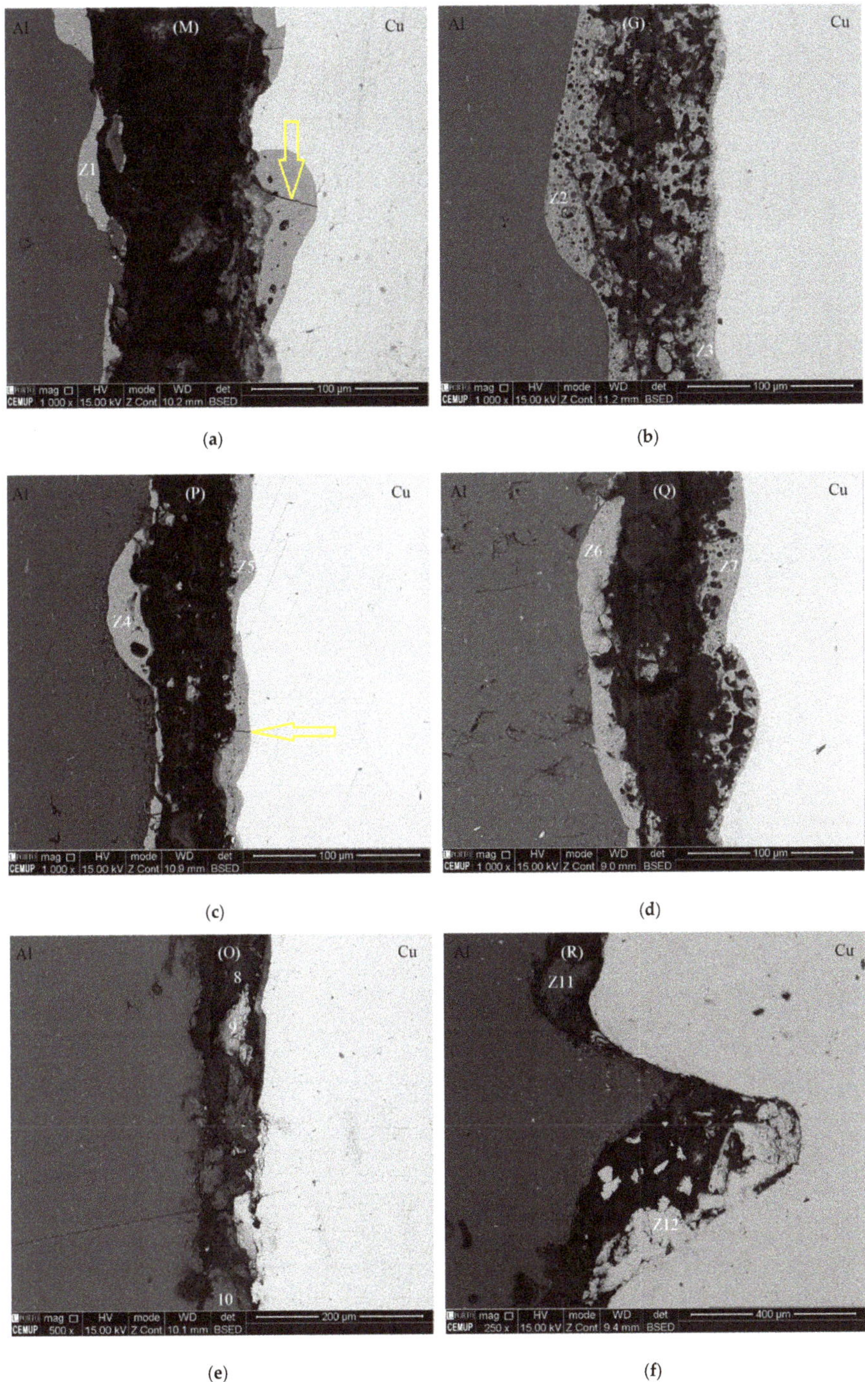

Figure 5. SEM/BSE images used for the chemical analysis of the interested zones at the interfaces of the (**a**) Al/Cu-M, (**b**) Al/Cu-G, (**c**) Al/Cu-P, (**d**) Al/Cu-Q, (**e**) Al/Cu-O, and (**f**) Al/Cu-R joints produced by MPW technique.

3.2. Tensile Test

Table 8 presents the load values registered at the failure points of the welds, performed in this study, obtained from uniaxial tensile tests. All joints failed in the welded zones except for the specimen prepared by threading (Al/Cu-R), in this specimen failure occurred at the Al tube base as illustrated in Figure 6. It means that the resistance of the Al/Cu-R joint was superior to the base material though there was not any metallurgical bonding. The occurrence of a failure in the base material is an indication of a sound joint.

Table 8. The result of the tensile test performed for the joints.

Joint	Al/Cu-M	Al/Cu-G	Al/Cu-P	Al/Cu-Q	Al/Cu-O	Al/Cu-R
Load at fracture (kN)	12.1	12.1	13.1	14.3	9.6	19.7

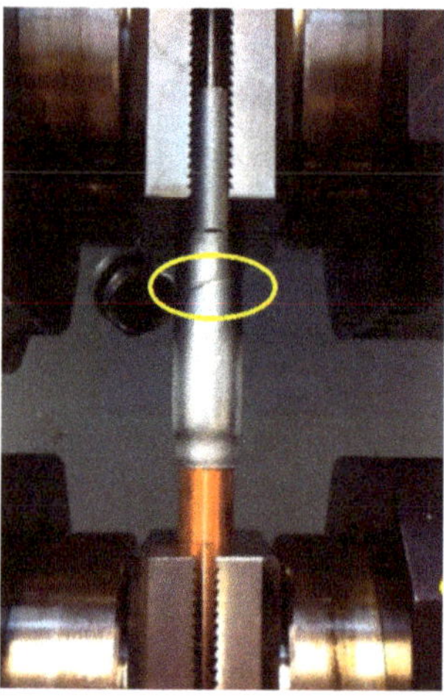

Figure 6. The Al/Cu-R tensile specimen illustrating the fractured zone (seen in yellow insert).

The smallest tensile load registered in this study belongs to the joint produced by the use of lubricated specimen (Al/Cu-O), this weak bonding is attributed to the lack of welding due to the excess of contamination and poor efficiency of the jet phenomenon. Nonetheless, the joint produced by the use of threaded target (Al/Cu-R) showed the highest load at the failure point. This achievement is attributed to the occurrence of the crimping joint rather than welding (Figures 4f and 5f). Regarding the load values, Table 8 also depicts that the Al/Cu-Q joint follows the Al/Cu-R, the microstructure of the joint interface (Figures 4d and 5d) contributed to this level of load, and the waviness pattern, which is a strong characteristic of MPW process, was more pronounced in the Al/Cu-Q joint.

In fact, the impact of surface roughness has been influenced by the welding windows. However, a moderate roughness and a perfectly clean surface are required, which is consistent with similar studies [19].

4. Conclusions

A high velocity plastic deformation joining technique was performed on dissimilar Al/Cu tubular assemblies through the application of magnetic pulse forces. This study evaluated the influence of the surface roughness of the copper target component on the quality of the joint considering the microstructure of the interface and the fracture load obtained from the uniaxial tensile test.

The surface condition of the target component influences the welding quality. However, it is not possible to correlate the surface roughness of the target with the waviness interface and also with the tensile load. Nonetheless, the formation of a waviness pattern at the joint interface requires a moderate roughness since neither the smoothest surface (polished copper) nor the least smooth surface (sandblasted surface) established a waviness pattern, and the best pattern was formed in the joint produced by the use of chemically treated surface. The surface roughness should have influenced the jet formation and consequently the pattern at the joint interface. Though the MPW is called a solid-state process, the occurrence of localized melting was verified. The formation of the intermetallic compounds (mainly Al_2Cu phase) at the Al/Cu joint interfaces which found wavy patterns is confirmed. Moreover, some microvoids and a few microcracks were also formed in these reacted zones.

The strongest joint was obtained from the use of a threaded target, and it was achieved without the formation of any intermetallic or waviness at the interface. This joint failed at the Al tube base during the tensile test, indicating a sound joint though it was not a metallurgical type. Thus, in the MPW process, it seems very challenging to achieve an effective metallurgical bond.

The joint assembly prepared by lubrication on the target surface failed during the MPW process, meaning that the preparation of a clean surface appears a prerequisite for this process.

For this study, the parameters such as the standoff distance, the voltage level, and LWZ distance were kept constant, and this arrangement was implemented to study the influence of the surface conditions for the same working conditions. However, it seems that the samples' length may influence joint formation. Thus, for any design, a new welding window should be optimized.

Author Contributions: Contributed to the conceptualization of this study, A.R., I.V.d.O., and G.A.T.; contributed to the experiments, A.M.R.; contributed to the conceptualization and writing the manuscript, O.E.; provided the revision, O.E. and A.R. All authors have read and agreed to the published version of the manuscript.

Funding: This research received no external funding.

Conflicts of Interest: The authors declare no conflict of interest.

References

1. *Properties and Selection: Nonferrous Alloys and Special Purpose Materials*, 10th ed.; ASM International: Almere, The Netherlands, 1990; Volume 2.
2. Kah, P.; Vimalraj, C.; Martikainen, J.; Suoranta, R. Factors influencing Al-Cu weld properties by intermetallic compound formation. *Int. J. Mech. Mater. Eng.* **2015**, *10*, 10. [CrossRef]
3. Mori, K.I.; Bay, N.; Fratini, L.; Micari, F.; Tekkaya, A.E. Joining by plastic deformation. *CIRP Ann.-Manuf. Technol.* **2013**, *62*, 673–694. [CrossRef]
4. Wang, H.; Wang, Y. High-velocity impact welding process: A review. *Metals* **2019**, *9*, 144. [CrossRef]
5. Kang, B.-Y. Review of magnetic pulse welding. *J. Weld. Join.* **2015**, *33*, 7–13. [CrossRef]
6. Psyk, V.; Risch, D.; Kinsey, B.L.; Tekkaya, A.E.; Kleiner, M. Electromagnetic forming—A review. *J. Mater. Process. Technol.* **2011**, *211*, 787–829. [CrossRef]
7. Cai, W.; Daehn, G.; Vivek, A.; Li, J.; Khan, H.; Mishra, R.S.; Komarasamy, M. A state-of-the-art review on solid-state metal joining. *J. Manuf. Sci. Eng.* **2019**, *141*, 031012. [CrossRef]
8. Watanabe, M.; Kumai, S. High-speed deformation and collision behavior of pure aluminum plates in magnetic pulse welding. *Mater. Trans.* **2009**, 0906290826. [CrossRef]
9. Kapil, A.; Sharma, A. Magnetic pulse welding: An efficient and environmentally friendly multi-material joining technique. *J. Clean. Prod.* **2015**, *100*, 35–58. [CrossRef]
10. Ben-Artzy, A.; Stern, A.; Frage, N.; Shribman, V.; Sadot, O. Wave formation mechanism in magnetic pulse welding. *Int. J. Impact Eng.* **2010**, *37*, 397–404. [CrossRef]
11. Stern, A.; Shribman, V.; Ben-Artzy, A.; Aizenshtein, M. Interface phenomena and bonding mechanism in magnetic pulse welding. *J. Mater. Eng. Perform.* **2014**, *23*, 3449–3458. [CrossRef]
12. Wu, X.; Shang, J. An investigation of magnetic pulse welding of Al/Cu and interface characterization. *J. Manuf. Sci. Eng.* **2014**, *136*, 051002. [CrossRef]
13. Rebensdorf, A.; Boehm, S. Increase of the reproducibility of joints welded with magnetic pulse technology using graded surface topographies. In Proceedings of the 7th International Conference on High Speed Forming, Dortmund, Germany, 27–28 April 2016; pp. 125–136.

14. Cui, J.; Sun, T.; Geng, H.; Yuan, W.; Li, G.; Zhang, X. Effect of surface treatment on the mechanical properties and microstructures of Al-Fe single-lap joint by magnetic pulse welding. *Int. J. Adv. Manuf. Technol.* **2018**, *98*, 1081–1092. [CrossRef]

15. Rebensdorf, A.; Böhm, S. Magnetic pulse welding—Investigation on the welding of high-strength aluminum alloys and steels as well as the influence of fluctuations in the production on the welding results for thin metal sheets. *Weld. World* **2018**, *62*, 855–868. [CrossRef]

16. Lee, W.-B.; Bang, K.-S.; Jung, S.-B. Effects of intermetallic compound on the electrical and mechanical properties of friction welded Cu/Al bimetallic joints during annealing. *J. Alloy. Compd.* **2005**, *390*, 212–219. [CrossRef]

17. Oliveira, I.; Cavaleiro, A.; Taber, G.; Reis, A. Magnetic Pulse Welding of Dissimilar Materials: Aluminum-Copper. In *Materials Design and Applications*; Silva, L.F.M.d., Ed.; Springer: Berlin/Heidelberg, Germany, 2017; pp. 419–431. [CrossRef]

18. Inês Vieira de Oliveira, I.V. *Pulse Technology as a Tool for Multi-Material Joining*; Faculty of Engineering University of Porto: Porto, Portugal, 2016.

19. Marre, M.; Weddeling, C.; Hammers, T.; Merzkirch, M.; Rautenberg, J.; Tekkaya, A.; Schulze, V.; Biermann, D.; Zabel, A. Innovative joining methods in lightweight designs, Part II. *Alum. Int. J. Ind. Res. Appl.* **2010**, *86*, 55–59.

 *metals*

Article

Particle Ejection by Jetting and Related Effects in Impact Welding Processes

Jörg Bellmann [1,2,*], **Jörn Lueg-Althoff** [3], **Benedikt Niessen** [4], **Marcus Böhme** [5], **Eugen Schumacher** [6], **Eckhard Beyer** [1], **Christoph Leyens** [2,7], **A. Erman Tekkaya** [3], **Peter Groche** [4], **Martin Franz-Xaver Wagner** [5] and **Stefan Böhm** [6]

[1] Institute of Manufacturing Science and Engineering, Technische Universität Dresden, George-Baehr-Str. 3c, 01062 Dresden, Germany; eckhard.beyer@tu-dresden.de

[2] Fraunhofer IWS Dresden, Winterbergstr. 28, 01277 Dresden, Germany; christoph.leyens@tu-dresden.de

[3] Institute of Forming Technology and Lightweight Components, TU Dortmund University, Baroper Str. 303, 44227 Dortmund, Germany; joern.lueg-althoff@iul.tu-dortmund.de (J.L.-A.); erman.tekkaya@iul.tu-dortmund.de (A.E.T.)

[4] Institute for Production Engineering and Forming Machines—PtU, The Technical University (TU) of Darmstadt, Otto-Berndt-Straße 2, 64287 Darmstadt, Germany; niessen@ptu.tu-darmstadt.de (B.N.); groche@ptu.tu-darmstadt.de (P.G.)

[5] Institute of Materials Science an Engineering, Chemnitz University of Technology, Erfenschlager Straße 73, 09125 Chemnitz, Germany; marcus.boehme@mb.tu-chemnitz.de (M.B.); martin.wagner@mb.tu-chemnitz.de (M.F.-X.W.)

[6] Department for Cutting and Joining Manufacturing Processes, The University of Kassel, Kurt-Wolters-Str. 3, 34125 Kassel, Germany; e.schumacher@uni-kassel.de (E.S.); s.boehm@uni-kassel.de (S.B.)

[7] Institute of Materials Science, Technische Universität Dresden, Helmholtzstr. 7, 01069 Dresden, Germany

[*] Correspondence: joerg.bellmann@tu-dresden.de; Tel.: +49-351-83391-3716

Received: 31 July 2020; Accepted: 16 August 2020; Published: 18 August 2020

Abstract: Collision welding processes are accompanied by the ejection of a metal jet, a cloud of particles (CoP), or both phenomena, respectively. The purpose of this study is to investigate the formation, the characteristics as well as the influence of the CoP on weld formation. Impact welding experiments on three different setups in normal ambient atmosphere and under vacuum-like conditions are performed and monitored using a high-speed camera, accompanied by long-term exposures, recordings of the emission spectrum, and an evaluation of the CoP interaction with witness pins made of different materials. It was found that the CoP formed during the collision of the joining partners is compressed by the closing joining gap and particularly at small collision angles it can reach temperatures sufficient to melt the surfaces to be joined. This effect was proved using a tracer material that is detectable on the witness pins after welding. The formation of the CoP is reduced with increasing yield strength of the material and the escape of the CoP is hindered with increasing surface roughness. Both effects make welding with low-impact velocities difficult, whereas weld formation is facilitated using smooth surfaces and a reduced ambient pressure under vacuum-like conditions. Furthermore, the absence of surrounding air eases the process observation since exothermic oxidation reactions and shock compression of the gas are avoided. This also enables an estimation of the temperature in the joining gap, which was found to be more than 5600 K under normal ambient pressure.

Keywords: impact welding; collision welding; pressure welding; process glare; jet; cloud of particles; shock compression; surface roughness; collision conditions

---

1. Introduction

The use of the ideal material for every single component of a structure can contribute to a reduction of the total mass, a lower consumption of resources, or to cost-saving. This development intensifies the requirement of modern joining technologies to be able to join dissimilar metals with strongly different mechanical, physical, or chemical properties. Many conventional fusion-based welding processes cannot be used for this purpose because of the heat-induced formation of brittle intermetallic phases lowering the weld quality [1].

Collision welding processes such as explosive welding (EXW), magnetic pulse welding (MPW), vaporizing foil actuator welding (VFAW) and laser impact welding (LIW) are based on the oblique collision of metallic surfaces at high velocities [2,3]. They are often described as "cold" solid-state processes since the mechanism for the bond formation is dominated by the prevalent pressure instead of fusion. The creation of intermetallic phases is therefore limited to an uncritical extent. Flat and tubular joints in overlap configuration can be manufactured [4]. In EXW, which was patented in 1962, the required pressure to accelerate one of the joining partners is generated via a controlled detonation [5]. EXW is mainly used for large scale cladding operations [6]. Compared to EXW, MPW processes are easier to control via the electrical charging energy stored in a capacitor bank. The precise energy input allows joining of smaller and thinner parts. So-called welding windows are used to plot the welding result as a function of the collision kinetics [7], characterized by the collision angle β and the impact velocity v_{imp} or axial collision velocity v_c at the point of collision, respectively. A crucial influencing factor is the exposure of the joining partner surfaces to the material flow phenomenon, called "jetting". Jetting is often described as a necessary criterion for the bond formation. There are several approaches to investigate this phenomenon. Some analytical theories describe the material flow due to the fluid-like behavior of the metals in the collision zone during EXW [8]. This assumption is based on the occurrence of high strain rates up to 10^6 s^{-1} and high pressures up to several gigapascals at the collision point [9,10]. Different numerical approaches used mesh-less methods to simulate the jetting [10–12]. Experimentally, the jet was observed by high-speed imaging [13–15] or collected after leaving the gap using so-called witness plates that enabled a quantitative analysis of the composition of the ejected material [16–18]. Another imaging method that has been used for the analysis of EXW is pulsed radiography [19]. Comparing the findings of the afore mentioned literature reveals that "jetting" is not necessarily a cumulative stream but can occur as a "cloud of particles", too. This differentiation was introduced by Deribas et al. [20] and supported by Groche et al. using high-speed imaging and a subsequent investigation of the microstructure [14]. Both phenomena are depicted in Figure 1. The stream can either remain in a cumulative shape or disperse in particles during the further progression of the collision, depending on the collision conditions and involved materials. This is defined as a "jet" [3]. Additionally, brittle oxide layers and other surface contaminations spall from the surfaces due to high strain rates at the point of collision. This "cloud of particles" is the result of the dispersed material stream, the spalled surface layers, or both phenomena. If a cumulative metal stream is formed and sustained during the collision, it might partly be hidden by the cloud of particles [13]. In case of a comparable low energy input, it might even be too small to be visually detectable during the high-speed observation [14].

Taking a closer look at the microstructure at the bond interface of Al$_3$Cu welded by EXW, a thin layer of ultrafine grained (UFG) microstructure has been observed within the newly formed interface, which might be the result of melting and rapid solidification [21]. In contrast, adjacent regions showed strongly elongated grains in welding direction without any macroscopic contraction of flyer and parent. Interestingly, UFG and columnar grown grains, which indicate local melting, were found at the interface for aluminum welds produced by MPW and at comparatively low energies, too [22,23]. In comparison, copper joints with similar weld parameters exhibited an UFG microstructure, but no columnar solidification grains. This indicates that copper did not melt at the collision point and the UFG microstructure was a result of dynamic recrystallization due to the high strain rates at the collision point [23].

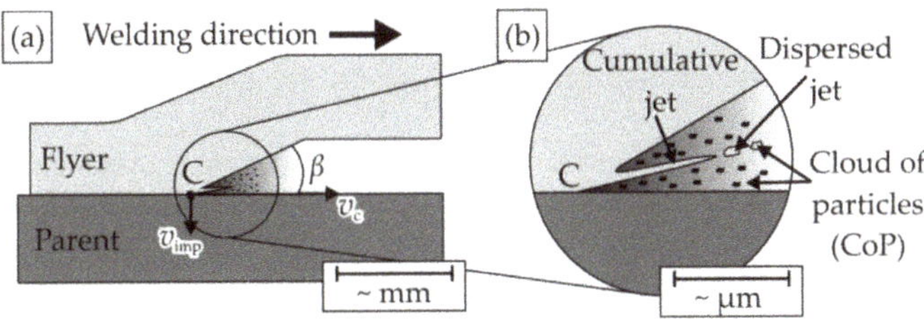

Figure 1. (**a**) Macroscopic scheme of the oblique high-speed collision of a metal pair and (**b**) microscopic view at the collision point C with differentiation between the jet as a cumulative or dispersed stream and the cloud of particles (CoP) that can consist of the dispersed jet, too.

A special process model test rig, which operates at relatively low-impact energies, was applied for the determination of the welding window for aluminum and copper. It allowed to differentiate between the macroscopic cloud of particles (CoP), visible in the high-speed images, and a jet in the shape of a metal stream with a thickness of a few microns [14]. Both phenomena were used to describe the boundaries of the welding window in terms of the impact velocity and the collision angle [24]. Bond formation required the metal stream or the CoP to be formed as a result of the plastic deformation of boundary layers close to the colliding surfaces. Due to the high strain rates, oxide layers and contaminations were removed from the surfaces and then ejected. With increasing impact velocity and, thus, induced energy, bonds could also be formed at larger collision angles. The investigation showed that for aluminum and copper, in addition to process-related variables, the yield strength of the material played an important role in the description of this highly dynamic process. Thus, the plastic deformation of the boundary layers had an influence on the formation of joints, even at low energies. In contrast, the lower boundary angle, which was almost constant for different impact velocities, was limited by the possible ejection of the CoP. If the gap was too small and hindered the outflow of the CoP material, it was incorporated into the newly formed interface and prevented bond formation or led to a porous microstructure [25].

Furthermore, the shape of the CoP was determined by fluid dynamic effects depending on the collision angle [14]. For small angles, the CoP appeared as a turbulent flow due to the wall friction at the surfaces of the joining partners. At larger collision angles, flow separations occurred due to boundary layer friction and widening of the joining gap. Therefore, the central flow part moved faster than the ones at the edges. These findings can also be transferred to MPW [15], where the collision angle was identified as a crucial parameter for a thermal surface activation prior to the contact of the surfaces and a successful weld formation [26].

The collision was observed by high-speed imaging and the shape, velocity, and vertical thickness of the CoP were investigated and related to the geometrical bond properties and its strength. A certain velocity of the CoP was required to achieve the targeted geometry and high strength of the bond. At this point it should be mentioned that the increase of the CoP velocity was related to a higher energy induction in the process zone. In contrast, an increased vertical thickness of the CoP reduced the bond strength due to an enhanced formation of intermediate layers, cracks, and voids. It was concluded that a CoP with higher thickness can be trapped more easily in the closing gap and may disturb the bond formation [14,15].

Experiments in a varied gas atmosphere confirmed these results concerning the role of the CoP during bond formation. Pabst and Groche used gases with different densities and found that lower gas densities led to higher CoP velocities and to a successful weld formation even at lower charging energies [27]. This effect became even more apparent in high vacuum where sound welds were achieved already at lower charging energies, probably due to the escape of the ejected particles which was facilitated by the absence of the surrounding medium [28].

Collision welding processes are also accompanied by a bright and characteristic impact flash ("process glare"). Bellmann et al. [29] showed that this impact flash is a necessary, but not a sufficient criterion for the weld formation in ambient atmosphere. The light emission can be a key parameter for an assessment of the temperature conditions in the joining gap [30]. First results indicated that for a few microseconds the local temperatures can exceed the melting temperatures of most metals, which appears to contradict the theory of collision welding as a "cold" process. These findings are supported by Khaustov et al. [31], who identified the shock-compressed gas in the joining gap and the plastic deformation of the joining partners as potential heat sources.

Thus, there are many hints and proofs for very high temperatures before or during the surface contact in collision welding processes. Nevertheless, the influences of certain input parameters on the properties of the CoP are still not completely understood, neither its effect on weld formation. This paper aims for a deeper understanding of collision welding processes and addresses the following questions:

1. How do material properties, surface properties, the collision environment, and the collision kinetics influence the characteristics of the CoP?
2. What is the temperature in the joining gap due to the CoP or compressed air?
3. Is the process glare a multiple superposition of different effects, depending on the process environment?
4. Under which conditions can the process glare be used as a sufficient welding criterion?

2. Materials and Methods

2.1. Nomenclature and Overview

Table 1 lists all symbols that are used within this paper to shorten the captions of figures and tables. The questions mentioned above cannot be answered by a single experiment due to multiple influencing factors. Thus, three different joining setups were used; their specific characteristics are summarized in Table 2. After the welding experiments, the result in terms of weld quality was checked by manual peel testing, as presented by Bellmann et al. [32]. Peel testing of test rig samples was not necessary since non-welded samples were easily identifiable due to the separation of flyer and parent part right after the collision. Some of the welded samples were further prepared for different types of microscopy, see Table 2.

Table 1. Nomenclature for experimental setup.

Symbol	Parameter	Symbol	Parameter
E	Charging energy	R_i	Inner resistance of the pulse generator
g	Initial joining gap	s	Thickness of the flyer
I	Discharge current	S	High voltage switch
I_{max}	Maximum discharge current	T	Time
L_i	Inner inductance of the pulse generator	$t_{f,start}$	Flash starting time
l_w	Working length (axial overlap between the workpiece and the tool coil)	$U_{f,max}$	Voltage equivalent to maximum intensity of the impact flash
p	Ambient pressure	v_{imp}	Impact velocity
p_m	Magnetic pressure	v_c	Axial collision velocity
R_a	Mean roughness index	β	Collision angle

Table 2. Characteristics of the deployed experimental setups.

Experimental Setup	Test Rig [1]	MPW for Sheets [2]	MPW for Tubes [3]
Manufacturer	*PtU*[4]	*PST products*[5]	*Bmax*[6]
Pulse generator	Not applicable	PS48-16	MPW50/25
Acceleration	Purely mechanical	Electromagnetic pulse technology	Electromagnetic pulse technology
Investigated geometry	Sheets 12×12 mm^2	Sheets 100×40 mm^2	Tubes $\varnothing$ 40 mm
Maximum impact velocity [7]	262 m/s	245 m/s	270 m/s [8]
Impact velocity adjustable	By rotational speed	By discharge current	By discharge current
Collision angle adjustable	By bending	By acceleration gap	By acceleration gap and working length
Ambient pressure	Normal	Normal/1 mbar	Normal
High-speed camera	hsfc pro by *PCO* [9]	hsfc pro by *PCO* [9]	No
Photonic Doppler Velocimetry (PDV)	No	Yes	No
Digital single-lens reflex (DSLR) camera for long time exposures	*Canon*[10] 5D with a 100 mm macro lens	*Canon*[10] 5D with a 100 mm macro lens	*Canon*[10] EOS 700D
Measurement of the impact time	Electrical contact between flyer and parent	Time-resolved flash detection with phototransistor	Time-resolved flash detection with phototransistor
Collision angle accessible	By high-speed camera and image acquisition	By high-speed camera (limited)	By modified parent parts, see [7]
Number of trials for each parameter set	1	1 for the lowest velocity level, otherwise <3	1
Characterization of weld quality		Peel test [32]	Peel test [32]
Microstructural characterization	Optical microscopy (GX-51 by *Olympus*, Tokyo, Japan), scanning electron microscopy (SEM, Neon 40 EsB by *ZEISS AG*, Oberkochen, Germany) and electron backscatter diffraction (EBSD, DigiView IV camera controlled by TEAM v4.5, *EDAX*, USA) for samples joined in vacuum-like and normal ambient atmosphere		Scanning electron microscopy, energy dispersive X-ray spectroscopy (EDS) and 3D microscopy (see Section 2.3 for details)

[1] described in detail in [33], [2] see also [15], [3] see also [26], [4] Institute for Production Engineering and Forming Machines—PtU, The Technical University (TU) of Darmstadt, Germany, [5] Alzenau, Germany, [6] Toulouse, France, [7] for the experiments presented here, [8] based on the flash appearance time $t_{f,start}$ determined in [34], [9] Kelheim, Germany, [10] Tokyo, Japan.

2.2. Test Rig

The first part of the experiments was executed with a special test rig performing the collision process in a purely mechanical approach [33]. The joining partners, two material strips, were mounted at the end of two rotors, see Figure 2a. The rotors rotated in the same turning direction, but started with a phase offset of 45°. When both rotors reached half of the intended impact velocity, the phase offset was eliminated and the joining partners collided with high accuracy and high repeatability in the center between the rotor points, Figure 2b,c. By bending of one joining partner prior to the collision experiment, the oblique collision angle β could be adjusted to a desired value. In the applied configuration the test rig allowed experiments with a maximal absolute impact velocity of 262 m/s; the sheet thickness was kept constant at 2 mm.

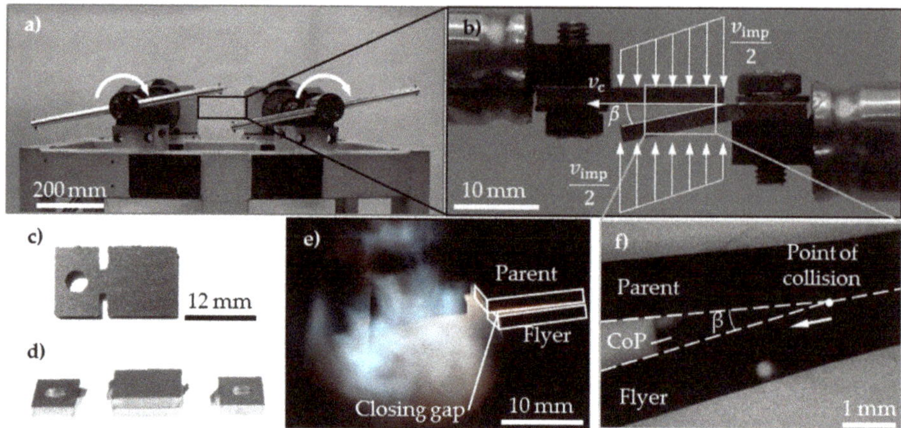

Figure 2. (a) Test rig assembly [35], (b) shown in detail: mounted joining partners and the resulting process parameters [35], (c) geometry of the specimen in the test rig and welded specimens [14], (d) welded specimens [35], (e) process glare with two seconds exposure time, (f) high-speed image of the collision welding process with 20 ns exposure time and measured collision angle β [14]. (a,b,d) are reproduced from [35], with permission from Elsevier, 2017; (c,f) are reproduced from [14], with permission from John Wiley and Sons, 2019.

Table 2 were taken with a digital single-lens reflex camera (DSLR) *Canon* 5D with a 100 mm macro lens that was positioned at an angle of approximately 45° to the rotor center axis. The exposure time was set to two seconds, the aperture fixed at F13, and the light sensitivity ISO 100 was applied. An image intensifier camera hsfc pro from *PCO* working at a dynamic range of 12 bit was used to observe the closing gap between the colliding joining partners [35,36]. Due to the highly dynamic process a short exposure time of 20 ns was necessary. Sufficient lighting was ensured by the laser (CAVILUX Smart by *Cavitar*, Tampere, Finland) with a wavelength of 640 nm, which was used in a transmitted light configuration. Overexposures of the high-speed images due to the process glare were suppressed by an optical band pass filter.

During the initial experiment the aluminum alloys EN AW-1050 H14 and EN AW-6060 in T4 condition with their chemical composition given in Table 3 served as flyer and parent material, respectively. Both joining surfaces were laser ablated to R_a ~ 0.5 with the laser system CL50 (Nd:YAG laser, 1064 nm wavelength by *Clean-Lasersysteme*, Herzogenrath, Germany) prior to joining to remove debris. Within this study, the influence of an increased surface roughness of the parent sheet was investigated. The surface topography was generated via laser ablation, too, and exhibited a wavy structure perpendicular to the welding direction with heights of approx. 15 μm (R_a ~ 3.2) and 30 μm (R_a ~ 7.8), see Figure 3.

Table 3. Composition and quasi-static yield strength of aluminum EN AW-1050 alloy, EN AW-6060 alloy, and C45.

Material	EN AW-1050 H14 [1]	EN AW-6060 [2] T4 [3]	C45 (1.0503) [4], Normalized, Surface Polished ($R_a = 1$)	
Element	Weight%	Weight%	Element	Weight%
Si	0.25	0.3–0.6	C	0.42–0.5
Fe	0.4	0.1–0.3	Mn	0.5–0.8
Cu	0.05	≤0.1	P	<0.045
Mn	0.05	≤0.1	S	<0.045
Mg	0.05	0.35–0.6	Si	<0.4
Cr	-	≤0.05	Ni	<0.4
Ni	-	-	Cr	<0.4
Zn	0.07	≤0.15	Mo	<0.1
Ti	0.05	≤0.1		
Quasi-static yield strength approx.	102 MPa [5]	91 MPa [5]/60 MPa [6]	490 MPa [4]	

[1] adapted from [37], [2] adapted from [38], [3] T4 temper: solution annealing for 1 h at 500 °C, sheets quenched in water and aged at room temperature, tubes cooled by air and naturally aged, [4] adapted from [39], [5] determined by tensile test, [6] determined by tube tensile test.

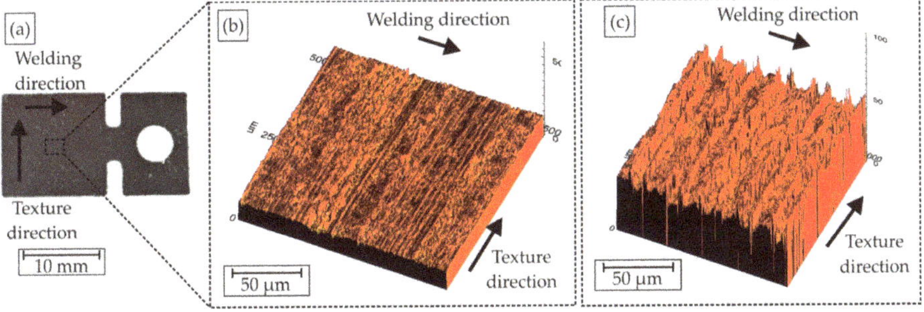

Figure 3. (a) Texture and welding directions at the specimen and detailed surface structures after (b) laser ablation and (c) laser structuring.

In the next step, the parent material's strength was successively increased by replacing it with oxygen-free high conductivity copper (OFHC-Cu) with a purity of more than 99.99 weight percent in two different conditions: as delivered and ultrafine grained after severe plastic deformation by equal channel angular pressing (ECAP) with average grain sizes of ~176 µm and 1.2 µm and yield strengths of 70 MPa and 520 MPa, respectively [40].

2.3. MPW Process

The second part of the impact welding experiments was performed on two commercial pulse generators. In contrast to the test rig, the acceleration of the flyer part was driven by the magnetic pressure p_m between a coil and the electrically conductive flyer workpiece. Higher impact velocities were achievable compared to the test rig, but the adjustment and measurement of the collision conditions was more challenging. The sheet welding setup offered the possibility to execute impact welding experiments under vacuum-like conditions and, thus, suppressed oxidation effects and shock compression of the surrounding air [41]. The process analysis in vacuum via long-term exposures [26] or high-speed camera was simplified due to the lowered intensity of the impact flash. The velocity of the jet or of the CoP, respectively, could be estimated by dividing the progressing distance between two photos by the known time shift. Furthermore, the flash starting time $t_{f,start}$ of the process glare and the emission spectrum of the process glare could be analyzed with the setup depicted in Figure 4 by means of a time-resolved flash measurement and diffraction grating, respectively. This procedure allowed an estimation of the temperature in the joining gap while the material combination EN AW-1050 and EN AW-6060 in T4 temper in the sheet configuration was welded. A constant acceleration gap g of 1.5 mm was chosen, and the charging energy E was varied between 9 and 16 kJ. The flat coil B80/10 by *PST products* was used in combination with the pulse generator PS48-16 manufactured in 2011 by *PST products*, resulting in a discharging frequency of approx. 18 kHz.

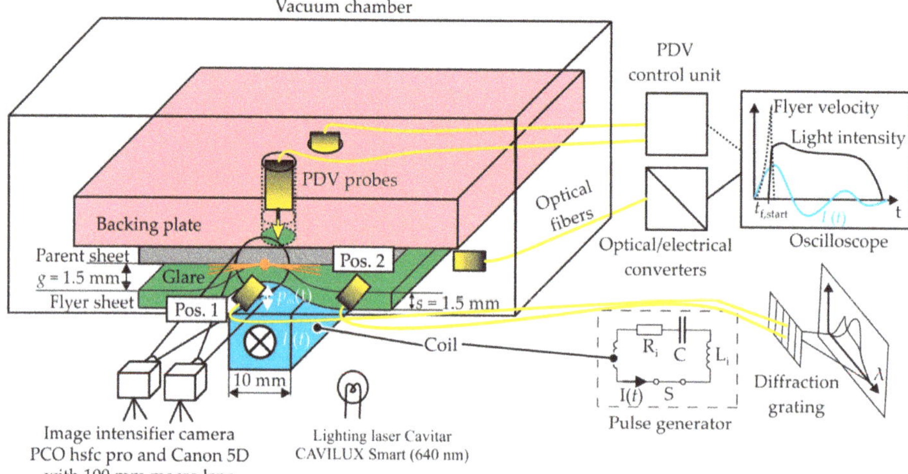

Figure 4. Setup for the analysis of the flyer movement, the formation of the CoP, the flash occurrence, and the emission spectrum in the MPW sheet welding setup.

For a spectral analysis of the process glare at the sheet welding configuration, a simple setup consisting of two polymer optical fibers, the transmission diffraction grating GT50-08 by *thorlabs* (Newton, MA, USA) (830 lines/mm), and the same DSLR used for the long time exposure shots of the welding process was applied, see Figure 4. The first optical fiber was aligned parallel to the coil wire and monitored a position in the welding gap close to the initial contact point (position 1). The second fiber monitored the outer region of the welding gap, which was not affected by the welding process (position 2). The wavelength and sensitivity calibrations of this setup were done with two known wavelengths (green laser pointer $\lambda = 532$ nm, bandpass filter $\lambda = 640 \pm 12$ nm) and a 60 W incandescent bulb modeled as a blackbody emitter at 2700 K.

The movement of the flyer part was monitored and recorded using Photonic Doppler Velocimetry (PDV). This laser-based technology is capable of measuring velocities up to several kilometers per second [32]. The laser source was a 1 W fiber laser module by *Redfern Integrated Optics Inc* (Santa Clara, CA, USA). The wavelength of the emitted laser beam was 1550 nm; with this setup part velocities of up to approx. 1200 m/s can be recorded. Two PDV probes were applied, which recorded the movement of the flyer through small holes in the parent part, see Figure 4. For instance, the impact velocity v_{imp} could be determined.

The tube welding setup shown in Figure 5 included witness pins with a diameter of 2 mm close to the joining zone. They were made of different materials (St-steel, W-tungsten, and C-graphite) and vaporized by the CoP. A tracer copper coating was placed on the parent surface, beginning at a distance of five millimeters from the initial collision point. The surfaces of the steel and tungsten pins were investigated after MPW using scanning electron microscopy (JSM-6610LV by *Jeol*, Tokyo, Japan) and energy dispersive X-ray spectroscopy (X-Max 80 mm^2, Model 51-XMX0002 by *Oxford instruments*, Abingdon, UK). The surface topography of the graphite pins was scanned with a microscope (VHX-5000 and universal zoom lens VH-Z100UR by *Keyence*, Osaka, Japan) The collision angle between flyer and parent workpiece could be modified via the working length l_w, as described in [26]. It is defined as the axial overlap between the flyer part and the concentration zone of the working coil, see Figure 5. The charging energy of the pulse generator MPW 50/25 by *Bmax* was set to 4.5 kJ and led to a maximum tool coil current of approx. 377 kA at a discharge frequency of approx. 20 kHz. The chemical composition and selected mechanical properties of the flyer material aluminum EN AW-6060 and the parent material steel C45 are also listed in Table 3.

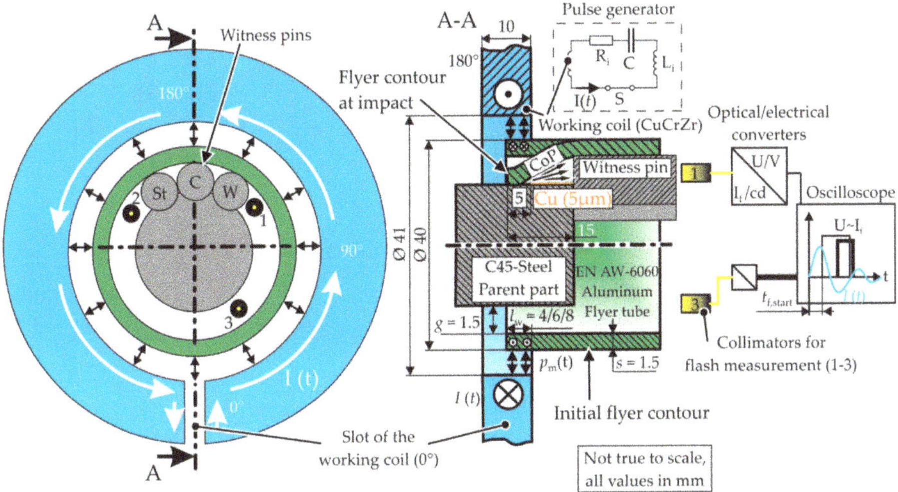

Figure 5. Setup for flash detection, analysis of the CoP composition and properties in the MPW tube welding setup.

3. Results

3.1. Test Rig

The experiments with the initial material combination of the aluminum alloys EN AW-1050 and EN AW-6060 in T4 temper revealed a welding window for collision angles β between 4.5° and 6.6° at a constant impact velocity v_{imp} of 262 m/s. Increasing the roughness of the parent surface to 15 µm narrowed and shifted the range for successful welds to a collision angle of 9.1°. The amount of ejected material for both roughness conditions within their specific welding windows was similar, see Figure 6c,d, while the intensity of the process glare was reduced, see Figure 6f,g. Furthermore, polished cross sections revealed that the waviness of the joining zone increased and exhibited pockets for the laser structured surface (b) instead of a continuous layer (a) along the interface. A roughness of 30 µm inhibited the CoP formation, the process glare and welding completely at the given impact velocity, see Figure 6e,h, respectively.

Replacing the parent material with the copper alloy Cu-OFHC enabled welding for collision angles between 6.6° and 7.5°, while welding was not achievable in this range with the ultrafine grained copper as parent material. Although the shadows in the joining gap and, thus, the CoP densities seemed to be comparable, the process glare was significantly reduced, see Figure 7

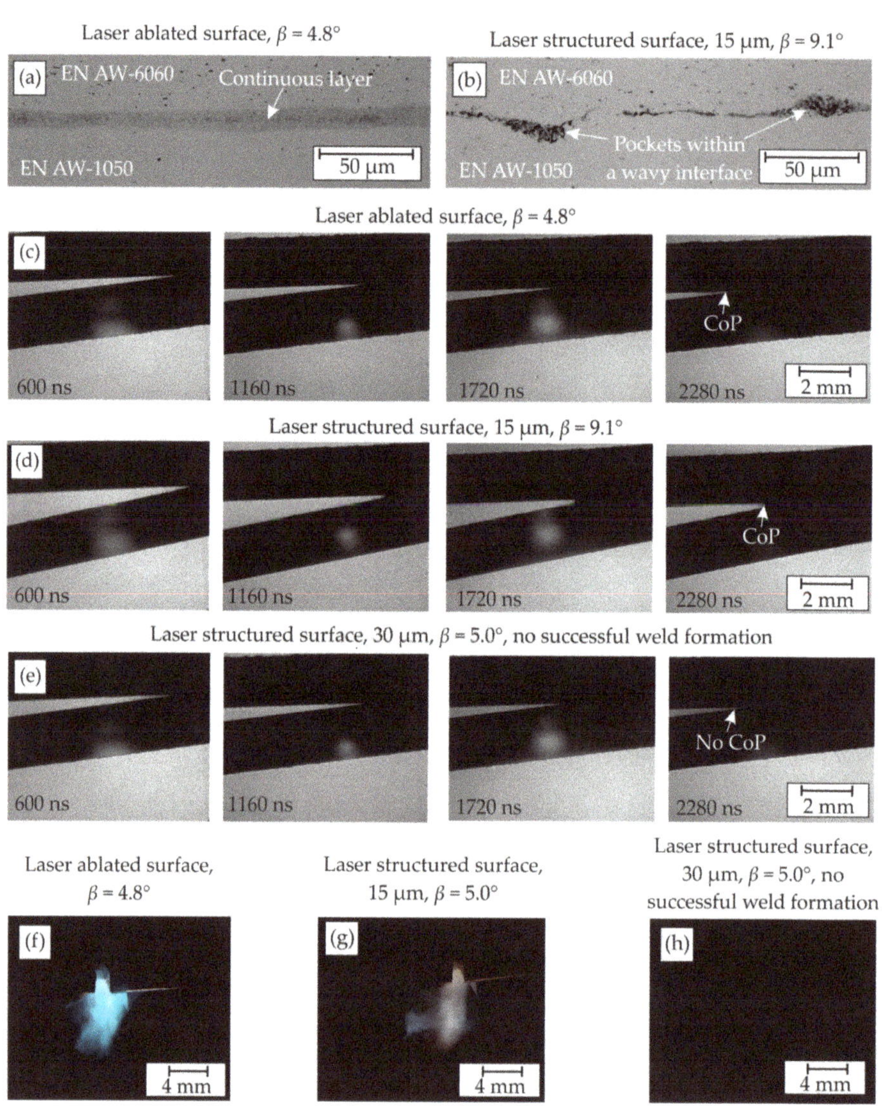

Figure 6. Influence of the surface structure on (**a,b**) the joining zone in polished cross sections, (**c–e**) the formation of a cloud of particles, and (**f–h**) the process glare at v_{imp} = 262 m/s. Welding direction to the left from the initial collision point.

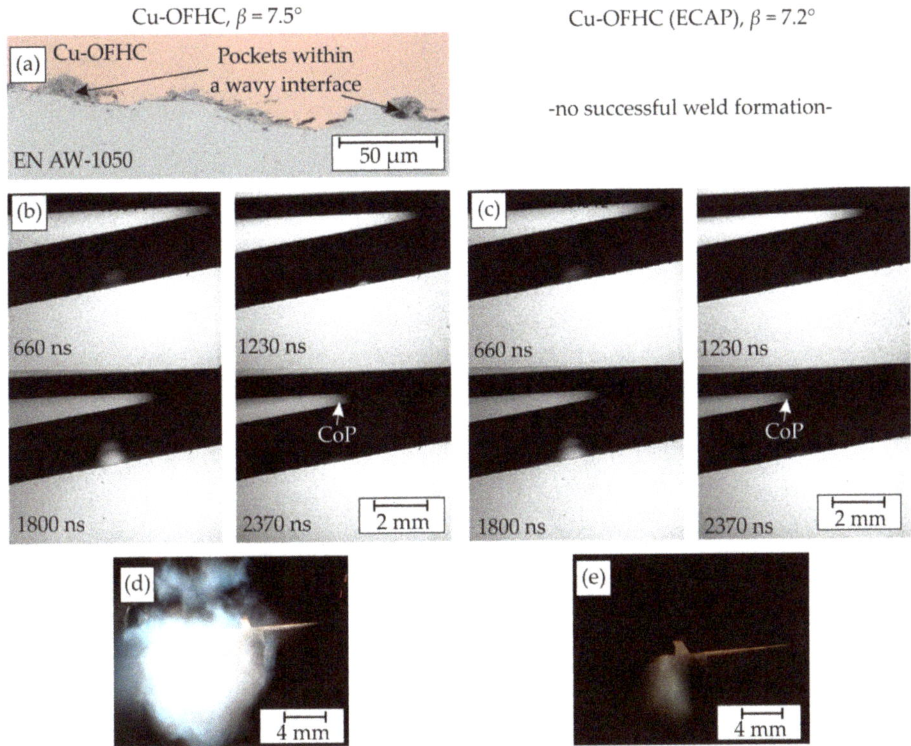

Figure 7. Influence of the parent properties on (**a**) the joining zone in a polished cross section, (**b**,**c**) the formation of a cloud of particles and (**d**,**e**) the process glare. Welding direction to the left from the initial collision point.

3.2. MPW Process for Sheets

MPW experiments with the same material EN AW-1050 for both the flyer and the parent sheet were performed at ambient pressure first. Due to missing PDV measurements, the non-welded samples processed with 18 kJ charging energy are not listed in Table 4. Nevertheless, they enabled the definition of the lower process boundary at 19 kJ charging energy since welding occurred at this energy level with an impact velocity of 224 m/s. Reducing the ambient pressure in the joining gap to approx. 100 Pa enabled welding even with lower impact velocities of 203 m/s. At this point it should be mentioned that due to the design of the vacuum chamber, the distance between the coil wire and the flyer sheet in the active zone increased during the evacuation process of the vacuum chamber, leading to a reduction of the acceleration distance between the joining partners. This effect was compensated by adjusting the charging energy E in 30 preliminary tests to ensure the targeted level of impact velocity v_{imp} and monitor it via PDV.

The reduced ambient pressure in the joining gap not only decreased the lower process boundary, but also influenced the process glare and the interfacial microstructures. Compared with the experiment in ambient atmosphere at an impact velocity of 243 m/s, the process glare under reduced pressure decreased significantly and appeared in an orange to red color, see Figure 8. In addition, the welded area increased, see Table 5. The interface did not contain a porous layer close to the central gap, which resulted in a smooth transition from the non-welded zone in the middle to the adjacent welded regions in Figure 9.

Table 4. Influence of the impact velocity and ambient pressure on the welding result.

Parameter	Impact Velocity [1]	Ambient Pressure	Charging Energy	Max. Discharge Current	Flash Appearance Time [1]	Welding Result [2]
Symbol	v_{imp}	p	E	I_{max}	$t_{f,start}$	
Unit	m/s	Pa	kJ	kA	µs	
	~190	100,000	15	352	19.7	Not welded
	~203	100	19	403	21.6	Welded
	~224	100,000	19	400	17.9	Welded
	~225	100	21	416	18.7	Welded
	~243	100,000	21	420	17.1	Welded
	~245	100	24	451	18.0	Welded

[1] measured by PDV, [2] checked with manual peel test.

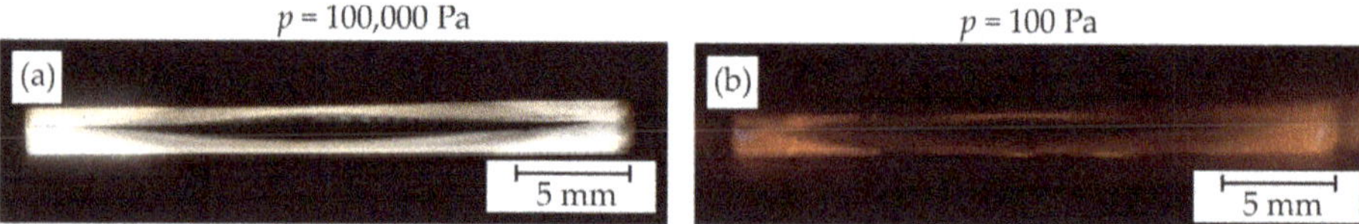

Figure 8. (a) Bright process glare at 100,000 Pa ambient pressure and (b) reduced process glare at 100 Pa ambient pressure in the joining gap at constant impact velocity $v_{imp} \approx 243$ m/s. Welding directions to the left and right from the central initial collision point.

Table 5. Influence of the ambient pressure in the joining gap on the length of the weld seams at constant impact velocity $v_{imp} \approx 243$ m/s.

Ambient Pressure	p = 100,000 Pa	p = 100 Pa
Weld length, left [mm]	0.9	2.4
Not welded central gap [mm]	3.7	2.5
Weld length, right [mm]	1.7	2.4

The combined inverse pole figure (IPF) and Band Contrast/Image Quality (IQ) maps in Figure 9 reveal some interesting differences of the weld interfaces caused by the change in ambient pressure. Most notably, the central gap at 100,000 Pa is wider close to the initial weld compared to the sample welded at 100 Pa ambient pressure. The detail map at 100,000 Pa ambient pressure in (b) reveals a slightly porous layer in the weld interface with many very small grains, although to a smaller extend compared to previous reports [22,25]. In comparison, the initial weld interface produced at 100 Pa ambient pressure in (d) is smoother and without a nano-crystalline interlayer.

The high-speed images reveal a CoP that dispersed in the joining gap and glowed brightly in normal ambient atmosphere. If the ambient pressure was reduced, the appearance of the CoP changed to a dark and tongue-like shape, see Figure 10. In both cases no band pass filter was used. The CoP velocity was estimated by dividing the propagation distance between two time steps. It was approx. 4 to 6 km/s in normal ambient atmosphere and about 10 km/s at reduced ambient pressure.

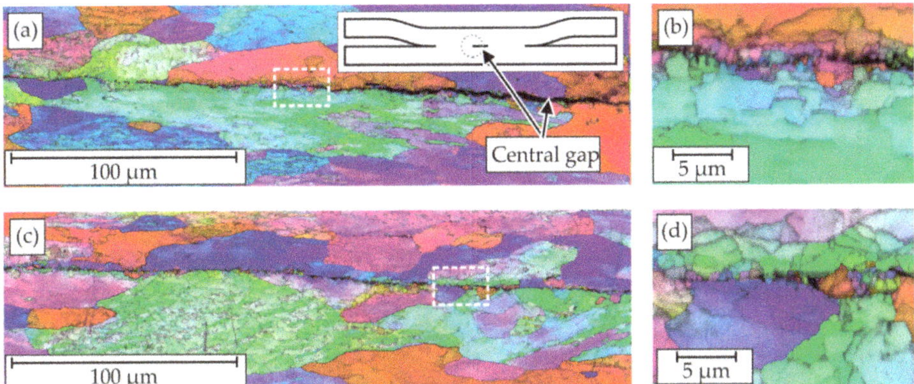

Figure 9. Combined inverse pole figure (IPF) and image quality (IQ) maps of the interface region obtained at constant impact velocity $v_{\mathrm{imp}} \approx 243$ m/s at (**a,b**) 100,000 Pa and (**c,d**) 100 Pa ambient pressure: the inset in (**a**) shows the sample positions where the maps were obtained using EBSD measurements. The white outlines in (**a,c**) mark the positions of the detailed maps in (**b,d**). Welding direction to the left from the central initial collision point.

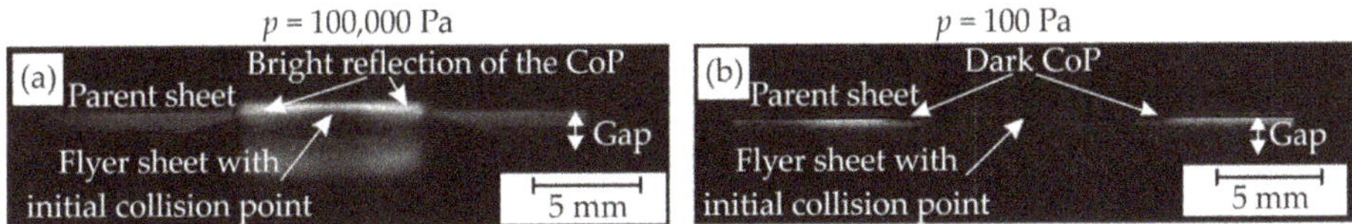

Figure 10. High-speed images of the MPW process at (**a**) normal ambient pressure and (**b**) at reduced ambient pressure with $v_{\mathrm{imp}} \approx 243$ m/s shortly after the initial collision. Welding directions to the left and right from the central initial collision point.

All optical spectra of the process glare in Figure 11 show a prominent peak at short wavelengths, which can be assigned to the intense characteristic aluminum emission line at approximately 396 nm [42]. Although the overall shape of the spectra is unchanged between the two different positions under reduced ambient pressure, the intensity is increased over the whole spectral range at normal ambient pressure. Due to the limited spectral resolution, it is impossible to distinguish bundles of individual emission lines and the continuous blackbody radiation. However, the different process glare colors in Figure 8 and the shift of the intensity maxima to longer wavelengths in normal ambient pressure from the position close to the initial collision point (Pos. 1) to the end of the joining gap (Pos.'2) allow for two main conclusions:

1. The temperature of the light-emitting medium is much higher at normal ambient pressure compared to the reduced pressure conditions.
2. At normal ambient pressure, the temperature rises further during the propagation of the collision point. Using Wien's displacement law (see Appendix A for discussion) and an upper boundary for the photon emission maximum of $\lambda = 650$ nm, the temperature of the process glare under normal ambient pressure can be expected to exceed 5600 K.

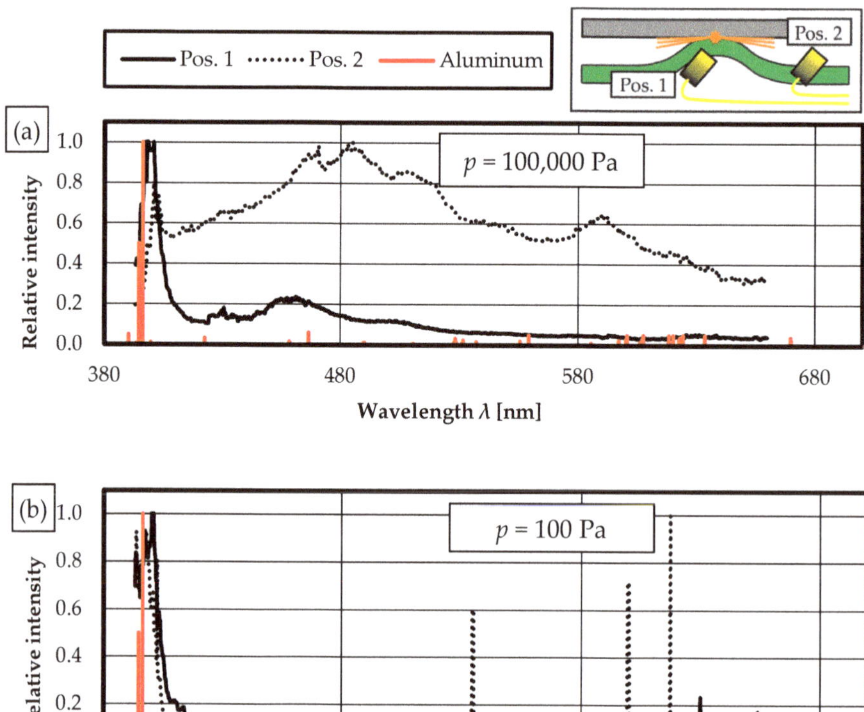

Figure 11. Relative spectral intensities obtained at constant impact velocity $v_{\mathrm{imp}} \approx 243$ m/s at two different positions under (**a**) normal ambient pressure and (**b**) reduced pressure 100 Pa in comparison with the characteristic aluminum emission lines [42].

3.3. MPW Process for Tubes

The collision angle during MPW can be controlled by the working length l_{w} [34]. It influences both the appearance of the process glare and the welding result [26]. A working length of 4 mm led to an initial collision angle of approx. 9° where welding was not achievable in the given setup. Increasing the working length to 8 mm decreased the collision angle to approx. 3° and allowed for a weld formation between the aluminum flyer and the steel parent [34]. Moreover, an evaluation of the process glare in the joining gap under vacuum-like conditions revealed a significant temperature increase far above the vaporization temperature of the involved materials. Now, additional MPW experiments were performed at normal ambient pressure to gain deeper insights into the relation between the collision angle and the formation of the CoP as well as its interaction with the witness pins that consisted of three different materials, see Figure 12a. The witness pins were placed side by side at the 180° position of the tool coil. The pin diameter of 2 mm was small compared to the inner circumference of the flyer tube with 116 mm. Thus, the influence of the different radial positions can be neglected. The results are summarized in Table 6. The CoP penetrated into the soft graphite pin, see Figure 12b, while it was deposited on the surfaces of the tungsten and steel pins, as shown in Figures 13 and 14, respectively. In both figures, sections (a) and (b) show the pin surfaces after the cutting procedure before the MPW experiments. Obviously, the deposited layer in (e) and (f) consisted mainly of the aluminum flyer material as well as the dominating iron parent material. Furthermore, the increased oxygen content indicated a partial oxidation of the CoP during MPW.

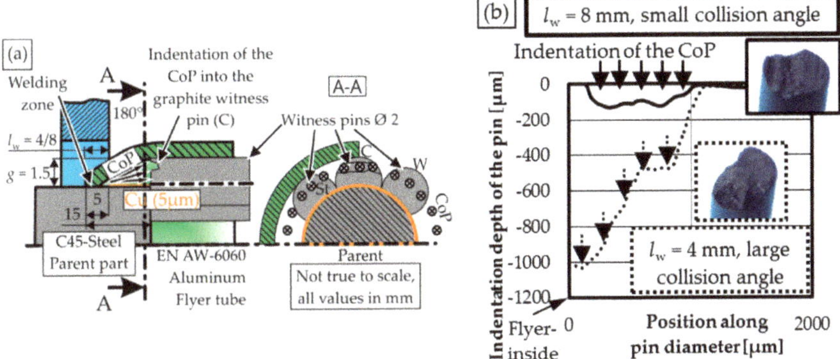

Figure 12. (**a**) Detailed sketch of the joining setup and witness pins to study (**b**) the influence of the collision angle on the indentation depth of the CoP into the graphite witness pin (0 µm-position of the pin is in contact with the inside of the flyer tube during MPW).

Table 6. Influence of the collision angle on the process glare and interaction between the CoP and the witness pins.

Working Length l_W	4 mm	8 mm
Collision Angle β [1]	"Large" (9.5°)	"Low" (3.4°)
Welding result [2]	Not welded	welded
Voltage equivalent to maximum light intensity $U_{f,max}$ [3]	6.6 V	5.8 V
Depth and shape of the penetration zone in the graphite witness pin [4]	~1000 µm, large area	~150 µm, line-shaped
Surface characteristic of the tungsten and steel witness pin	Many coarse particles	Homogenous aluminum cover layer with a few coarse particles
Content of copper on the tungsten witness pin	0 wt %	1.6 wt %
Content of copper on the steel witness pin	0 wt %	2.3 wt %

[1] by analogy with [34], [2] checked with manual peel test, [3] defined in [43], [4] see Figure 12b.

The collision angle had a big impact on the penetration of the CoP in the graphite witness pin and the structure of the vaporized surfaces on the steel and tungsten witness pins, respectively. For high collision angles the penetration depth in the graphite witness pins was approx. 1 mm, see Figure 12b. There were many single particles deposited on the tungsten pin, leading to a coarse and ragged structure. Energy dispersive X-ray spectroscopy (EDX) revealed no copper, neither on the tungsten pin nor on the steel pin, see Figures 13f and 14f, respectively. In contrast, small collision angles led to a line-shaped penetration area in the graphite pin. Based on the distance of 15 mm to the initial collision point, the angle of ejection could be calculated. It is within a range of 1.9° to 5°, which is in good agreement with the simulated collision angle of 3.4° [34]. At this small angle, a homogeneous aluminum layer with a copper content of ~2 weight percent from the tracer coating and with only a few single aluminum particles was deposited. Compared to large collision angles, the thickness of the layer was lower since the subjacent tungsten was detectable during EDX analysis, too.

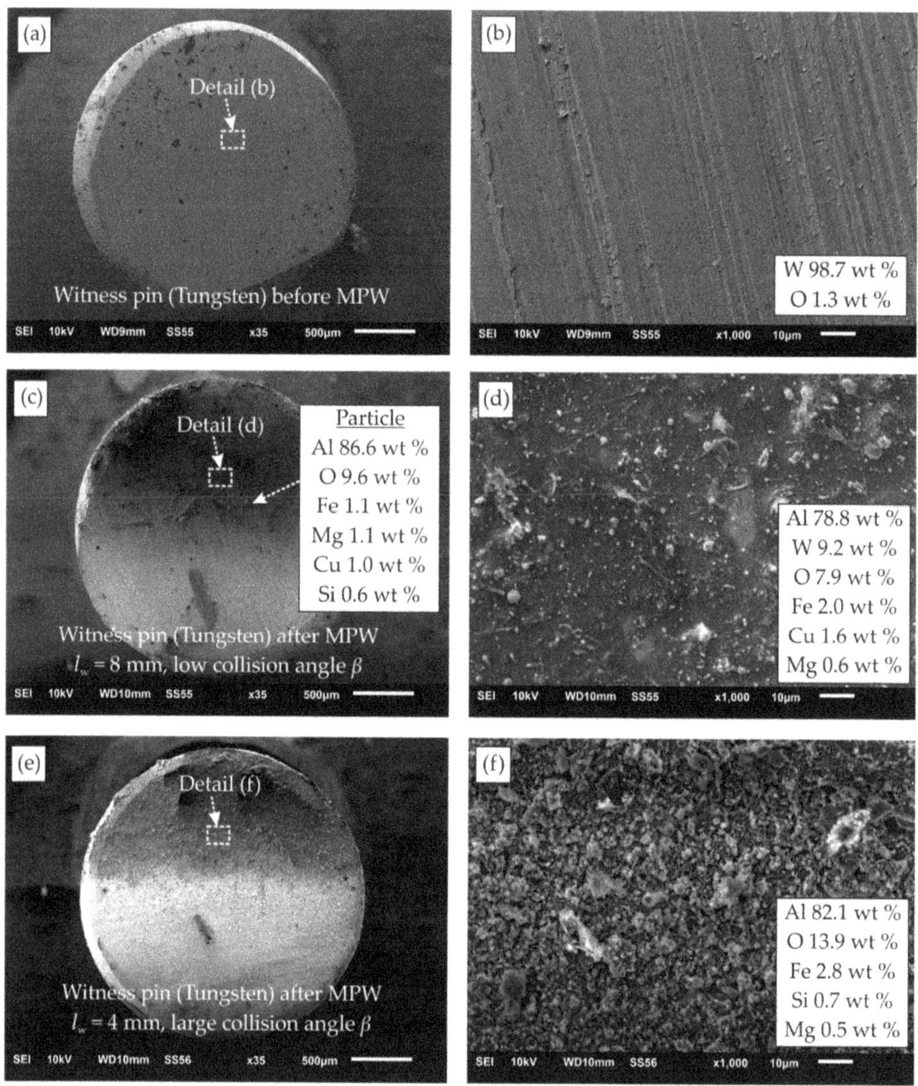

Figure 13. Surface topography and chemical composition of the tungsten witness pins (**a**,**b**) before MPW and after MPW with (**c**,**d**) low collision angle and (**e**,**f**) large collision angle.

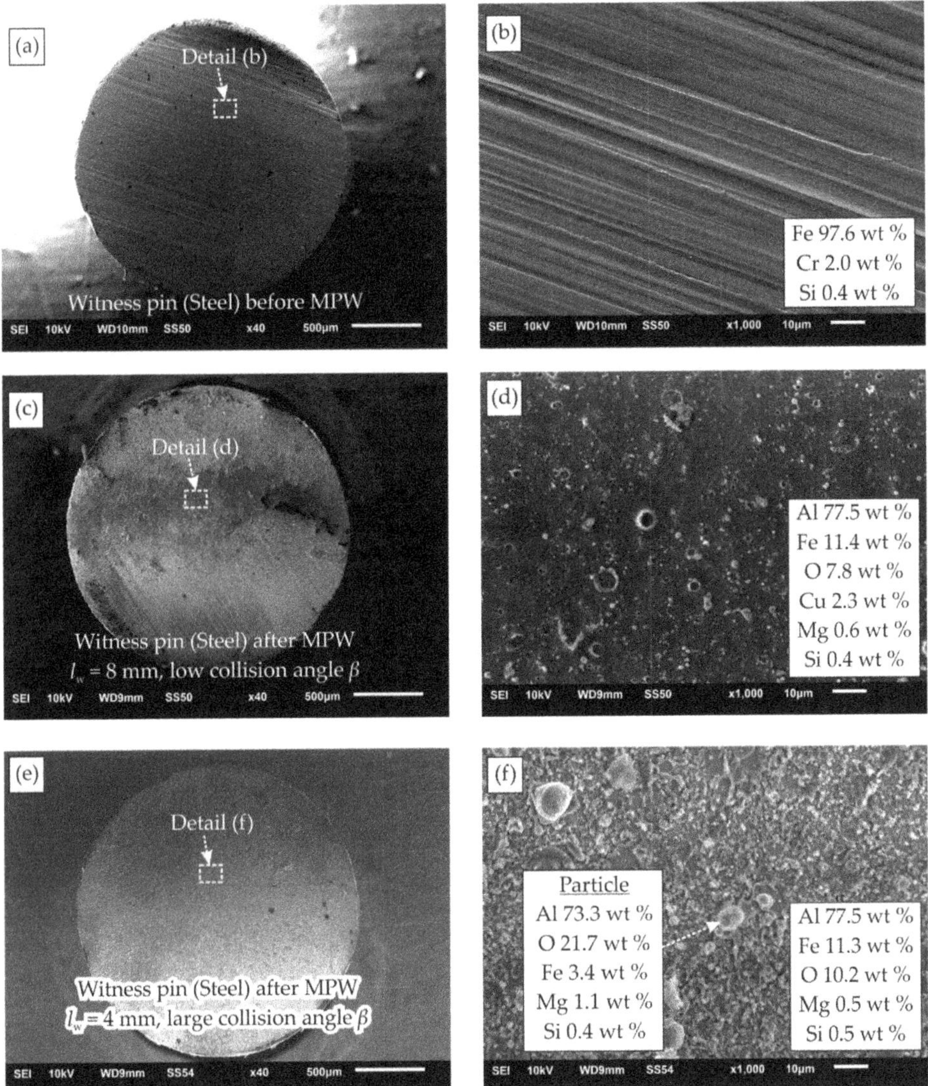

Figure 14. Surface topography and chemical composition of the steel witness pins (**a**,**b**) before MPW and after MPW with (**c**,**d**) low collision angle and (**e**,**f**) large collision angle.

4. Discussion

The experimental results presented in this study clearly confirm the hypothesis that high temperatures occur within the joining gap during collision welding processes. The experiments at reduced ambient pressure showed that the temperature increase also takes place in the absence of the surrounding air. Thus, it seems reasonable to relate the temperature increase to the material that is ejected from the process zone. Depending on the collision kinetics and the involved materials, material ejection can either occur as a dense material flow as a jet and/or as a CoP in case of lower impact energies and depending on the collision angle. The thermal energy of the CoP enables and supports all three bonding types that have been reported for collision welding processes so far. Depending on the impact kinetics at the propagating collision point either solid-state bonding, solid–liquid coexistence state bonding, or liquid-state bonding can occur [44]. The influence of the material properties, surface properties, the collision environment, and the collision kinetics on the characteristic features of the CoP were investigated in three different experimental setups, as discussed and listed below:

a. Test rig experiments: Decreasing the grain size of the parent material led to an increase in the material's strength and hindered plastic deformation in the joining zone. This makes it difficult to generate a sufficient CoP for the heating of the surfaces and to form a solid jet that uncovers the base materials.

b. Test rig experiments: The escape of the CoP was hindered by an increased surface roughness, which also led to a reduced process glare. The high-speed picture frames as well as the melting pockets in the cross section point to the conclusion that the CoP was partly entrapped in the wavy surface during the movement of the collision point. This effect can be detrimental for the weld formation if the kinetic or thermal energy of the CoP is too high and leads to extensive melting along the interface. Furthermore, it weakens or hinders bond formation due to a lack of direct contact of the activated surfaces of the base material.

c. Sheet welding setup: The surrounding gas lowered the velocity of the escaping CoP, as also reported in [27]. This led to a porous microstructure if the CoP was partially enclosed in the joining zone. In contrast, the microstructure revealed no pores when MPW was performed under vacuum-like conditions, see also [25]. Similar to the observations reported in [28], the necessary impact velocity for a successful weld seam can be reduced in vacuum compared to normal atmosphere. Moreover, the process glare is reduced significantly, either due to the absence of chemical interactions with the surrounding oxygen or because of shock compression of the gas in the joining gap, as described in [45].

d. Tube welding setup: The collision kinetics have a big impact on the characteristics of the CoP. The deep and voluminous craters that are formed by the CoP in soft graphite pins indicated that the kinetic energy of the CoP was higher for larger collision angles, whereas smaller collision angles only led to a single line in the graphite witness pin with significantly reduced depth. The structures of the vaporized surfaces of the witness pins made of steel and tungsten provided insights into the thermal energy of the CoP: Small collision angles led to a higher degree of compression of the CoP in the joining gap and, thus, pronounced heating of the surfaces in the joining zone. The copper tracer that was placed at the end of the actually welded area was detectable on the surface of the witness pins. Since the impact velocity, the plastic deformation, and, thus, the heating due to forming were reduced compared to the initial collision point, the thermal and kinetic energy of the *compressed CoP must have been responsible for the melting and transportation of the copper tracer* towards the witness pins. The finely dispersed structure of the vaporized pin surfaces supports the hypothesis that the high temperatures in the joining gap resulted from the thermal energy of the CoP. In contrast, *compression and heating were reduced for large collision angles*, leading to the ragged surface of the witness pins. The surface contained larger aluminum particles and no signs of copper from the tracer surface coating. The chemical composition of the CoP was dominated by approx. 80 weight percent aluminum due to its lower melting temperature and lower strength compared to steel. Consequently, the amount of aluminum that was plastically deformed, melted and finally contributed to the CoP is higher.

One of the key results of the present study is that the temperature in the joining gap at ambient pressure was found to exceed 5600 K. Due to the arrangement of the spectral measurements it can be concluded that this temperature occurred in the rapidly closing welding gap *in front* of the actual point of collision. This effect can be suppressed at reduced ambient pressure. However, to determine the temperature of the CoP under vacuum-like conditions, a different setup with an increased spectral sensitivity in the infrared range would be needed.

Furthermore, the findings of the present study point to the conclusion that the process glare results from a superposition of multiple, different effects that depend on the process environment:

(a) In ambient atmosphere, the shock compression of the surrounding air led to a thermal glow. Furthermore, the remarkably high oxygen content on the witness pins provided evidence of an exothermic reaction of the metal vapor with the surrounding oxygen. This reaction contributed to the flash effects, too.

(b) Under vacuum-like conditions, the intensity of the light emission was reduced, and its appearance depended on the involved materials. In [13], a bright appearance was reported for magnesium (which has a lower boiling temperature than copper), where only a dark metal jet was observed. Thus, it can be concluded that even in the absence of the surrounding air, particles of the involved joining partners with temperatures above the vaporization temperature emitted light and contributed to the process glare. Depending on the local temperatures and pressures, the formation of plasma is also conceivable, but the present study does not provide direct evidence for this phenomenon.

To conclude, it should be noted that the process glare alone cannot be used as a sufficient welding criterion because multiple parameters contribute to the light emission, weld formation, and the corresponding side effects. In a conventional ambient atmosphere, the lightning effect is dominated by the interaction with the surrounding air. Thus, the effect of the collision angle, which significantly influences the welding result, is not directly accessible. Nevertheless, the process glare can be seen as a necessary criterion while additional conditions must be fulfilled to ensure a good weld quality. For example, the thermal properties of the involved materials must be suitable to ensure the cooling of the materials after the contact, as described in [26].

5. Conclusions, Research Highlights, and Outlook

The experimental results at normal ambient pressure indicate temperatures more than 5600 K in the joining gap that enable not only solid-state bonding, but also solid–liquid coexistence state bonding or liquid-state bonding. The process glare consists of different components and depends on certain factors. It occurs if the kinetic energy of the moving joining partner is sufficient to extract a certain number of particles from the surfaces by plastic deformation during the collision. The particles accumulate and then form a CoP. The CoP is compressed in the closing joining gap and heats up until it glows. This effect can be intensified by small collision angles where high temperatures are reached by a higher compression rate and comparatively more effective wall friction, sufficient to melt the surfaces of the joining partners. To form a sound weld, the CoP must leave the joining zone before the joining partners come into contact. Thus, smooth surfaces and vacuum-like conditions are preferable for collision welding. The absence of surrounding air eases the process observation since exothermic oxidation reactions and shock compression of the gas are avoided. Nevertheless, the occurrence of the process glare is not a sufficient welding criterion, but just a necessary condition.

From these findings, technological guidelines can be derived for collision welding processes. For example, the formation of the CoP is facilitated by soft materials; a small collision angle increases the temperature and surface activation; a low surface roughness supports the escape of the CoP.

Nevertheless, an important question that remains is which mechanism is ultimately responsible for the activation of the joining surfaces. On the one hand, it might be the kinetic energy of the flowing CoP that rubs intensively against the surfaces. On the other hand, surface activation could also be attributed to the heat transfer between the compressed CoP and the surfaces. Furthermore, the possible plasma state might as well play an important role for surface activation, since plasma activation is a well-established technology, see [46]. Although plasma formation during collision welding in normal ambient atmosphere was only attributed to the shock-compressed air in [23], it may also occur under vacuum-like conditions. The CoP itself could be transferred into plasma due to the sudden compression and heating in the joining gap. Future investigations should focus on the time-resolved measurement of the temperature in the joining gap to identify the dominating mechanisms during collision welding. Furthermore, to determine the temperature of the CoP under vacuum-like conditions, a different setup with sufficient spectral sensitivity in the IR range would be needed.

6. Patents

The flash measurement system enables the identification of suitable collision conditions and can be used for quality assurance during production. It is patented for different impact welding processes [47,48].

Author Contributions: Conceptualization, J.B., J.L.-A., B.N., M.B., E.S., E.B., A.E.T., P.G., M.F.-X.W., S.B.; methodology, data analysis, J.B. (MPW of tubes, design of vacuum chamber for MPW of tubes, flash measurements, witness pins, SEM), J.L.-A. (MPW of tubes, PDV measurements), B.N. (Test rig, high-speed imaging, long time exposures), M.B. (spectroscopic measurements, optical microscopy, EBSD), E.S. (MPW of sheets, design of vacuum chamber for MPW of sheets, SEM, High-speed imaging, long time exposures); Design of experiments, investigation, validation, J.B., J.L.-A., B.N., M.B., E.S.; writing—original draft preparation, visualization, J.B., J.L.-A. (1., 2.3.), B.N. (1., 2.2., 3.1., 4., 5.), M.B. (2.2., 2.3., 2.4., 3.2.), E.S. (2.3., 3.2.); writing—review and editing, supervision, project administration, funding acquisition, resources, E.B., C.L., A.E.T., P.G., M.F.-X.W., S.B. All authors have read and agreed to the published version of the manuscript.

Funding: This research was funded by the Deutsche Forschungsgemeinschaft (DFG, German Research Foundation), grant number BE 1875/30-3, TE 508/39-3, GR 1818/49-3, WA 2602/5-3, BO 1980/23-1. This work is based on the results of the working group "high-speed joining" of the priority program 1640 ("joining by plastic deformation"). It consists of the subprojects A1, A5, A8, and A9. We acknowledge support by the Open Access Publication Funds of the SLUB/TU Dresden.

Acknowledgments: We would like to acknowledge the effort for the sample preparation, SEM and EDS analysis at Fraunhofer IWS Dresden. As well, we would like to thank Walter Tutsch of PCO AG for the support in setting up the image intensifier camera. The authors also greatly appreciate the help of Stephan Ditscher of Baumüller who supported the programming of the electronical control system of the test rig. We thank Jeanette Brandt for her efforts in proofreading the manuscript.

Conflicts of Interest: The authors declare no conflict of interest. The funders had no role in the design of the study; in the collection, analyses, or interpretation of data; in the writing of the manuscript, or in the decision to publish the results.

Appendix A Blackbody Model for Temperature Estimation

According to Planck's law, power density M_λ^O in the range between λ and $\lambda + d\lambda$ for thermal emission can be calculated as

$$M_\lambda^O(\lambda, T)dA\,d\lambda = \frac{2\pi h c^2}{\lambda^5} \frac{1}{e^{\left(\frac{hc}{\lambda k_B T}\right)} - 1} dA\,d\lambda \tag{A1}$$

However, the sensitivity of imaging sensors (complementary metal-oxide-semiconductor, CMOS) is proportional to the photon flux, which can be calculated from the power density distribution as

$$J_\lambda^{ph}(\lambda, T)dA\,d\lambda = \frac{2\pi c}{\lambda^4} \frac{1}{e^{\left(\frac{hc}{\lambda k_B T}\right)} - 1} dA\,d\lambda \tag{A2}$$

As can be seen in Figure A1 that these two expressions yield different shapes of the thermal emission spectra and subsequently different maxima.

A simple method for estimating the temperature of a thermal emitter is the tracking of those spectral maxima. According to Wien's displacement law, higher temperatures cause a shift of emission maxima to shorter wavelengths in both distributions.

$$\lambda_{max}^M = \frac{2897.8\ \mu m \cdot K}{T} \quad \lambda_{max}^J = \frac{3669.7\ \mu m \cdot K}{T} \tag{A3}$$

However, due to the limited spectral range and low resolution of the setup used in this work, tracking of those maxima is only feasible for temperatures above 5400 K. Furthermore, the position of those maxima is difficult to track in the noise-afflicted experimental spectra given the low slope of the theoretical spectra for temperatures above 5000 K (Figure A2).

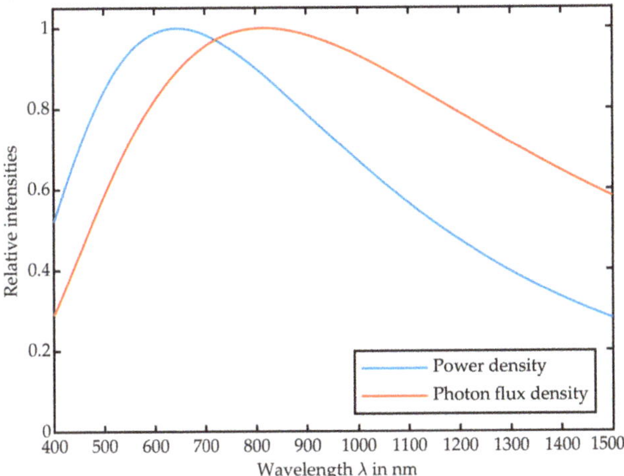

Figure A1. Spectral power density and photon flux density for thermal emission at T = 4500 K.

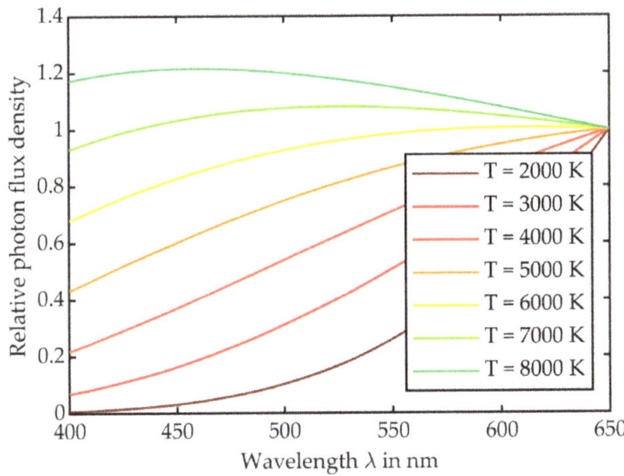

Figure A2. Modeled thermal emission spectra in the investigated spectral range for temperatures between 2000 K and 8000 K.

References

1. Kapil, A.; Sharma, A. Magnetic Pulse Welding: An efficient and environmentally friendly multi-material joining technique. *J. Clean. Prod.* **2015**, 35–58. [CrossRef]
2. Wang, H.; Wang, Y. High-Velocity Impact Welding Process: A Review. *Metals* **2019**, *9*, 144. [CrossRef]
3. Sadeh, S.; Gleason, G.H.; Hatamleh, M.I.; Sunny, S.F.; Yu, H.; Malik, A.S.; Qian, D. Simulation and Experimental Comparison of Laser Impact Welding with a Plasma Pressure Model. *Metals* **2019**, *9*, 1196. [CrossRef]
4. Mori, K.; Bay, N.; Fratini, L.; Micari, F.; Tekkaya, A.E. Joining By Plastic Deformation. *CIRP Ann. Manuf. Technol.* **2013**, *62*, 673–694. [CrossRef]
5. Philipchuk, V.; Scituate, N.; Roy, L.F. Explosive Welding. US Patent US 3,024,526, 13 March 1962.

6. Crossland, B. *Explosive Welding of Metals and Its Application*; Clarendon Press: Oxford, UK, 1982.
7. Lueg-Althoff, J.; Bellmann, J.; Gies, S.; Schulze, S.; Tekkaya, A.E.; Beyer, E. Influence of the Flyer Kinetics on Magnetic Pulse Welding of Tubes. *J. Mater. Process. Technol.* **2018**, 189–203. [CrossRef]
8. Cowan, G.R.; Holtzmann, A.H. Flow Configurations in Colliding Plates: Explosive Bonding. *J. Appl. Phys.* **1963**, *34*, 928–939. [CrossRef]
9. Stern, A.; Shribman, V.; Ben-Artzy, A.; Aizenshtein, M. Interface Phenomena and Bonding Mechanism in Magnetic Pulse Welding. *J. Mater. Eng. Perform.* **2014**, *23*, 3449–3458. [CrossRef]
10. Akbari Mousavi, A.A.; Al-Hassani, S.T.S. Numerical and experimental studies of the mechanism of the wavy interface formations in explosive/impact welding. *J. Mech. Phys. Solids* **2005**, *53*, 2501–2528. [CrossRef]
11. Kakizaki, S.; Watanabe, M.; Kumaji, S. Simulation and experimental analysis of metal jet emission and weld interface morphology in impact welding. *Mater. Trans.* **2011**, *52*, 1003–1008. [CrossRef]
12. Gleason, G.; Sunny, S.; Sadeh, S.; Yu, H.; Malik, A. Eulerian Modeling of Plasma-Pressure Driven Laser Impact Weld Processes. *Procedia Manuf.* **2020**, *48*, 204–214. [CrossRef]
13. Mori, A.; Tanaka, S.; Hokamoto, K. Observation for the High-Speed Oblique Collision of Metals. In *Explosion Shock Waves and High Strain Rate Phenomena*; Hokamoto, K., Raghukandan, K., Eds.; Materials Research Forum LLC: Puducherry, India, 2019; pp. 74–78. ISBN 978-1-64490-032-1.
14. Groche, P.; Niessen, B.; Pabst, C. Process boundaries of collision welding at low energies. *Mater. Werkst.* **2019**, *50*, 940–948. [CrossRef]
15. Schumacher, E.; Rebensdorf, A.; Böhm, S. Influence of the jet velocity on the weld quality of magnetic pulse welded dissimilar sheet joints of aluminum and steel. *Mater. Werkst.* **2019**, *50*, 965–972. [CrossRef]
16. Watanabe, M.; Kumai, S. High-Speed Deformation and Collision Behavior of Pure Aluminum Plates in Magnetic Pulse Welding. *Mater. Trans. Jpn. Inst. Light Met.* **2009**, *50*, 2035–2042.
17. Hassani-Gangaraj, M.; Veysset, D.; Nelson, K.A.; Schuh, C.A. In-situ observations of single micro-particle impact bonding. *Scr. Mater.* **2018**, *145*, 9–13. [CrossRef]
18. Bergmann, O.R.; Cowan, G.R.; Holtzmann, A.H. Experimental Evidence of Jet Formation During Explosion Gladding. *Trans. Metall. Soc. AIME* **1966**, *236*, 646–653.
19. Drennov, O.B. Structure of a Shaped Jet Formed in an Oblique Collision of Flat Metal Plates. *MSF* **2004**, *465*, 409–414. [CrossRef]
20. Deribas, A.A.; Zakharenko, I.D. Surface effects with oblique collisions between metallic plates. Translated from Fisika Goreniya, I. *Vzyva* **1974**, *10*, 409–421.
21. Hammerschmidt, M.; Kreye, H. Microstructure and bonding mechanism in explosive welding. In *Shock Waves and High-Strain-Rate Phenomena in Metals*; Springer: Berlin/Heidelberg, Germany, 1981; pp. 961–973.
22. Sharafiev, S.; Pabst, C.; Wagner, M.F.-X.; Groche, P. Microstructural characterisation of interfaces in magnetic pulse welded aluminum/aluminum joints. *IOP Conf. Ser. Mater. Sci. Eng.* **2016**, *118*, 12016. [CrossRef]
23. Pabst, C. *Ursachen, Beeinflussung, Auswirkungen sowie Quantifizierung der Temperaturentwicklung in der Fügezone beim Kollisionsschweißen*; Technische Universität Darmstadt: Darmstadt, Germany, 2019.
24. Niessen, B.; Groche, P. Weld interface characteristics of copper in collision welding. In Proceedings of the 22nd International ESAFORM Conference on Material Forming (ESAFORM 2019), Vitoria-Gasteiz, Spain, 8–10 May 2019; Galdos, L., Arrazola, P., Saenz de Argandoña, E., Otegi, N., Mendiguren, J., Madariaga, A., Saez de Buruaga, M., Eds.; AIP Publishing: Melville, NY, USA, 2019; p. 50018, ISBN 978-0-7354-1847-9.
25. Böhme, M.; Sharafiev, S.; Schumacher, E.; Böhm, S.; Wagner, M.F.X. On the microstructure and the origin of intermetallic phase seams in magnetic pulse welding of aluminum and steel. *Mater. Werkst.* **2019**, *50*, 958–964. [CrossRef]
26. Bellmann, J.; Lueg-Althoff, J.; Schulze, S.; Hahn, M.; Gies, S.; Beyer, E.; Tekkaya, A.E. Thermal Effects in Dissimilar Magnetic Pulse Welding. *Metals* **2019**, *9*, 348. [CrossRef]
27. Pabst, C.; Groche, P. Identification of process parameters in electromagnetic pulse welding and their utilisation to expand the process window. *Int. J. Mater. Mech. Manuf.* **2018**, *6*, 69–73.
28. Kümper, S.; Schumacher, E.; Böhm, S. Influence of the Ambient Pressure on the Weld Quality for Magnetic Pulse Welded Sheet Joints. In Proceedings of the 8th International Conference on High Speed Forming, Columbus, OH, USA, 13–16 May 2018; Daehn, G.S., Tekkaya, A.E., Eds.; The Ohio State University: Columbus, OH, USA, 2018.
29. Bellmann, J.; Lueg-Althoff, J.; Schulze, S.; Gies, S.; Beyer, E.; Tekkaya, A.E. Parameter Identification for Magnetic Pulse Welding Applications. *Key Eng. Mater.* **2018**, *767*, 431–438. [CrossRef]

30. Ishutkin, S.N.; Kirko, V.I.; Simonov, V.A. Thermal action of shock-compressed gas on the surface of colliding plates. *Fiz. Goreniya/Vzryva* **1979**, *16*, 69–73. [CrossRef]

31. Khaustov, S.V.; Kuz'min, S.V.; Lysak, V.I.; Pai, V.V. Thermal processes in explosive welding. *Combust. Explos. Shock Waves* **2014**, *50*, 732–738. [CrossRef]

32. Bellmann, J.; Lueg-Althoff, J.; Schulze, S.; Gies, S.; Beyer, E.; Tekkaya, A.E. Measurement and Analysis Technologies for Magnetic Pulse Welding: Established Methods and New Strategies. *Adv. Manuf.* **2016**, 322–339. [CrossRef]

33. Groche, P.; Wagner, M.F.-X.; Pabst, C.; Sharafiev, S. Development of a novel test rig to investigate the fundamentals of impact welding. *J. Mater. Process. Technol.* **2014**, *214*, 2009–2017. [CrossRef]

34. Bellmann, J.; Ueberschär, F.; Lueg-Althoff, J.; Schulze, S.; Hahn, M.; Beyer, E.; Tekkaya, A.E. Effect of the Forming Behavior on the Impact Flash during Magnetic Pulse Welding of Tubes. In Proceedings of the 13th International Conference on Numerical Methods in Industrial Forming Processes, NUMIFORM 2019, Portsmouth, NH, USA, 23–27 June 2019; Korkolis, Y.P., Kinsey, B.L., Knezevic, M., Padhye, N., Eds.; The Minerals, Metals & Materials Society (TMS): Pittsburgh, PA, USA, 2019; pp. 651–654, ISBN 978-0-87339-769-8.

35. Groche, P.; Becker, M.; Pabst, C. Process window acquisition for impact welding processes. *Mater. Des.* **2017**, *118*, 286–293. [CrossRef]

36. Pabst, C.; Sharafiev, S.; Groche, P.; Wagner, M.F.X. A Novel Method to Investigate the Principles of Impact Welding: Development and Enhancement of a Test Rig, Experimental and Numerical Results. *AMR* **2014**, *966*, 500–509. [CrossRef]

37. Seeberger. Datasheet 3.0255 (EN AW-1050A). Available online: https://seeberger.net/_assets/pdf/werkstoffe/aluminium/de/3.0255.pdf (accessed on 23 January 2020).

38. Seeberger. Datasheet AlMgSi (EN AW-6060). Available online: http://www.seeberger.net/_assets/pdf/werkstoffe/aluminium/de/AlMgSi.pdf (accessed on 18 February 2020).

39. Deutsche Edelstahlwerke. Unlegierter Vergütungsstahl 1.1191/1.1201: C45E/C45R. Available online: https://www.dew-stahl.com/fileadmin/files/dew-stahl.com/documents/Publikationen/Werkstoffdatenblaetter/Baustahl/1.1191_1.1201_de.pdf (accessed on 21 February 2019).

40. Frint, S.; Hockauf, M.; Frint, P.; Wagner, M.F.-X. Scaling up Segal's principle of Equal-Channel Angular Pressing. *Mater. Des.* **2016**, *97*, 502–511. [CrossRef]

41. Pabst, C.; Pasquale, P. Identification of additional process parameters for impact welding and their influence on the process window. In Proceedings of the 8th International Conference on High Speed Forming, Columbus, OH, USA, 13–16 May 2018; Daehn, G.S., Tekkaya, A.E., Eds.; The Ohio State University: Columbus, OH, USA, 2018.

42. Tucker, N. Spectra V1.0. Available online: https://www.mathworks.com/matlabcentral/fileexchange/27796-spectra-v1-0 (accessed on 29 June 2020).

43. Bellmann, J.; Lueg-Althoff, J.; Schulze, S.; Gies, S.; Beyer, E.; Tekkaya, A.E. Measurement of Collision Conditions in Magnetic Pulse Welding Processes. *J. Phys. Sci. Appl.* **2017**, *7*, 1–10. [CrossRef]

44. Cui, J.; Ye, L.; Zhu, C.; Geng, H.; Li, G. Mechanical and Microstructure Investigations on Magnetic Pulse Welded Dissimilar AA3003-TC4 Joints. *J. Mater. Eng. Perform.* **2020**, *29*, 712–722. [CrossRef]

45. Saravanan, S.; Raghukandan, K. Thermal kinetics in explosive cladding of dissimilar metals. *Sci. Technol. Weld. Join.* **2012**, *17*, 99–103. [CrossRef]

46. Kotte, L.; Mäder, G.; Roch, J.; Kaskel, S. Extended DC arc atmospheric pressure plasma source for large scale surface cleaning and functionalization. *Contrib. Plasma Phys.* **2018**, *58*, 327–336. [CrossRef]

47. Bellmann, J. Verfahren und Vorrichtung zur Prozessüberwachung bei einer mittels Kollisionsschweißen gebildeten Schweißnaht. German Patent DE 10 2016 217 758 B3, 16 September 2016.

48. Bellmann, J. Method and Device for Monitoring the Process for a Welding Seam Formed by Means of Collision Welding. U.S. Patent Application No. 16/333,917, 8 September 2017.

Article

Interface Formation during Collision Welding of Aluminum

Benedikt Niessen [1,*], Eugen Schumacher [2], Jörn Lueg-Althoff [3], Jörg Bellmann [4,5], Marcus Böhme [6], Stefan Böhm [2], A. Erman Tekkaya [3], Eckhard Beyer [4], Christoph Leyens [5,7], Martin Franz-Xaver Wagner [6] and Peter Groche [1]

[1]   Institute for Production Engineering and Forming Machines—PtU, The Technical University (TU) of Darmstadt, Otto-Berndt-Strasse 2, 64287 Darmstadt, Germany; groche@ptu.tu-darmstadt.de

[2]   Department for Cutting and Joining Manufacturing Processes—tff, The University of Kassel, Kurt-Wolters-Str. 3, 34125 Kassel, Germany; e.schumacher@uni-kassel.de (E.S.); s.boehm@uni-kassel.de (S.B.)

[3]   Institute of Forming Technology and Lightweight Components, TU Dortmund University, Baroper Str. 303, 44227 Dortmund, Germany; joern.lueg-althoff@iul.tu-dortmund.de (J.L.-A.); erman.tekkaya@iul.tu-dortmund.de (A.E.T.)

[4]   Institute of Manufacturing Science and Engineering, Technische Universität Dresden, George-Baehr-Str. 3c, 01062 Dresden, Germany; joerg.bellmann@tu-dresden.de (J.B.); eckhard.beyer@tu-dresden.de (E.B.)

[5]   Fraunhofer IWS Dresden, Winterbergstr. 28, 01277 Dresden, Germany; christoph.leyens@tu-dresden.de

[6]   Institute of Materials Science and Engineering, Chemnitz University of Technology, Erfenschlager Straße 73, 09125 Chemnitz, Germany; marcus.boehme@mb.tu-chemnitz.de (M.B.); martin.wagner@mb.tu-chemnitz.de (M.F.-X.W.)

[7]   Institute of Materials Science, Technische Universität Dresden, Helmholtzstr. 7, 01069 Dresden, Germany

*   Correspondence: niessen@ptu.tu-darmstadt.de; Tel.: +49-6151-16-23148

Received: 3 August 2020; Accepted: 3 September 2020; Published: 8 September 2020

Abstract: Collision welding is a high-speed joining technology based on the plastic deformation of at least one of the joining partners. During the process, several phenomena like the formation of a so-called jet and a cloud of particles occur and enable bond formation. However, the interaction of these phenomena and how they are influenced by the amount of kinetic energy is still unclear. In this paper, the results of three series of experiments with two different setups to determine the influence of the process parameters on the fundamental phenomena and relevant mechanisms of bond formation are presented. The welding processes are monitored by different methods, like high-speed imaging, photonic Doppler velocimetry and light emission measurements. The weld interfaces are analyzed by ultrasonic investigations, metallographic analyses by optical and scanning electron microscopy, and characterized by tensile shear tests. The results provide detailed information on the influence of the different process parameters on the classical welding window and allow a prediction of the different bond mechanisms. They show that during a single magnetic pulse welding process aluminum both fusion-like and solid-state welding can occur. Furthermore, the findings allow predicting the formation of the weld interface with respect to location and shape as well as its mechanical strength.

Keywords: collision welding; impact welding; magnetic pulse welding; model test rig; welding window; jet; cloud of particles; welding mechanisms

1. Introduction

One of the biggest challenges today is climate change and its impact on the environment and human society. Driven by politics and self-motivation, the industry is increasingly striving for sustainable products and manufacturing processes, e.g., by introducing clean production methods without toxic components and with less energy and raw material consumption. Furthermore, new products have to be more environmentally friendly. For example, the consistent lightweight design in the transport sector is an important factor to reduce emissions. Reliable joining techniques are key to implement load-adapted material usage and to fulfill further operational functions. Conventional fusion-based joining processes, however, reach their technological limits when it comes to metallurgical joining between dissimilar metals. In contrast, solid-state welding techniques like magnetic pulse welding (MPW) can lead to advantageous properties like high bond strengths, no heat-affected zones and low electrical resistance, even between metals with differing thermomechanical and chemical properties [1,2]

MPW is based on the oblique collision between two joining partners at high relative velocities [3], thus belonging to the category of collision welding processes like explosion welding or laser impact welding. Usually, one of the joining partners, called flyer, is accelerated up to several hundred meters per second and collides with a stationary so-called target at an impact velocity v_{imp} under a collision angle β (see Figure 1). Due to this angle, a collision front (or in the two-dimensional case a point of collision (PoC)) moves along the colliding surfaces characterized by the collision point velocity v_c. High strain rates of up to 10^6 1/s and high pressures of up to several GPa occur at the collision point [4,5]. When the dynamic elastic limit of the material is exceeded, material flow results from the plastic deformation of the contact surfaces and a stream of material is pushed ahead of the collision point, see detail in Figure 1 [6–8]. This phenomenon is called jetting and is, besides other criteria, regarded as a necessary condition for bond formation. The jet can remain as a cumulative stream or can disperse in particles. Furthermore, the extensive local strains at the point of the collision lead to the removal of brittle oxide layers and surface contaminations from the surfaces, which are ejected either as a compact stream or as a dispersed cloud of particles (CoP; see Figure 1) [9]. Depending on the collision conditions, the CoP either results of the dispersed material stream, the spalled surface contamination and oxide layers or both phenomena, whereas the cumulative jet can be partly or completely hidden by the CoP [9,10]. This ejection is typically accompanied by a process glare in the form of a bright light or flash emission [11]. Due to the high pressure and temperature at the PoC, the clean surfaces are forced into intimate contact, which ultimately triggers the bonding mechanism. Cui et al. [12] identified three different joining mechanisms when joining aluminum and titanium by MPW, depending on the prevalent kinetic and thermal energies: solid-phase metallurgical bonding by diffusion, liquid-state bonding by melting and solid-liquid coexistence state bonding.

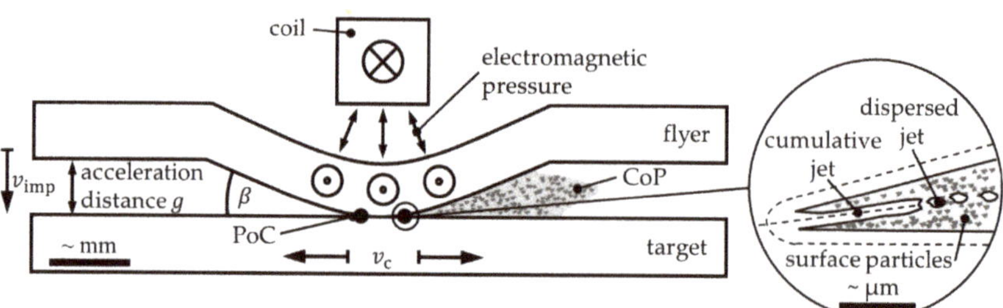

Figure 1. Schematic illustration of symmetric sheet magnetic pulse welding (MPW) with the formation of the jet as material flow at the point of collision according to [7,10] in detail. The cloud of particles (CoP) is either formed by dispersed jet particles, spalled contamination and oxide layers or both phenomena.

During collision welding, the process-related acceleration of the joining partners determines the provided kinetic energy. In the case of MPW, the mobile joining partner is accelerated by an induced electromagnetic pressure (see Figure 1) that is generated by closing an electrical circuit that consists of a charged capacitor bank and a coil actuator [3]. The energy input can be adjusted to the welding task by variation of the charging energy. Different coil designs allow welding of overlap joints of profiles, tubes and sheets, respectively. The high repetition rate and the ability to integrate the system into production lines are two main advantages for the usage of MPW in mass production [13]. However, the joining process is not yet widely applied. One reason for this is the necessary certification of the joining properties, which, in addition to characterization of joint strength, also requires the verification of fatigue strength, corrosion behavior or gas tightness, for instance. Recently, these issues have been objects of concentrated research, and some promising results have already been obtained [14–18].

In addition, a deeper understanding of the joint behavior under different loading scenarios during service is essential for an adequate design of the joining partners and the joint. Ideally, this would cover the prediction of the weld interface position, its shape and area as a function of the material and process parameters. Therefore, a wider comprehension of the bond formation and its influencing parameters is necessary. However, the collision parameters change along the propagating collision front due to the transient behavior during MPW and, thus, cannot be simply calculated or measured [19–21]. In addition, the nature of the process hinders a separate investigation of the influence of the velocity, the mass and the resulting effective energy, which result from the selected acceleration distance and the charging energy of the MPW setup [22,23]. Thus, no direct correlation between the energy input and the welding result can be drawn. Therefore, a purely mechanical model test rig (described in Section 2.1) was designed that allows an independent adjustment of impact velocity and collision angle [24]. A study with copper-copper joints on weld interface formation depending on these parameters showed that after reaching a minimum velocity, the shape and size of the weld interface was influenced particularly by the collision angle [25]. Higher impact velocities increase the region where large weld interfaces can be produced and shift this region towards larger collision angles [25].

The role of the energy input for bond formation is, however, still unclear. This includes additional effects which occur when the kinetic energy is varied either by changing the accelerated mass or the impact velocity. Former investigations mostly varied the kinetic energy of the flyer via the impact velocity at a constant flyer mass. However, this also changes the collision conditions. For explosion welding, Lysak et al. [26] related the collision conditions to the energy part for the metal's plastic deformation by adding the averaged mass to the classical welding window as a third dimension. Thereby the hydrodynamic processes of the collision are linked to the metallophysical processes of bond formation.

To transfer these findings to the MPW process at comparatively low energy input, the influence of the energy input is examined in this paper separately from the collision conditions by changing the flyer mass and keeping the impact velocity constant. For this purpose, several experimental series with flyers of three different thicknesses are carried out using two different experimental setups. The experiments on the mechanical model test rig serve to change the particular parameters individually and to determine their influence on the weld interface, they are supplemented by experiments with different impact velocities. For this purpose, the different phenomena occurring at collision welding can be related to the formation of the weld interface for different points in the welding window. Subsequently, the results are validated by experiments on an MPW setup with different flyer thicknesses, for which the necessary impact velocity parameter settings are determined with an adapted measurement setup. This allows the determination of the influence of different energy inputs at otherwise comparable collision conditions. Based on the experimental results, the following questions are addressed in the present paper:

(1) How does the kinetic impact energy of the flyer influence the collision process and the formation of the weld interface and how is the welding window affected?

(2) Do the collision kinetics influence the governing phenomena, the resulting bond mechanism and thus the weld interface's properties, e.g., mechanical strength?

(3) Can the formation and properties of the weld interface be predicted with respect to location, shape and strength, and how can these properties be controlled by the process parameters?

2. Materials and Methods

2.1. Description of the Series of Experiments

Three series of experiments, performed in two different setups, a model test rig (see Section 2.2) and MPW setup (see Section 2.3), were carried out with aluminum sheets (EN AW-1050A Hx4, yield strength: 99 MPa, tensile strength: 105 MPa) with an initial thickness of $s = 2$ mm for the target as well as the flyer. The specimens for the test rig were produced by laser cutting, while specimens for the MPW setup were cut to size (40 mm × 100 mm) by plate shears. To study the influence of the flyer thickness while keeping the material properties constant, the sheet thickness for both setups was reduced by milling to 1 and 1.5 mm, respectively. The main experiments were carried out with an impact velocity of 262 m/s (see Table 1, Series 1.1 and 2.1., 2.2, 2.3). Furthermore, a second series of complementary experiments were carried out in the test rig with a target and flyer thickness of 2 mm and varied impact velocities of 220 m/s and 240 m/s, respectively (see Table 1, Series 1.2).

Table 1. Summary of series of experiments in applied setups with varied and constant parameters.

Series of Experiment	Applied Setup	Varied Key Parameter	Constant Parameter
1.1	Test rig	Flyer thickness ($s = 1$ mm, 1.5 mm, 2 mm) Collision angle ($\beta = 3$–$9°$)	$v_{imp} = 262$ m/s
1.2	Test rig	Impact velocity $\left(v_{imp} = 220$ m/s, 240 m/s$\right)$ Collision= 3–9°)	$S = 2$ mm
2.1	MPW	Flyer thickness ($s = 1$ mm, 1.5 mm, 2 mm)	$v_{imp} = 262$ m/s, g = 1.5 mm
2.2	MPW	Flyer thickness ($s = 1$ mm, 1.5 mm, 2 mm)	$v_{imp} = 262$ m/s, g = 2.0 mm
2.3	MPW	Flyer thickness ($s = 1$ mm, 1.5 mm, 2 mm)	$v_{imp} = 262$ m/s, g = 2.5 mm

In the test rig, the collision angle was varied to define the weldable region of the welding window. From earlier investigations, it was known that collision angles that led to welding are in the range of 3° to 9° for 2 mm thick joining partners of aluminum at an impact velocity of 262 m/s. Based on this, the weldable range was determined for each flyer thickness and impact velocity. The joints are considered welded if they cannot be separated manually after the experiments.

The initial impact velocity was adjusted close to 262 m/s using the MPW setup by means of photonic Doppler velocimetry (see Section 2.4.3 and Table 1). As mentioned above, the further progression of impact velocity and collision angle was unsteady and difficult to measure in this setup. However, the initial collision angle and its rate of change along the propagating collision front was varied by different acceleration distances of 1.5 mm, 2 mm and 2.5 mm.

2.2. Model Test Rig

Besides the individual and precise adjustment of the collision parameters at stationary process conditions, the model test rig was built up (by the PtU Institute, Darmstadt, Germany) with the intention to provide good observability, which was realized by a purely mechanical concept (see Figure 2a). The main components are two rotors with a diameter of 500 mm, each one driven by a synchronous motor. As a joining partner, specimens with a collision area of 12 mm × 12.5 mm were mounted and prebent with a certain angle at one end of each rotor (see Figure 2b,c). In order to start the collision welding operation, both rotors rotated in the same turning direction but with a phase offset of 45°. As the rotational speed reached half of the desired impact velocity, the phase offset was compensated within one revolution. Thus, the two specimens collided with high accuracy and repeatability in the center between the two turning points. After the collision and the accompanying welding process, the specimens were torn off at the predetermined breaking point since the rotors could not be stopped instantaneously; see Figure 2d. The applied configuration of the test rig for these experiments led to a maximum absolute impact velocity v_{imp} of 262 m/s [27].

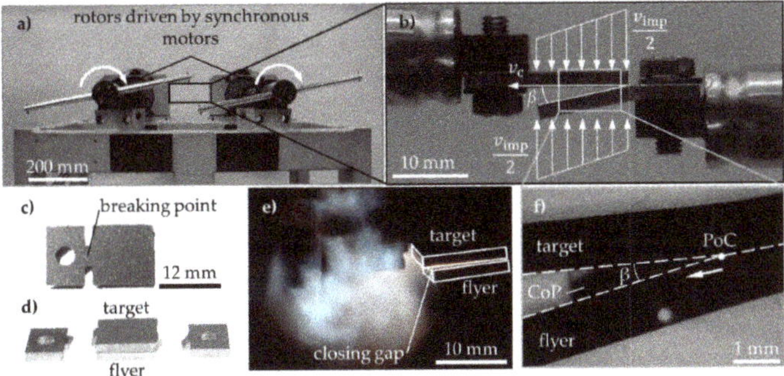

Figure 2. (**a**) Test rig assembly [28], (**b**) shown in detail: mounted joining partners and the resulting process parameters at the moment of the initial impact [28]. (**c**) Geometry of the specimen in the test rig [24] and (**d**) welded specimens [28]. (**e**) A long-term exposure of process glare (2 s exposure time) and (**f**) high-speed image [24] of the collision welding process with 20 ns exposure time were recorded. (a,b,d) are reproduced from [28], with permission from Elsevier, 2017; (c,f) are reproduced from [24], with permission from John Wiley and Sons, 2019. The entire figure is also published in the companion paper [10].

2.3. MPW Setup

The MPW experiments were carried out with the pulse generator BlueWave PS48–16 from PSTproducts GmbH, Alzenau, Germany in combination with the sheet welding tool coil B80/10. The pulse generator provides a maximum charging energy of 48 kJ and a maximum charging voltage of 16 kV. The effective part of the tool coil had a width of 10 mm, a length of 80 mm, and a thickness of 5 mm. It can be operated up to a maximum peak current of 500 kA. During the welding experiments, the flyer and target sheets were positioned in the center above the coil with an overlap of 30 mm and were fixed by a steel backing plate (see Figure 3).

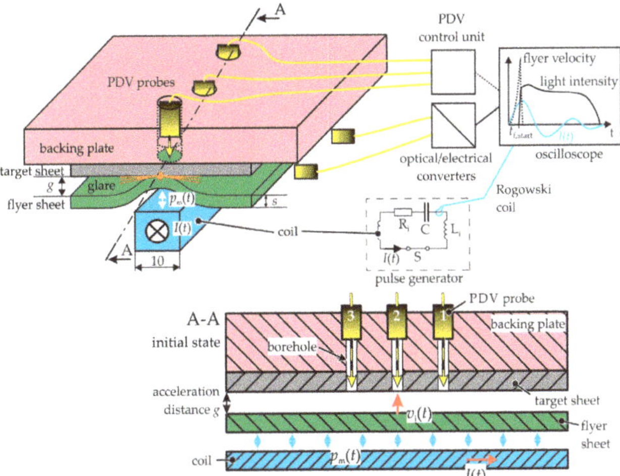

Figure 3. MPW setup for sheet welding: the acceleration gap g and the flyer sheet thickness s were varied. Section A-A shows the photonic Doppler velocimetry (PDV) measurement configuration at the initial state of the welding process (no deformation of the flyer). The oscilloscope recorded the signals of the PDV system, Rogowski coil and measured intensities of the process glare.

2.4. Methods of Process Observation

2.4.1. Process Observation in the Model Test Rig

Two observation methods were implemented in the test rig. First, the collision welding process was observed by an image intensifier camera hsfc pro (by PCO, Kelheim, Germany) with a long-distance microscope lens. This system allowed taking up to eight images per experiment. Due to the high-velocity collision an exposure time of 20 ns was applied (Figure 2f). During the experiments, a CAVILUX Smart lighting laser (by Cavitar, Tampere, Finland) with a power of 400 W and a wavelength of 640 nm provided sufficient brightness. In combination with an optical bandpass filter placed in front of the camera lens, the bright process glare was suppressed, which would otherwise outshine the phenomena in the closing gap. A script in MATLAB (by MathWorks, Natick, Massachusetts, MA, USA) was used to measure the collision angle β by edge detection in each high-speed image, as shown exemplarily in Figure 2f [24,28,29].

Second, a qualitative examination of the process glare was realized by long-term exposures (Figure 2e), with a single-lens reflex camera 5D (by Canon, Ōta, Tokio, Japan) and a 100 mm macro lens, which was positioned at an angle of about 45° to the rotor center axis. The image acquisition settings were set to an exposure time of 2 s, an aperture of F13 and a light sensitivity of ISO 100. The results are shown in Appendix A.

2.4.2. Rogowski Coil

The recording of the discharge current was performed via a Rogowski pickup coil type *CWR* 3000 B (by Power Electronic Measurements Ltd., Long Eaton, Nottingham, UK) in combination with a high-resolution oscilloscope (see Figure 3). Rogowski coils are especially suitable for the measurement of oscillating currents with high amplitudes and frequencies like those occurring during MPW. The discharge current curve $I(t)$ contains important information for the evaluation of the temporal evolution of the MPW process and is shown in Figure 4, together with the recorded light intensity.

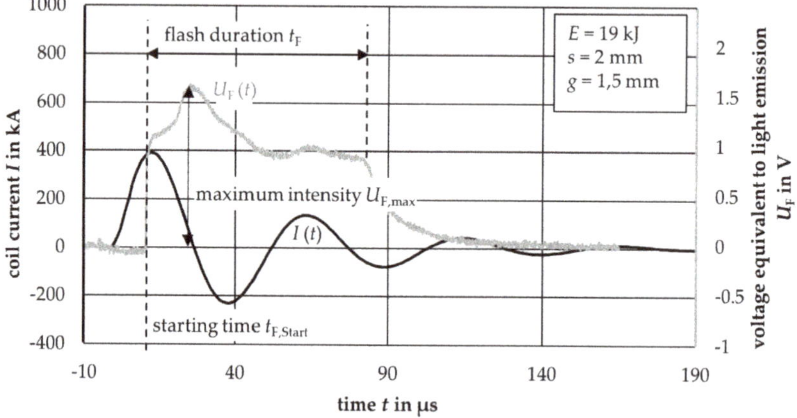

Figure 4. Signals of the Rogowski pickup coil and the flash measurement system showing the time-dependent evolution of the tool coil current and of the light emission, respectively, in the setup with the pulse generator BlueWave PS48-16.

2.4.3. Photonic Doppler Velocimetry

An accurate, quantitative determination of the flyer velocity at the moment of impact is crucial for the setting of similar impact velocities at the different experimental setups. Compared to the test rig, the accessibility of the collision zone during the MPW process is very limited for cameras due to the small acceleration distance and the progression of the collision front. The initial impact velocity at a fixed acceleration distance can be adjusted via the charging energy, but the analytical or numerical determination is elaborate and the MPW process is very sensitive to small disturbances like variations of material properties or flyer dimensions. Therefore, photonic Doppler velocimetry was applied for the measurement of the impact velocity v_{imp} during the MPW experiments. This robust and accurate method, developed by Strand et al [30], is based on the laser Doppler effect. It is an established measurement technology in the field of high-velocity forming and joining [31]. With the applied PDV system, velocities of up to 1.1 km/s can be measured using three parallel channels. The characteristics of the applied system are described by Lueg-Althoff in [32].

The application of PDV measurements requires the direct accessibility of the surface of the moving object, i.e., the flyer part in MPW. Therefore, three focuser probes were positioned normal to the setup (see Figure 3). Small boreholes were drilled into dummy target parts and the backing plate in order to allow the laser beams to directly illuminate the moving flyer surface. This allows measuring the temporal evolution of the flyer velocity from the beginning of the movement until the impact, and to adjust the impact velocity v_{imp} for all three flyer thicknesses and acceleration distances by modifying the charging energy of the pulse generator and the corresponding current flow in the tool coil, respectively. In Table 2, the determined charging energies are listed for several combinations of flyer thickness and acceleration gap (Experimental Series 2.1, 2.2, 2.3), as well as the averaged measurements of maximum current, discharge frequency and impact velocity (including minimum and maximum values). The boreholes in the target plate inhibited welding between the flyer and the dummy target parts. Nevertheless, the evolution of the flyer impact velocity was not affected by the presence of the boreholes until the collision because the magnetic field is completely shielded by the conductive flyer part. Therefore, it was assumed that the normal impact velocity v_{imp} determined at the three measuring points was identical in the experiments with a solid target for the real welding experiments.

Table 2. Determined charging energies for combinations of flyer thickness and acceleration gap and averaged measurements of maximum current, discharge frequency and impact velocity.

Flyer Thickness s in mm	Acceleration Gap g in mm	Charging Energy in kJ	Ø max. Current in kA	Ø Discharge Frequency in kHz	Min. Impact Velocity in m/s	Max. Impact Velocity in m/s	Ø Impact Velocity in m/s
2.0	1.5	19.0	392.4	19.7	255	261	257.0
2.0	2.0	18.0	381.3	19.7	255	258	256.3
2.0	2.5	17.5	376.0	19.7	259	262	260.7
1.5	1.5	14.5	340.5	19.7	251	257	254.3
1.5	2.0	13.7	330.4	19.7	251	256	253.0
1.5	2.5	13.3	325.3	19.7	249	255	252.3
1.0	1.5	9.7	274.6	19.7	261	269	263.7
1.0	2.0	9.3	268.7	19.7	260	261	260.3
1.0	2.5	8.7	259.8	19.7	249	250	249.3

2.4.4. Flash Detection

High-speed collision processes are accompanied by a characteristic flash, which is called impact flash [33]. During the MPW experiments, the time-dependent evolution of the light emission was measured with the flash measurement system explained in [34]. Figure 4 shows an example of the time-resolved light intensity as well as the derived characteristic values—the starting time of the flash, its duration and maximum intensity. According to previous studies, the flash duration was defined as the duration between the initial increase of the light intensity and its steep decrease [34]. These parameters were taken with two independent sensors at the same distance to the welding zone of approximately 15 mm and were then averaged for each series of experiments with a specific flyer thickness and acceleration gap, respectively. The results are shown in Appendix A.

2.5. Analysis of the Weld Interface

A nondestructive analysis method of the welded area was carried out using a 2D ultrasonic measurement with the MiniScanner (by Amsterdam Technology, Zwinderen, The Netherlands) for experiments on the model test rig. This device scans an area of 12 mm × 25 mm during a single scan run in pulse-echo mode with a local resolution of approximately 0.1 mm × 0.1 mm spots. For each spot, a so-called A-scan echo was recorded with its two-dimensional coordinates. This scan information was analyzed using a MathWorks MATLAB script. Depending on the signal, a differentiation was made between "bond", "no bond" or "no information" for each scanned point. This approach delivered both a qualitative and quantitative result in terms of the welded area [24]. Since the joining partners tear off after their collision in the model test rig, subsequent collisions with the still rotating rotors and the housing can occur. These joints were post-treated by flattening prior to ultrasonic examination. In addition, scratches and burrs were removed from the surfaces by grinding to improve the signal quality. If the deformation exceeds a certain limit, the evaluation of the joint was affected and hindered locally or completely.

The joints produced at the MPW setup were tested for their bond strength by tensile shear testing in a Z100 testing machine (by Zwick, Ulm, Germany) with three repetitions for each parameter set at a testing velocity of 10 mm/min. Additionally, for one joint of each series, a cross-section was prepared parallel to the central plane in the welding direction as shown in Figure 5. Due to the symmetric collision of the flyer with the target, there were two symmetric points of collision that moved in opposite directions and formed two weld interfaces; see also Figure 1. To characterize the welding result, the widths of the two welding interfaces and the gap between them were measured in the cross-section using an optical microscope (OM) DM2700 (by Leica Microsystems, Wetzlar, Germany).

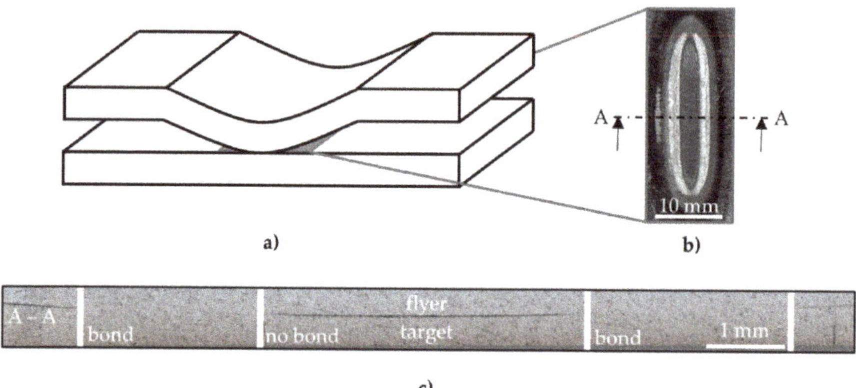

Figure 5. (**a**) Schematic sketch of a welded MPW joint (**b**) with a fracture image of the welding interface with typical elliptical ring-shape after tensile shear test. The location of the cross-section is marked by the dot-dashed line (A–A) and was analyzed by OM and SEM. An exemplary OM-image of a cross-section (A–A) is shown in (**c**).

The microstructures of the weld interfaces of joints made with both setups were further analyzed with a Ultra Plus (by Zeiss, Jena, Germany) scanning electron microscope (SEM), using the secondary electron detector to validate the results of the ultrasonic examination and also to determine how the different weld interface types have been formed.

3. Results

3.1. Model Test Rig

Figure 6 shows the welding windows for the series of Experiments 1.1 and 1.2 performed in the test rig. Although in both cases, the lower boundary collision angle, below which bond formation is inhibited, did not differ strongly, the upper boundary collision angle increased both with higher flyer thickness and increased impact velocity.

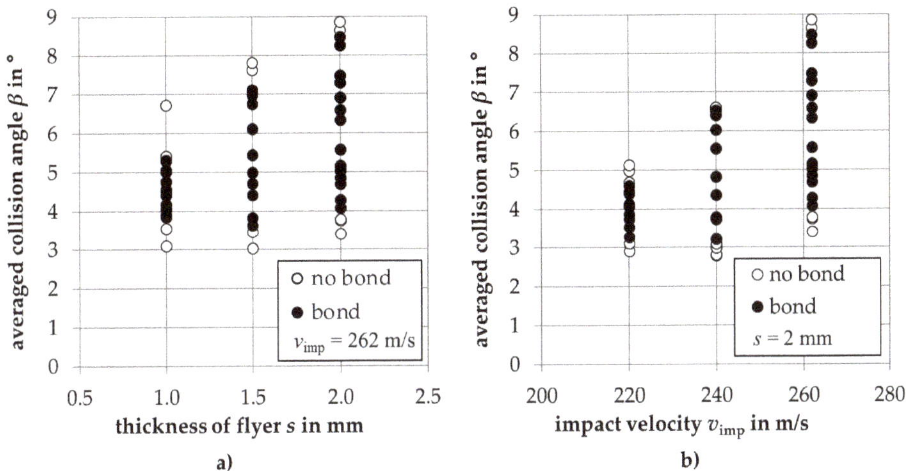

Figure 6. Welding windows (**a**) for different flyer thicknesses and collision angles and (**b**) for different impact velocities and collision angles, each point represents one experiment. The upper boundary angle increased in both cases, while the lower boundary varied only slightly.

For Series 1.2, the findings correspond to those already obtained in welding window investigations for another batch of the same material. In this context, the lower boundary angle was related to the suppressed ejection of the CoP, which inhibits bond formation by the reinclusion of the CoP particles at the PoC. In contrast, the upper boundary angle defines the process parameters up to which jet formation can be initiated and sustained [23].

The results of the ultrasonic analysis of the weld interface of Series 1.1 are shown in Figure 7. A similar curve of the ratio between the welded and overlapping areas over the collision angle was achieved for all three flyer thicknesses (see Figure 7a). Just a small amount of the overlapping area was welded close to the lower and upper boundary angles. In between, there is a region where large area welds can be formed. However, none of the specimens was completely welded. The range of this region, its maximum value and the corresponding collision angle increased with increasing flyer thickness and was shifted to larger collision angles. This behavior had also been observed in experiments where copper was welded at different impact velocities [24] and was further validated by the results of Experimental Series 1.2 with lower impact velocities using a flyer thickness of 2.0 mm. The types of weld interfaces described in [25] can also be found here and support the described phenomena at the upper and lower boundary angle. For small collision angles and without inhibiting the bond formation, the CoP could only escape sufficiently in the lateral regions and at the end of the closing gap. At large collision angles, the jet formation was initiated after an entry region, but then broke down while the flyer continued to deform on top of the target (see Figure 7b). However, this weld interface type was not as pronounced regarding the termination of the weld interface formation as in the previous experiments with copper. Furthermore, no completely welded interface could be obtained, which might be due to the fact that the investigations were carried out close to the lower limit of the welding process with respect to the energy input.

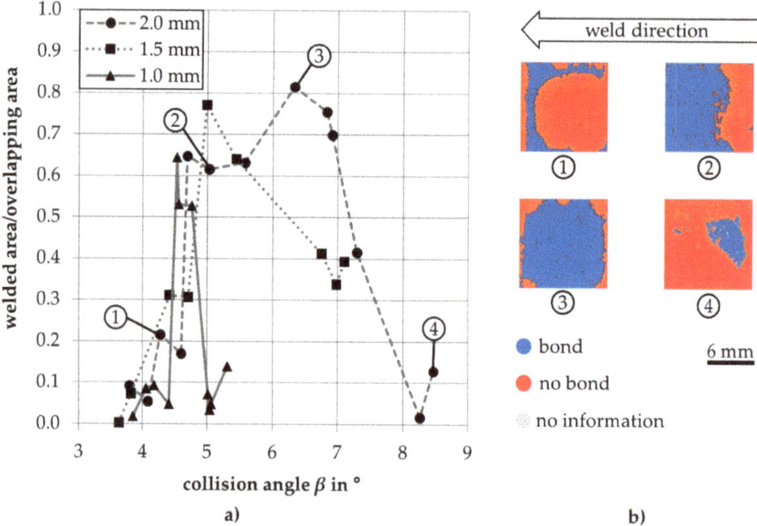

Figure 7. (**a**) Progression of welded to the overlapping area over different collision angles for the three flyer thicknesses (target thickness: 2 mm, v_{imp} = 262 m/s); (**b**) two-dimensional representation of the weld interface obtained from the ultrasonic analysis.

Figure 8 summarizes the SEM analysis of the interfaces of Experimental Series 1.1. The welded interfaces are mostly straight and only single instances of wavy patterns can be found. Furthermore, the findings of the nonwelded interfaces and the transition regions support the hypothesis regarding the boundaries by collision angle and the related mechanisms. In Figure 8a the collision at an angle close to the lower boundary angle for 2 mm flyer thickness started without visible interaction of the surfaces (1). Shortly afterwards, the surfaces were contaminated by the enclosed CoP (2) whose amount increased along the joining gap (3). At a certain stage, the conditions in the gap changed in a way that local melting and resolidification occurred at the surfaces and the surfaces got continuously closer, until the formation of the weld interface began (4, 5). At the end of the weld interface a continuous melted and resolidified interlayer was found (6).

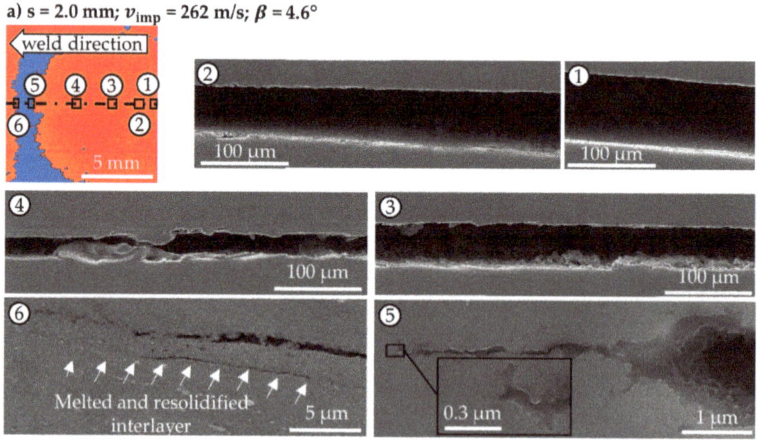

Figure 8. *Cont.*

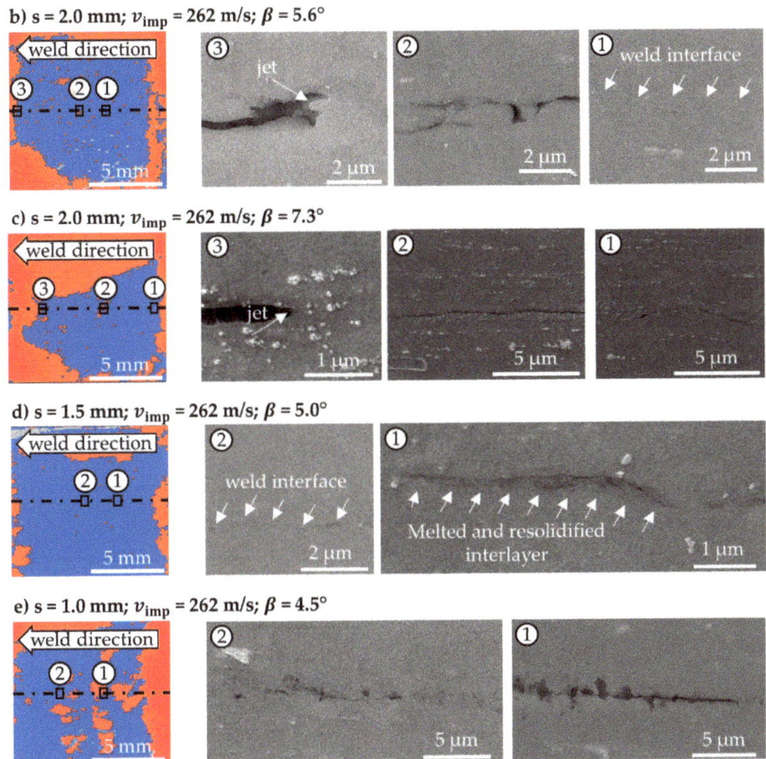

Figure 8. SEM analysis of test rig joints at different locations of the weld interface. Blue coloring indicates bond, red indicates no bond. (**a–c**): *s* = 2 mm: (**a**) Close to the lower boundary collision angle (4.6°); (**b**) in the region with a large welded interface (5.6°), (**c**) close to the upper boundary collision angle (7.3°); (**d**) largest welded area (5.0°) for *s* = 1.5 mm; (**e**) largest welded area (4.5°) for *s* = 1 mm (note that different magnifications are used in the SEM micrographs to highlight relevant features).

In the region with a large welded area for 2 mm flyer thickness, the weld interface could hardly be recognized in the SEM micrograph (see Figure 8b) (1); however, nonwelded areas could be clearly identified. Only at the beginning of the weld interface, some melted structures and a porous interface were observed. Later on, the few nonwelded regions in the center did not contain porous material from the CoP but obviously parts of the jet stream which were torn off and rolled over by the PoC and hindered the bond formation (2). Moreover, at the end of the weld interface, the spilled jet was clearly visible (3—arrow).

In joints with 2 mm flyer thickness, produced close to the upper boundary angle, the welded regions were not properly formed and contained several imperfections (Figure 8c) (1, 2). The jet at the end of the weld interface was significantly thinner (3) than in the region described above.

Looking at the joints with 1.5 mm and 1 mm flyer thickness, large welded areas were found. While for *s* = 1.5 mm the sound weld interface was mostly hardly visible in the SEM micrographs (Figure 8d) (2), other parts contained locally melted and resolidified interlayers (1). The weld interface for s = 1.0 mm exhibited partially porous regions, melting structures and cracks (see Figure 8e), which were partly declared as sound weld by the ultrasonic analysis (2). Nonwelded regions showed larger melting defects (1).

3.2. MPW *Setup*

Figure 9 shows the results of the weld interface formation in the MPW setup for the different flyer thicknesses and acceleration distances at selected positions of the weld interfaces (compare Table 1, Series 2.1). Considering the parameters separately, the start of the weld interface was not influenced by the flyer thickness. The end position of the weld interface increased with increasing flyer thickness. When the acceleration distance was enlarged, weld interface formation started earlier, but also ended earlier.

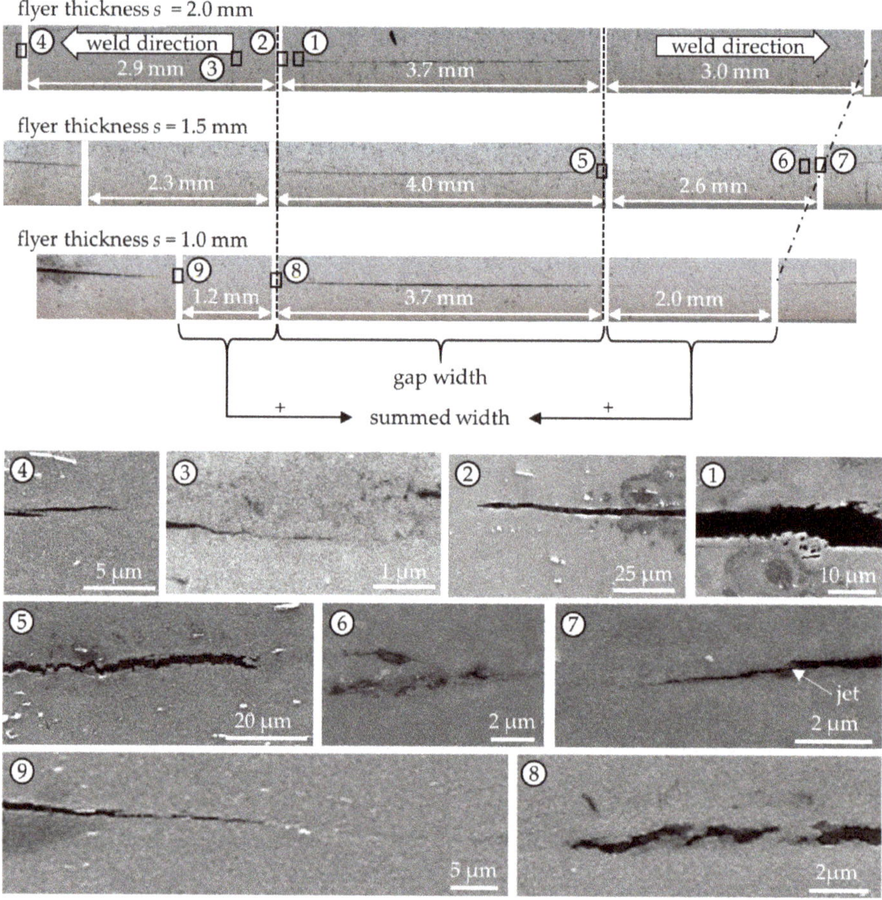

Figure 9. Microsections along the central plane parallel to the welding direction of weld interfaces in the MPW setup for an acceleration gap of $g = 1.5$ mm (Series 2.1): While weld interface began at the same position (dashed line), it ends later with increasing flyer thickness (dot-dashed line). Below, the results of the SEM investigation are shown in detail (1–9).

The summed width of both weld interfaces tended to decrease with increasing acceleration distance, especially for smaller flyer thicknesses, see also Figure 10. Furthermore, the flyer thickness of 1 mm exhibited an asymmetric image of the weld interfaces for 1.5 mm and 2.5 mm acceleration distances (Series 2.2, 2.3). The latter was only welded on one side in the sectioned joint (see cross-section in Figure 10), which is visible in the diagram by the smaller total weld interface width and thus, no gap width. This was either a result of the comparably low energy input or of an asymmetric rolling movement of the flyer due to the clamping situation in the weld setup.

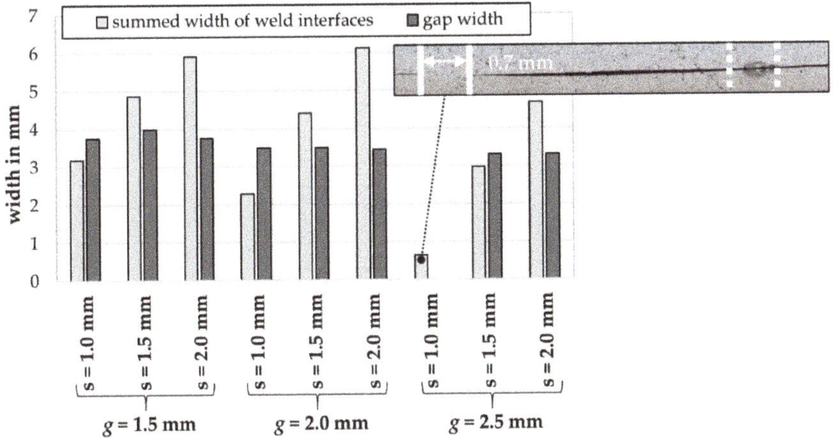

Figure 10. Summed width of both magnetic pulse welded interfaces and width of the gap in between for different flyer thicknesses and acceleration gaps at constant impact velocity of 262 m/s. All configurations show the same trend: while the gap width varied slightly for the acceleration distances *g*, the summed width of the weld interface increased with increased flyer thickness *s*. Due to the asymmetric weld formation (see cross-section), no gap width could be determined for the configuration *s* = 1.0 mm and *g* = 2.5 mm.

The SEM analysis of the weld interface in Figure 9 revealed similarities with the results that were produced in the test rig, see Section 3.1. Pores and partly melted and resolidified structures were found in front of (1) and at the beginning of the weld interfaces (2, 5, 8). Such weld defects were related to the heating and/or entrapment of the CoP and indicated a collision angle close to the lower boundary. Similar defects were also located in further sections along the weld interfaces (6). The ends of the weld interfaces (4, 7, 9) were similar to the ones produced in the test rig that were welded at a collision angle close to the upper boundary. A thin jet at the end of the weld was identified for *s* = 1.5 mm (7).

The bearable tensile forces of the different weld configurations (Series 2.1, 2.2, 2.3) are shown in Figure 11. The comparison of the tensile forces revealed that for an acceleration distance of 1.5 mm all joints achieved the bearable tensile force calculated from the tensile strength of the base material and thus, all failed in the base material except for two joints. These two parts showed a nonuniform weld interface formation in the fracture pattern. For joints with a flyer thickness of 2 mm and an acceleration distance of 2 mm, failure occurred in the base material, which was also apparent in the achieved tensile force. In this case, the fracture occurred in the neck region of the flyer close to the welded area, where the cross-sectional area was reduced due to plastic deformation during the welding process.

All other joints failed in the weld interface. Relating the tensile force to the total weld interface width resulted in a ratio of approximately 1.5 kN/mm for all configurations. Only the configurations of 2 mm and 2.5 mm acceleration distance with 1 mm flyer thickness varied to lower values due to the incomplete global weld interface formation. The fracture images of these samples revealed that the weld interface was characterized as two parallel lines instead of a complete elliptical ring (see fracture images in Figure 11). Furthermore, all fracture surfaces showed a symmetric weld interface in contrast to the cross-section in Figure 10. Therefore, the width value of this configuration was corrected by the multiplication by a factor of two to calculate the ratio of tensile force to the width in Figure 11 to represent a symmetric weld interface.

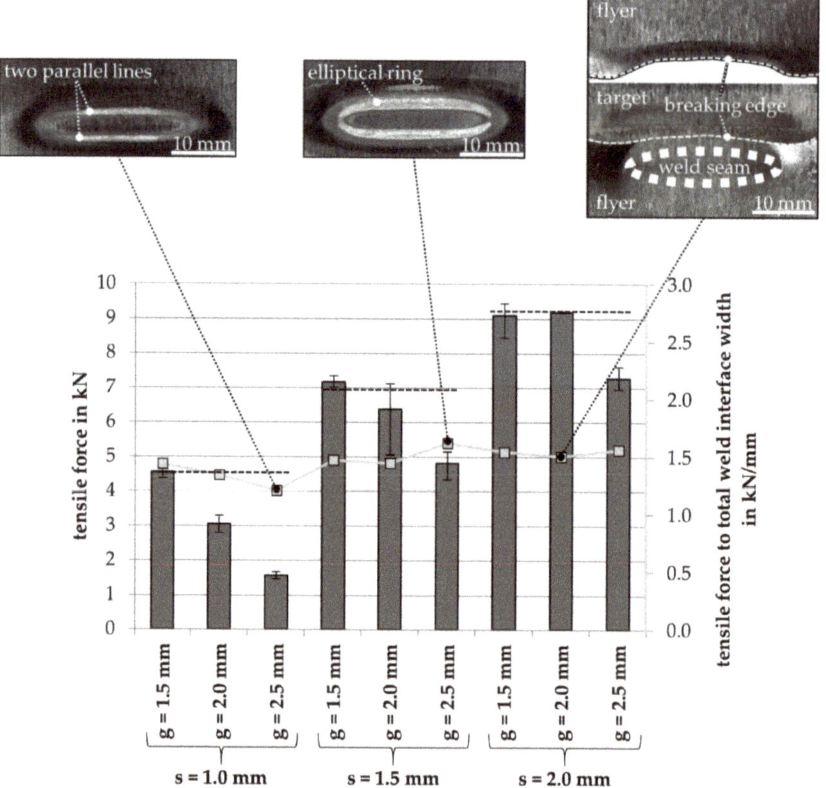

Figure 11. Averaged tensile force with minimal and maximal deviation for different flyer thicknesses *s* and acceleration distances *g* (left axis, vertical columns). The dashed lines represent the theoretical bearable tensile force value for the particular flyer thickness calculated by the tensile strength of the base material. The ratio of the tensile force to total weld interface width is represented by rectangles and grey line and the right axis scale. The images of the different fracture surface types in top view are shown to explain the variation of the ratio tensile force to total weld interface width.

4. Discussion

4.1. Influence of Kinetic Energy Input on Weld Interface and Welding Window

For the comparison of the results of the model test rig and the MPW setup, it is important to recall that the collision conditions continuously change during the MPW process [18–20]. Hence, the impact starts at a small collision angle close to 0° and at the maximum impact velocity. This results in a high collision point velocity. Subsequently, the collision angle increases and the impact velocity decreases until the weldable region of the welding window is entered and then, after further progression of the collision conditions, left again. Depending on the process parameters, the path through the material-specific welding window differs. In the test rig, in contrast, the collision angle and the impact velocity stay constant during the collision. Therefore, each experiment represents a single point in the welding window. Performing a series of experiments with different collision conditions allows defining welding windows for different process and material parameters. This in turn provides an extended understanding of the governing welding mechanisms and delivers additional information for the design of the MPW process with industrial relevance.

Considering the determined welding windows for different thicknesses in the test rig (Series 1.2), it was possible to analyze the formation of the weld interface in the MPW setup (Series 2.1, 2.2, 2.3); compare Figures 6 and 9. The progression of the collision angle was similar for different thicknesses, until the start of the bond formation. In addition, higher thicknesses resulted in longer weld interfaces. For MPW it was not possible to identify, whether the bond formation stopped (i.e. the collision kinetics left the weldable region of the welding window) due to an increase of the collision angle or due to a combined change of collision angle and impact velocity. The latter depends on the different thicknesses, the resulting different stiffness values and thus a varied forming behavior of the flyer during the rolling movement on the target.

When the acceleration distance was increased, the location of the weld interface was shifted closer to the initial point of collision. This was due to the higher progression rate of the collision angle leading to a flyer rolling movement that the lower boundary angle was exceeded earlier and the upper boundary angle was reached faster. This is in good agreement with the findings of Sarvari et al. [21] who investigated different acceleration distances in an MPW setup.

To understand the influence of the mass induced kinetic energy on the bond formation, the results of the test rig experiments with varied thickness (Series 1.1) and varied impact velocity (Series 1.2) are plotted together in the classical β-v_c-welding window in Figure 12. It can be observed that the increase in energy input by changing both the flyer mass and a higher impact velocity led to an expansion of the weldable region towards lower collision point velocities. The change of the impact velocity to achieve a certain kinetic energy also changed the collision conditions while they remained equal when the flyer mass was increased. Moreover, the weldable area of the welding window was affected by the increase in energy. Similar findings were reported by Lysak and Kuzmin in [26] for explosion welding. To explain the solid-state welding mechanism, they related the impact velocity, the collision point velocity and the involved mass to three physical parameters, the pressure at the PoC, the duration of the applied pressure and the temperature in the weld interface zone. In the next section, it is explained how these parameters influenced the phenomena in the joining gap and the governing bond mechanisms in the case of changed energy input.

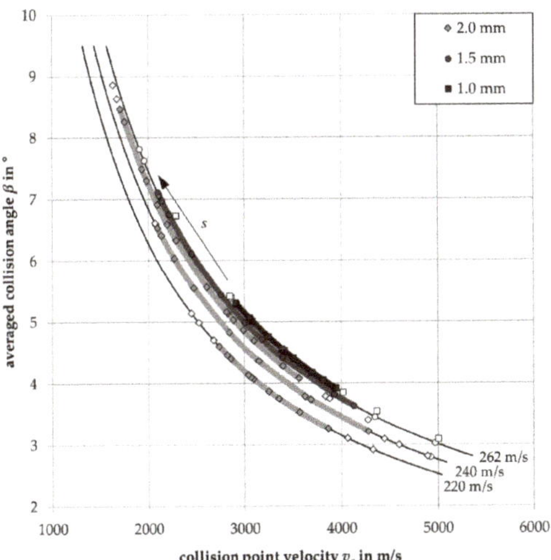

Figure 12. Welding window by collision angle β over collision point velocity v_c for different flyer thicknesses s and impact velocities v_{imp} measured in the test rig; symbols in shaded graphs represent experiments with bond, symbols outside of the graphs represent experiments with no bond. The different flyer thicknesses at $v_{imp} = 262$ m/s are plotted at slightly different impact velocities to improve the visibility. Arrow indicates the influence of the increasing value on the upper left boundary.

4.2. Influence of the Collision Kinetics on the Phenomena and the Bond Mechanisms

The findings of the test rig experiments (Series 1.1 and 1.2) and the analysis of the weld interface revealed that the acting phenomena, especially the role of the CoP, at different points of the welding window have a pronounced influence on the occurring bond formation mechanisms. This has not been taken into account previously. Hence, due to the transient propagation through the welding window during MPW, it can be assumed that different mechanisms of bond formation apply in certain sections depending on the process parameters and the formed CoP.

During test rig experiments with small collision angles, maximum flyer mass and high impact velocity, the CoP had a strong impact on the bond formation. The SEM images in Figure 8a (1–3) for $s = 2$ mm indicate that at first, due to the small angle and resulting high flow resistance, the CoP could not be ejected out of the gap, was entrapped and hindered the contact of the activated base materials. It was found by the analysis of the associated process glare that the CoP can reach temperatures of several thousand Kelvin (see the accompanying paper [10] and compare it with Figure A1). This might also result in the nonwelded regions due to excessive melting of the surfaces (4). Later on, during the collision front propagation towards the free end of the flyer, the resistance to eject the CoP was low enough with the result that the interfaces came into contact and formed a bond (5). In this case, the mechanism of bond formation was attributed to fusion-like bonding, regarding the estimated high-temperature development and no visible signs of a jet in terms of a metal stream. This hypothesis is in good agreement with the findings of Bellmann et al. [35], who observed no deformation of the surface-near layers in the form of a jet at small collision angles but melting close to and in the weld interface.

In contrast, the formation of a jet occurred at larger collision angles (Figure 8b). In the case of large welded areas (ratio: 0.82 to the total overlapping area) in the middle of the estimated welding window for $s = 2$ mm (Figure 8b), most parts of the weld interface exhibited only a few defects (2) and were almost not distinguishable from the base material in the SEM. A determination of the governing bond mechanism was thus not possible. The absence of porous structures at the weld interface may indicate solid-state welding. At this point, it should be noted that the present investigation focused on joints between similar metals. No additional aspects were considered that would occur during the welding of dissimilar metals, like the formation of intermetallic phases.

The weld interface close to the upper boundary angle for $s = 2$ mm exhibited a less distinct jet and several areas without bond (see Figure 8c). This indicates that the deformation of the surfaces was considerably smaller and less adjacent base material was deformed compared to smaller collision angles. Two approaches provide possible explanations for these findings: First, according to Manikandan et al. [36], the depth of deformation of the adjacent base material by the colliding surfaces depends on the induced kinetic energy. Therefore, it can be argued that for large collision angles less induced energy was available for the microscopic deformation close to the point of collision. This can be explained by the fact that at larger collision angles more work was needed to close the joining gap by the continuous bending of the flyer. A second explanation may be the different interactions of the CoP with the surfaces in front of the point of collision. As mentioned above, the CoP can reach very high temperatures. Even if it did not melt the surfaces, the induced heat could cause a considerable reduction of the flow stress close to the point of collision, allowing more material to flow in the area of the contact surfaces, which was already described by Khaustov et al. [37] for explosion welding. At larger collision angles the temperature was lower due to the lower compression and less heat was transferred to the surfaces. Thus, the potential plastic deformation in the point of collision was reduced at increasing collision angles, which also explains the reduced intensity of the process glare (Figure A1). Moreover, the interaction of both phenomena is conceivable to explain the differences between the cases in Figure 8b,c.

Considering the influence of different flyer thicknesses, for $s = 1.5$ mm a comparable maximum welded area was obtained (ratio: 0.77), whereas for $s = 1.0$ mm the maximum value is smaller (ratio: 0.64). In both cases, the joint was produced in a region close to the lower collision angle boundary, where for $s = 2.0$ mm the bond formation (ratio: 0.62) has not reached its maximum value. Although some pores and melted interlayers were determined in sections of the weld interfaces of smaller flyer thicknesses, which likely resulted from the interaction with the hot CoP, large inclusions of the CoP could not be detected. Since the impact velocity, the collision angle and thus the air flow in the joining gap were comparable, this could not be attributed to a better ejection condition of the CoP. A possible explanation could be that due to the higher energy input more deformation around the point of collision occurred and, thus, also a larger jet was formed. As shown in Figure 8b (2), the jet can spall in particles. Furthermore, Pabst [38] determined that during the collision welding of aluminum, additional heat was induced into the CoP by an exothermic reaction of chipped particles of the base material. If more base material particles were chipped out by the impact, this in turn might cause more heating. The influence was lower for smaller flyer thicknesses and resulted in less spalled particles, heating and process glare (see Figure A1).

Concerning the governing bond formation mechanisms and considering the SEM images, the largest welded area for $s = 2$ mm was likely produced in the test rig experiments by solid-state welding, whereas excessive melting prevented welding at smaller collision angles. In contrast, the largest welded areas for the smaller flyer thicknesses were formed at least partly by fusion-like bonding. Both processes were strongly influenced by local interaction of the surfaces with the CoP.

These findings are transferable to the MPW interface (Series 2.1, 2.2, 2.3). A certain angle had to be exceeded to allow the ejection of the CoP and to initiate bond formation. The weld interface started with a partially melted but bonded interface where parts of the CoP were entrapped which is also supported by the findings in [6,39]. The further transient rolling movement of the flyer facilitated the ejection of the CoP, whereas the formation of the jet by plastic deformation occurred more intensively at first and then weakened again. This resulted in a jet behavior similar to that at large collision angles in the test rig (see Figure 9) (6). These sections were not clearly distinguishable in the SEM images. Considering the locations of weld defects by melting or CoP-entrapment, it is assumed that in the case of 1.5 mm, most of the weld interface was a result of fusion-like bonding, whereas the latter sections were bonded by solid-state welding due to the occurrence of the jet at the end. In the case of 2 mm flyer thickness, solid-state bonding dominated the weld interface due to the absence of such weld failures in the interface. On the other hand, a jet at the end of the weld interface was not clearly visible (Figure 9) (4).

Despite these different bond mechanisms and resulting weld interfaces, no significant dependence of the joint strength on bond mechanisms was identified (see Figure 11). The differences in the determined tensile forces mostly depended on the weld interface width and the question of whether a complete ring-elliptical weld interface was formed or not, which directly influences the size of the welded area (compare with Figure 11).

4.3. Prediction and Control of the Weld Interface's Formation and Properties

The results in Appendix A show that the process glare detected during the MPW experiments was brighter for an increased flyer thickness (Series 2.1, 2.2, 2.3). This is in good agreement with the experiments performed in the test rig where the kinetic collision conditions were kept constant. It is likely that the increased kinetic flyer energy led to an increased formation of CoP (or jet) as also described by Eichhorn [33]. Figure A2 shows the tendency of an increased flash intensity for smaller acceleration distances, which points to the conclusion that the collision angles were smaller and, thus, the compression and light emission of the CoP (or jet) were intensified. This finding can also be related to higher temperatures, as reported by Bellmann et al. [35].

The correlation of process glare parameters with the welding results revealed that the longest weld seams and the highest bond strengths were achieved in experiments with the longest and brightest impact flashes. This was the case for the highest flyer thickness and energy input, respectively, as well as for the smallest acceleration distance. Although the impact flash is only a *necessary* but not a sufficient welding criterion, this observation underlines the importance of a sufficient surface activation prior to surface contact, which comes along with a bright impact flash. At this point, it should be noted once more that only the impact velocity was adjusted directly during the MPW experiments, but not the collision angle. Thus, the reason for the increased weld length and strength obtained by using thicker flyers and smaller acceleration distances might also be attributed to the differences in the collision kinetics, especially the reduced collision angle. Starting with a small collision angle and assuming similar progression rates, a larger portion of the flyer collides under weldable conditions with the target compared to a high initial collision angle. This led to longer weld seams and brighter impact flashes. Hence, the correlation between the impact flash and the weld strength reported here is valid and appears to be a powerful tool for process development and quality assurance during MPW processes.

Considering the largest ratios of welded to the overlapping area at the test rig (Figure 7), it can be derived that there is a collision angle region leading to optimum collision welding conditions. This region increases and is shifted to larger angles when the kinetic energy input is increased. This kinetic energy input can be varied via the impact velocity [24] or the flyer thickness. The resulting adjustment of the progression rate of the collision angle allows controlling location and length of the weld interface during MPW.

5. Conclusions and Outlook

The collision kinetics influence both the CoP formation and its temperature (measured by analyzing the process glare in the companion paper [10]). It, therefore, determines the governing bond mechanism and thus, the reachable amount of welded area. The latter is, however, also influenced by other parameters like (initial) collision angle, its progression and the rolling movement.

Depending on the process conditions, the CoP can be useful for, or harmful to, the bond formation, it furthermore determines the predominant bonding mechanism.

The results of the test rig experiments confirm that the width of the weld interface can be increased by a smaller gradient of the collision angle, when the weldable area of the welding window is reached. Therefore, it could be useful to prepare the flyer geometry to influence its rolling movement on top of the target during MPW and, thus, to improve the weld interface formation. Together with the acquired knowledge about the different ways to increase the energy input, it is possible to adjust the size and location of the weld interface by setting the process parameters. In addition, monitoring of the process glare potentially enables quality assurance of high-strength joints.

Considering that the amount of the CoP increases continuously with the length of the colliding joining partners, the question arises in which areas it is still possible to consider a stationary welding process despite constant kinetic conditions. Therefore, it is of interest for further investigations if and how the local bond properties change and if the welding window can be extended by considering, as an additional factor, the quality of the weld interface.

Author Contributions: Conceptualization, B.N., E.S., J.L.-A., J.B. and M.B.; methodology, data analysis, B.N., E.S., J.L.-A. and J.B.; design of experiments, investigation, validation, B.N. (test rig experiments, ultrasonic investigation, long time exposures, SEM analysis), E.S. (MPW experiments, tensile shear tests, optical microscope and SEM analysis), J.L.-A. (PDV-measurement, flash measurement), J.B. (flash measurement) and M.B. (characterization of material's properties); writing—original draft preparation, visualization, B.N., E.S. (Sections 2.3, 2.5, 3.2, 4 and 5), J.L.-A. (Sections 2.3, 2.4.2, 2.4.3, 4 and 5) and J.B. (Sections 2.4.4, 4 and 5); writing—review and editing, S.B., A.E.T., C.L., M.F.-X.W. and P.G.; resources, supervision, project administration, funding acquisition, S.B., A.E.T., E.B., M.F.-X.W., and P.G. All authors have read and agreed to the published version of the manuscript.

Funding: This research was funded by the Deutsche Forschungsgemeinschaft (DFG, German Research Foundation), grant number BE 1875/30-3, TE 508/39-3, GR 1818/49-3, WA 2602/5-3, BO 1980/23-1 and is based on the results of the working group "high-speed joining" of the priority program 1640 ("joining by plastic deformation"). The working group consists of the subprojects A1, A5, A8 and A9. We also acknowledge support by the German Research Foundation and the Open Access Publishing Fund of the Technical University of Darmstadt for providing this publication as open access.

Acknowledgments: The experiments at the model test rig were conducted at the Bachelor Thesis of Yabing Wang under the supervision by Benedikt Niessen. Moreover, we would like to thank Walter Tutsch of PCO AG for the support in setting up the image intensifier camera. The authors also greatly appreciate the help of Stephan

Ditscher of Baumüller who supported the programming of the system control of the test rig. We thank Ammar Ahsan for his efforts in proofreading the manuscript.

Conflicts of Interest: The authors declare no conflict of interest. The funders had no role in the design of the study; in the collection, analyses, or interpretation of data; in the writing of the manuscript, or in the decision to publish the results.

Appendix A Analysis of the Process Glare

The qualitative examination of the process glare during test rig experiments is represented in Figure A1 for flyer thicknesses of 1 mm and 1.5 mm at different collision angles. It should be noted that the majority of process glare at the images was visible after leaving the joining gap and no temporal resolution is shown. Nevertheless, it can be seen that the intensity increases with the thickness of the flyer and the glare's color changes from red-orange to white-blue. Furthermore, the collision angle had a significant influence on the intensity, color and amount of the process glare. For angles lower than the lower boundary angle, the glare in the gap was not visible. Therefore, the glare was visible at the sides of the colliding joining partners with only weak intensity. With increasing collision angle its intensity and shape also increased up to a maximum. At larger collision angles the shape remained at the same level, but the intensity decreased, especially when the upper boundary angle was exceeded.

flyer thickness s = 1.0 mm

flyer thickness s = 1.5 mm

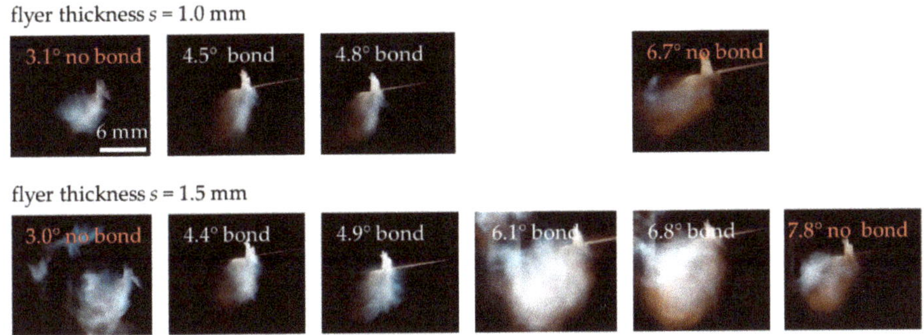

Figure A1. Comparison of the process glare for 1 mm and 1.5 mm flyer thickness at different collision angles, recorded by longtime exposure.

As mentioned in Section 2.4.4, the weld formation was accompanied by a bright flash when impact welding is performed in ambient atmosphere. The average values of the impact flash duration and its maximum intensities were evaluated for each parameter set (see Figure A2). The flash duration as well as the maximum intensity increased with higher flyer thickness. The influence of the acceleration distance on the light emission was not that clear due to the data spread. There is a slight tendency showing a decrease in the flash duration and intensity for increased acceleration distances.

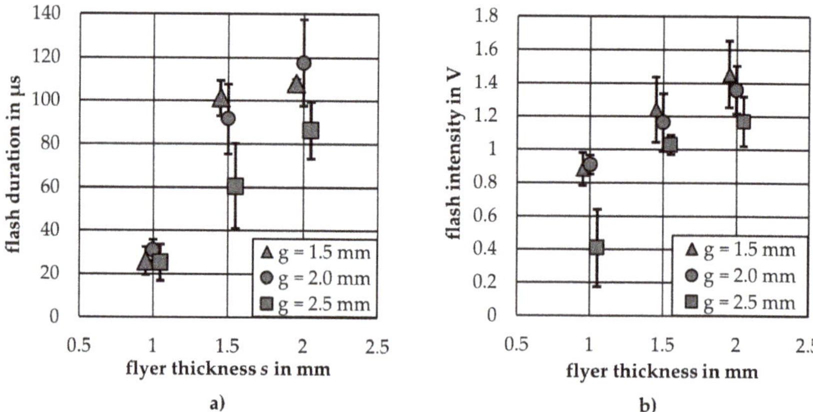

Figure A2. (**a**) Increasing flash durations and (**b**) increasing intensities for increasing flyer thicknesses (1.0 mm, 1.5 mm and 2.0 mm) at different acceleration distances *g*. Each point represents four or five experiments with the corresponding standard deviation.

References

1. Kapil, A.; Sharma, A. Magnetic pulse welding: An efficient and environmentally friendly multi-material joining technique. *J. Clean. Prod.* **2015**, *100*, 35–58. [CrossRef]
2. Jassim, A. Comparison of magnetic pulse welding with other welding methods. *J. Energy Power Eng.* **2011**, *5*, 1173–1178.
3. Wang, H.; Wang, Y. High-Velocity Impact Welding Process: A Review. *Metals* **2019**, *9*, 144. [CrossRef]
4. Stern, A.; Shribman, V.; Ben-Artzy, A.; Aizenshtein, M. Interface Phenomena and Bonding Mechanism in Magnetic Pulse Welding. *J. Mater. Eng. Perform.* **2014**, *23*, 3449–3458. [CrossRef]
5. Akbari Mousavi, A.A.; AL-Hassani, S.T.S. Numerical and experimental studies of the mechanism of the wavy interface formations in explosive/impact welding. *J. Mech. Phys. Sol.* **2005**, *53*, 2501–2528. [CrossRef]
6. Schumacher, E.; Rebensdorf, A.; Böhm, S. Influence of the jet velocity on the weld quality of magnetic pulse welded dissimilar sheet joints of aluminum and steel. *Materialwiss. Werkstofftech.* **2019**, *50*, 965–973. [CrossRef]
7. Carpenter, S.H.; Wittman, R.H. Explosion Welding. *Ann. Rev. Mater. Sci.* **1975**, *5*, 177–199. [CrossRef]
8. Cowan, G.R.; Holtzman, A.H. Flow Configurations in Colliding Plates: Explosive Bonding. *J. Appl. Phys.* **1963**, *34*, 928–939. [CrossRef]
9. Deribas, A.A.; Zakharenko, I.D. Surface effects with oblique collisions between metallic plates. *Combust. Explos. Shock Waves* **1974**, *10*, 358–367. [CrossRef]
10. Bellmann, J.; Lueg-Althoff, J.; Niessen, B.; Böhme, M.; Schumacher, E.; Beyer, E.; Leyens, C.; Tekkaya, A.E.; Groche, P.; Wagner, M.F.-X.; et al. Particle Ejection by Jetting and Related Effects in Impact Welding Processes. *Metals* **2020**, *10*, 1108. [CrossRef]
11. Bellmann, J.; Lueg-Althoff, J.; Schulze, S.; Gies, S.; Beyer, E.; Tekkaya, A.E. Magnetic Pulse Welding: Solutions for Process Monitoring within Pulsed Magnetic Fields. In Proceedings of the Euro-Asian Pulsed Power Conference & Conference on High-Power Particle Beams, Estoril, Portugal, 18–22 September 2016; pp. 18–22.
12. Cui, J.; Ye, L.; Zhu, C.; Geng, H.; Li, G. Mechanical and Microstructure Investigations on Magnetic Pulse Welded Dissimilar AA3003-TC4 Joints. *J. Mater. Eng. Perform.* **2020**, *212*, 8. [CrossRef]
13. Schäfer, R.; Pasquale, P.; Elsen, A. Material Hybrid Joining of Sheet Metals by Electromagnetic Pulse Technology. *Key Eng. Mater.* **2011**, *473*, 61–68. [CrossRef]
14. Baumgartner, J.; Schnabel, K.; Huberth, F. Fatigue assessment of EMPT-welded joints using the reference radius concept. *Procedia Eng.* **2018**, *213*, 418–425. [CrossRef]
15. Bellmann, J.; Schettler, S.; Dittrich, S.; Lueg-Althoff, J.; Schulze, S.; Hahn, M.; Beyer, E.; Tekkaya, A.E. Experimental study on the magnetic pulse welding process of large aluminum tubes on steel rods. *IOP Conf. Ser. Mater. Sci. Eng.* **2019**, *480*, 12033. [CrossRef]

16. Shribman, V.; Nahmany, M.; Levi, S.; Atiya, O.; Ashkenazi, D.; Stern, A. MP Welding of dissimilar materials: AM laser powder-bed fusion AlSi10Mg to wrought AA6060-T6. *Prog. Addit. Manuf.* **2019**, *11*, 2356. [CrossRef]

17. Mrzljak, S.; Gelinski, N.; Hülsbusch, D.; Schumacher, E.; Boehm, S.; Walther, F. Influence of Process Parameters, Surface Topography and Corrosion Condition on the Fatigue Behavior of Steel/Aluminum Hybrid Joints Produced by Magnetic Pulse Welding. *Key Eng. Mater.* **2019**, *809*, 197–202. [CrossRef]

18. Schumacher, E.; Kümper, S.; Kryukov, I.; Böhm, S. Analysis of the Weld Seam Area of Magnetic Pulse Welded Aluminium-Steel-Sheet-Connections on its Suitability as a Sign of Quality. In Proceedings of the 8th International Conference on High Speed Forming (ICHSF 2018), Columbus, OH, USA, 13–18 May 2018.

19. Groche, P.; Pabst, C. Numerical Simulation of Impact Welding Processes with LS-DYNA. In Proceedings of the 10th European LS-DYNA Conference, Wurzburg, Germany, 15–17 June 2015.

20. Cuq-Lelandais, J.-P.; Avrillaud, G.; Ferreira, S.; Mazars, G.; Nottebaert, A.; Teilla, G.; Shribman, V. 3D Impacts Modeling of the Magnetic Pulse Welding Process and Comparison to Experimental Data. In Proceedings of the 7th International Conference on High Speed Forming, Dortmund, Germany, 27–28 April 2016.

21. Sarvari, M.; Abdollah-zadeh, A.; Naffakh-Moosavy, H.; Rahimi, A.; Parsaeyan, H. Investigation of collision surfaces and weld interface in magnetic pulse welding of dissimilar Al/Cu sheets. *J. Manuf. Process.* **2019**, *45*, 356–367. [CrossRef]

22. Psyk, V.; Hofer, C.; Faes, K.; Scheffler, C.; Scherleitner, E. Testing of Magnetic Pulse Welded Joints—Destructive and Non-Destructive Methods. In Proceedings of the 22nd International ESAFORM Conference on Material Forming (ESAFORM 2019), Vitoria-Gasteiz, Spain, 8–10 May 2019; American Institute of Physics: Melville, NY, USA, 2019; p. 50010, ISBN 978-0-7354-1847-9.

23. Bellmann, J.; Lueg-Althoff, J.; Schulze, S.; Hahn, M.; Gies, S.; Beyer, E.; Tekkaya, A.E. Effect of the wall thickness on the forming behavior and welding result during magnetic pulse welding. *Materialwiss. Werkstofftech.* **2019**, *212*, 150. [CrossRef]

24. Groche, P.; Niessen, B.; Pabst, C. Process boundaries of collision welding at low energies. *Materialwiss. Werkstofftech.* **2019**, *50*, 940–948. [CrossRef]

25. Niessen, B.; Groche, P. Weld Interface Characteristics of Copper in Collision Welding. In Proceedings of the 22nd International ESAFORM Conference on Material Forming (ESAFORM 2019), Vitoria-Gasteiz, Spain, 8–10 May 2019; American Institute of Physics: Melville, NY, USA, 2019; p. 50010, ISBN 978-0-7354-1847-9.

26. Lysak, V.I.; Kuzmin, S.V. Lower boundary in metal explosive welding. Evolution of ideas. *J. Mater. Process. Technol.* **2012**, *212*, 150–156. [CrossRef]

27. Groche, P.; Wagner, M.F.-X.; Pabst, C.; Sharafiev, S. Development of a novel test rig to investigate the fundamentals of impact welding. *J. Mater. Process. Technol.* **2014**, *214*, 2009–2017. [CrossRef]

28. Groche, P.; Becker, M.; Pabst, C. Process window acquisition for impact welding processes. *Mater. Des.* **2017**, *118*, 286–293. [CrossRef]

29. Pabst, C.; Sharafiev, S.; Groche, P.; Wagner, M.F.X. A Novel Method to Investigate the Principles of Impact Welding: Development and Enhancement of a Test Rig, Experimental and Numerical Results. *Adv. Mater. Res.* **2014**, *966–967*, 500–509. [CrossRef]

30. Strand, O.T.; Goosman, D.R.; Martinez, C.; Whitworth, T.L.; Kuhlow, W.W. Compact system for high-speed velocimetry using heterodyne techniques. *Rev. Instrum.* **2006**, *77*, 83108. [CrossRef]

31. Johnson, J.R.; Taber, G.; Vivek, A.; Zhang, Y.; Golowin, S.; Banik, K.; Fenton, G.K.; Daehn, G.S. Coupling experiment and simulation in electromagnetic forming using photon doppler velocimetry. *Steel Res. Int.* **2009**, *80*, 359–365.

32. Lueg-Althoff, J. *Fügen von Rohren durch elektromagnetische Umformung—Magnetpulsschweißen*, 1st ed.; Shaker: Herzogenrath, Germany, 2019; ISBN 384406558X. (In German)

33. Eichhorn, G. Analysis of the hypervelocity impact process from impact flash measurements. *Planet. Space Sci.* **1976**, *24*, 771–781. [CrossRef]

34. Bellmann, J.; Beyer, E.; Lueg-Althoff, J.; Gies, S.; Tekkaya, A.E.; Schulze, S. Measurement of Collision Conditions in Magnetic Pulse Welding Processes. *J. Phys. Sci. Appl.* **2017**, *7*, 1–10. [CrossRef]

35. Bellmann, J.; Lueg-Althoff, J.; Schulze, S.; Hahn, M.; Gies, S.; Beyer, E.; Tekkaya, A. Thermal Effects in Dissimilar Magnetic Pulse Welding. *Metals* **2019**, *9*, 348. [CrossRef]

36. Manikandan, P.; Hokamoto, K.; Fujita, M.; Raghukandan, K.; Tomoshige, R. Control of energetic conditions by employing interlayer of different thickness for explosive welding of titanium/304 stainless steel. *J. Mater. Process. Technol.* **2008**, *195*, 232–240. [CrossRef]

37. Khaustov, S.V.; Kuz'min, S.V.; Lysak, V.I.; Pai, V.V. Thermal processes in explosive welding. *Combust. Explos. Shock Waves* **2014**, *50*, 732–738. [CrossRef]
38. Pabst, C. *Ursachen, Beeinflussung, Auswirkungen sowie Quantifizierung der Temperaturentwicklung in der Fügezone beim Kollisionsschweißen*; Shaker Verlag: Düren, Germany, 2019; ISBN 978-3-8440-7073-6. (In German)
39. Böhme, M.; Sharafiev, S.; Schumacher, E.; Böhm, S.; Wagner, M.F.-X. On the microstructure and the origin of intermetallic phase seams in magnetic pulse welding of aluminum and steel. *Materialwiss. Werkstofftech.* **2019**, *50*, 958–964. [CrossRef]

MDPI

St. Alban-Anlage 66

4052 Basel

Switzerland

Tel. +41 61 683 77 34

Fax +41 61 302 89 18

www.mdpi.com

Metals Editorial Office

E-mail: metals@mdpi.com

www.mdpi.com/journal/metals